OXFORD MONOGRAPHS IN
INTERNATIONAL LAW

General Editor: Professor Ian Brownlie QC, DCL, FBA
*Chichele Professor of Public International Law in the University of
Oxford and Fellow of All Souls College, Oxford.*

HUMAN RIGHTS IN STATES OF EMERGENCY
IN INTERNATIONAL LAW

OXFORD MONOGRAPHS IN INTERNATIONAL LAW

The aim of this series of monographs is to publish important and original pieces of research on all aspects of public international law. Topics which are given particular prominence are those which, while of interest to the academic lawyer, also have important bearing on issues which touch the actual conduct of international relations. None the less the series is wide in scope and includes monographs on the history and philosophical foundations of international law.

ALSO IN THIS SERIES

The Juridical Bay
GAYL WESTERMAN

The Exclusive Economic Zone in International Law
DAVID ATTARD

Judicial Remedies in International Law
CHRISTINE GRAY

The Legality of Non-Forcible Counter-Measures in
International Law
OMER ELAGAB

State Responsibility and the Marine Environment
BRIAN D. SMITH

The Shatt-al-Arab Boundary Question: A Legal
Reappraisal
KAIYAN HOMI KAIKOBAD

Relevant Circumstances and Maritime Delimitation
MALCOLM D. EVANS

Land-Locked and Geographically Disadvantaged
States in the International Law of the Sea
S. C. VASCIANNIE

Surrender, Occupation, and Private
Property in International Law
NISUKE ANDO

The Human Rights Committee:
Its Role in the Development of the
International Covenant on Civil and
Political Rights
DOMINIC McGOLDRICK

Human Rights in States of Emergency in International Law

JAIME ORAÁ

CLARENDON PRESS · OXFORD
1992

Oxford University Press, Walton Street, Oxford OX2 6DP
Oxford New York Toronto
Delhi Bombay Calcutta Madras Karachi
Petaling Jaya Singapore Hong Kong Tokyo
Nairobi Dar es Salaam Cape Town
Melbourne Auckland
and associated companies in
Berlin Ibadan

Oxford is a trade mark of Oxford University Press

Published in the United States
by Oxford University Press, New York

British Library Cataloguing in Publication Data
Data available

Library of Congress Cataloging in Publication Data
Oraá, Jaime.
Human rights in states of emergency in international law/
Jaime Oraá.
(Oxford monographs in international law)
"Revised version of a thesis submitted in January 1990 for the
degree of Doctor of Philosophy at Oxford University"—
Acknowledgements.
Includes bibliographical references (p.   ) and index.
1. Human rights.   2. War and emergency legislation.   3. Treaties.
4. International law.   I. Title.   II. Series.
K3240.4.O73   1992   341.4'81—dc20   91-27309
ISBN 0-19-825710-4

Typeset by Best-set Typesetter Ltd., Hong Kong
Printed in Great Britain by
Bookcraft Ltd, Midsomer Norton, Avon

To
my Parents
and to the
Society of Jesus

Editor's Preface

With this work the *Monographs* series acquires its second title in the sphere of human rights, and others will follow.

The significance of its subject cannot be overstated. Since the Second World War, states of emergency, however described in different legal systems, have been a familiar feature of government, and have been resorted to as much by States claiming adherence to the rule of law as by others.

Father Oraá's work is not the first in the field, but it is the impressive result of extensive research and the application of careful legal analysis. Among the particular virtues of the book is its teasing out of the emergent principles of general international law governing human rights in states of emergency. At the same time, the bulk of the study is devoted to the three most pertinent multilateral treaties each of which has been in force for a considerable period of time: the International Covenant on Civil and Political Rights, the European Convention on Human Rights, and the American Convention on Human Rights.

All Souls College,   IAN BROWNLIE
Oxford
12th September 1991

Acknowledgements

This study is a revised version of a thesis submitted in January 1990 for the degree of Doctor of Philosophy at Oxford University.

My principal debt of gratitude is to my supervisor, Professor Ian Brownlie, QC, FBA, Chichele Professor of International Law at Oxford, and Fellow of All Souls, From the beginning of my work in Oxford, he has been a major source of inspiration. I am especially grateful for his invaluable comments on the different drafts of this study, and the considerable kindness he has shown to me during my years at Oxford.

I wish also to record my debt of gratitude to the University of Deusto (Spain) for their continuing support since the beginning of this project, to the staff of the Bodleian Law Library for their helpful and kind assistance, and to the members of Campion Hall, whose encouragement throughout my years in Oxford has been very much appreciated.

J.O.

Contents

Abbreviations　　　　　　　　　　　　　　　　　　　　　　xvii
Table of Cases　　　　　　　　　　　　　　　　　　　　　　xix

Introduction　　　　　　　　　　　　　　　　　　　　　　　　I

PART I.　HUMAN RIGHTS STANDARDS IN STATES OF
EMERGENCY IN THE CONTEXT OF MULTILATERAL
TREATIES: THE LEGAL REGIME OF THE DEROGATION
CLAUSE　　　　　　　　　　　　　　　　　　　　　　　　5

Introduction　　　　　　　　　　　　　　　　　　　　　　　　7

**1. The Principle of Exceptional Threat: The Existence of the
Emergency as Envisaged in the Three Main Multilateral
Treaties**　　　　　　　　　　　　　　　　　　　　　　　　II

　1. Introduction　　　　　　　　　　　　　　　　　　　　II
　2. The Definition of Emergency in the Three Treaties　　　II
　　(*a*) The Concept in the Covenant and in the European
　　　　Convention　　　　　　　　　　　　　　　　　12
　　(*b*) The Wording of the American Convention　　　　14

　3. The Interpretation Given to the Concept of Emergency by
　　the International Bodies Entrusted with the Application of
　　the Three Treaties　　　　　　　　　　　　　　　　16
　　(*a*) The Jurisprudence under the European Convention　16
　　(*b*) The Practice of the UN Human Rights Committee　20
　　(*c*) The Practice of the IACHR　　　　　　　　　　22
　　　　(i) The Doctrine under the American Convention　23
　　　　(ii) The Principles Applicable to States Non-Parties to
　　　　　　the Convention　　　　　　　　　　　　26

　4. The Main Features of the Emergency Envisaged in the
　　Treaties　　　　　　　　　　　　　　　　　　　　27

(*a*) The Emergency Must be Actual or at Least Imminent   27
(*b*) Its Effects Must Involve the Whole Population   28
(*c*) The Threat Must be to the Very Existence of the
      Nation   29
(*d*) The Declaration of Emergency Must be a Last Resort   29
(*e*) The Declaration of Emergency as a Temporary
      Measure   30

5. Circumstances in which a Public Emergency can be
    Declared   30

6. Conclusion   32

2. The Principle of Proclamation of the State of Emergency   **34**

1. Introduction   34
(*a*) The Requirement of Official Proclamation in the
      Covenant   34
(*b*) The Absence of this Requirement in the European
      Convention: Its Effects   37

2. The Organ Competent to Proclaim the Emergency:
    The Legislature and the Executive   38

3. Judicial Control   40

4. International Control under the Human Rights Treaties   42
(*a*) Under the European Convention   42
    (i) The Competence of the European Organs to
        Control Compliance with the Principles of the
        Derogation Clause in Exceptional Situations   42
    (ii) The Construction of the Irish Government:
        The Principle of Good Faith   44
    (iii) The Concept of the Margin of Appreciation   45
    (iv) The Problem of the Burden of Proof   47
(*b*) Under the Covenant: The United Nations Human
      Rights Committe   47
    (i) The Reporting Procedure   48
    (ii) The Practice of the Committee in Certain Specific
        Cases   49
(*c*) Under the American Convention: The Control
      Exercised by the Inter-American Commission on
      Human Rights   51

5. Conclusion   55

3. The Principle of Notification   **58**

1. The Rationale of the Principle of Notification   58

2. The Contents of the Notice of Derogation or the Formal
Requirements of the Principle of Notification   60
   (*a*) The Element of Time   60
   (*b*) The Contents of the Notice of Derogation   61

3. The Effect of the Failure to Comply with the Notification
Requirement   64
   (*a*) Total Failure to Meet the Notification Requirement:
    The Question of the Application ex officio of the
    Derogation Clause in Cases in which no Notification
    has been Sent   66
     (i) The *Cyprus Case*   67
     (ii) The *Bolivian Case*   69
     (iii) The Attitude of the UN HR Committee   69
   (*b*) Partial Failure to Meet the Notification Requirement   70

4. The Distinction between the Obligation of Notifying and
the Obligation of Providing Information to International
Bodies under Other Procedures   72

5. The Practice of Notification under the Three Treaties   76
   (*a*) The European Convention   76
   (*b*) The ICCPR   77
   (*c*) The ACHR   81

6. Conclusion   83

4. The Principle of Non-Derogability of Fundamental Rights   **87**

1. Introduction   87
   (*a*) The *travaux préparatoires* of the Three Treaties   88
   (*b*) The Rationale behind the List of Non-Derogable
    Rights   94

2. The Four Common Non-Derogable Rights   96

3. The Other Non-Derogable Rights under the ICCPR and
the ACHR   97
   (*a*) Rights Unrelated to Public Emergencies   97
   (*b*) Rights which May have an Impact on Public
    Emergencies   100

4. Other Rights and Principles which are Held Non-Derogable
by Implication   101
   (*a*) Provisions Related to the Exercise of Non-Derogable
    Rights   102
   (*b*) Provisions which Contain General Exceptions   103
   (*c*) Provisions Related to the Machinery of
    Implementation of the Treaties   105

5. Study of Several Proposals to Make a priori
   Non-Derogable Some Minimum Guarantees against
   Arbitrary Detention and of Due Process of Law 106
   (*a*) Introduction 106
   (*b*) Guarantees against Arbitrary Detention in
       Emergencies 108
       (i) General Guarantees 108
       (ii) Minimum Rights of Detainees 108
   (*c*) Guarantees of Due Process of Law in Emergencies 114
       (i) The Minimum Rights of Due Process 115
       (ii) The Rights whose Derogation would be Justified
           in Principle 116
       (iii) The Right to be Tried by an Independent and
           Impartial Tribunal 118
       (iv) The Practice of the HR Committee 119
       (v) The Practice under the ACHR 121

6. Conclusion 124

Appendix: Reservations to Non-Derogable Rights in the
Three Human Rights Treaties 127

The Three Substantive Conditions for the Derogation of
Rights in States of Emergency 139

5. The Principle of Proportionality **140**

1. Introduction 140
   (*a*) The Legislative History of the Treaties 143
2. The Application of the Principle of Proportionality by the
   Supervisory Bodies under the Three Main Treaties 144
   (*a*) The Principle as Applied by the European Organs in
       Cases of Emergency 144
   (*b*) The Principle as Applied by the UN Human Rights
       Committee 152
       (i) The Review of States' Reports 152
       (ii) Under the Optional Protocol 157
   (*c*) The Application of the Principle of Proportionality
       under the ACHR 159
       (i) The General Principles Formulated by the
           Inter-American Court 159
       (ii) The Application of the Principle by the IACHR 161
3. Conclusion 168

6. The Principle of Non-Discrimination **171**

  1. Introduction   171
    (*a*) The Characteristics of the Provision   172
    (*b*) The Debate on the Principle of Non-Discrimination as
       Non-Derogable   172
    (*c*) The Introduction of the Principle as a Condition for
       Derogation   173
    (*d*) The Extent of the Provision   174
  2. The Application of the Principle by the Monitoring Organs
    under the Three Treaties   177
    (*a*) The European Organs: The *Ireland* v. *UK Case*   177
    (*b*) The Practice of the HR Committee   182
    (*c*) The Practice of the IACHR   185
  3. Conclusion   188

7. The Principle of Consistency **190**

  1. Introduction   190
  2. Legislative History of the Principle in the Three Treaties   191
  3. The Application of the Principle by the Monitoring Organs
    under the Three Treaties   193
    (*a*) The European Convention   193
    (*b*) The UN Human Rights Committee   196
    (*c*) The IACHR   197
  4. The Delimitation of the Content of the Expression 'Other
    Obligations under International Law'   200
    (*a*) The Possible Additional Obligations Arising from the
       Laws of War   201
       (i) Humane Treatment   202
       (ii) Fair Trial and Due Process of Law   204
       (iii) Safeguards for Internees in Convention IV   204
  5. Conclusion   205

PART II.   HUMAN RIGHTS IN STATES OF EMERGENCY IN
           GENERAL INTERNATIONAL LAW   207

Introduction   209

**8. Customary International Law and Human Rights: Two
   Preliminary Questions**   **214**

   1. The Existence of Human Rights Standards in Customary   214
      Law
   2. The Special Evidence Required to Prove the Existence of   216
      Customary Norms in the Area of Human Rights

**9. First Line of Inquiry: Human Rights in Emergencies: The
   Doctrine of State Necessity**   **220**

   1. Legal Doctrines in International Law which Justify the
      Non-Compliance with International Obligations in Cases of
      force majeure, Self-Defence, and Necessity   220
   2. The Doctrine of State Necessity   221
      (*a*) Characteristics   221
      (*b*) Conditions of Application   222
      (*c*) Decisions of the ILO Organs: the *Greek* and *Polish*
            Cases   223

**10. Second Line of Inquiry: The Emergence of Some of
    the Principles of the Derogation Clause as Principles of
    General International Law**   **227**

    1. The Legal Nature of the Derogation Clause   227
    2. Three Types of Principles   228
    3. The Derogation Clause and Customary International Law:
       The 'Norm-Creating Character' of the ICCPR and the
       Derogation Clause   229
    4. Evidence of the Emergence of Some of the Principles of the
       Derogation Clause as Customary Law   234
       (*a*) The Probative Value of Law-Making Treaties   234
       (*b*) The Repetition of the Same Norm in Several Human
             Rights Treaties   234
       (*c*) The Practice of the Organs of International
             Organizations and Similar Agencies   235
             (i) The Practice of the IACHR   236
             (ii) the Practice of the UN Organs   242
       (*d*) The Acceptance of these Principles by States
             Non-Parties to the Human Rights Treaties   253
    5. The Application of the Principles of the Derogation Clause
       to Other Areas of International Law   254

(*a*) State Responsibility for Injury to Aliens (Human Rights of Aliens in States of Emergency in General International Law) 254

(*b*) The Law of Belligerent Occupation: The Case of Israel 256

6. The Principles of the Derogation Clause as 'General Principles of Law' 257

Conclusion of the Inquiry: The Principles which Constitute Emergent Principles of General International Law Governing Human Rights in States of Emergency 260

1. The Principle of Exceptional Threat 261
2. The Principle of Proportionality 262
3. The Principle of Non-Discrimination 263
4. The Principle of Non-Derogability of Fundamental Rights 264

Conclusions of Part II 266

Concluding Remarks 270

Bibliography 273

Index 287

Abbreviations

ACHR	American Convention on Human Rights
AFDI	*Annuaire français de droit international*
AJIL	*American Journal of International Law*
Australian YBIL	*Australian Year Book of International Law*
BYBIL	*British Year Book of International Law*
Can. YBIL	*Canadian Yearbook of International Law*
CLP	*Current Legal Problems*
ECHR	European Convention on Human Rights
ECOSOC	Economic and Social Council
EHRR	*European Human Rights Reports*
German YBIL	*German Year Book of International Law*
HR	Human Rights
HRJ	*Human Rights Journal* (*Revue des droits de l'homme*)
HRLJ	*Human Rights Law Journal*
HR Quarterly	*Human Rights Quarterly*
IACHR	Inter-American Commission on Human Rights
IA Court HR	Inter-American Court of Human Rights
ICCPR	International Covenant on Civil and Political Rights
ICESCR	International Covenant on Economic, Social, and Cultural Rights
ICJ	International Court of Justice
ICJ Reports	Reports of Judgments, Advisory Opinions, and Orders of the International Court of Justice
ICLQ	*International and Comparative Law Quarterly*
ILA	International Law Association
ILC	International Law Commission
ILM	*International Legal Materials*
Int. CJ Review	International Commission of Jurists Review
Int. Commiss. of Jurists	International Commission of Jurists
Israel YBHR	*Israel Year Book on Human Rights*
NYIL	*Netherlands Year Book of International Law*
OAS	Organization of American States
Recueil	*Recueil des Cours: Academie de droit international de la Haye*
UDHR	Universal Declaration of Human Rights
UN	United Nations

UNGA	United Nations General Assembly
UNGAOR	United Nations General Assembly, Official Records
YBECHR	*Year Book of the European Convention on Human Rights*
YBILC	*Year Book of the International Law Commission*

Table of Cases[1]

Baboeran v. *Suriname* 26, 97

Baby Boy case (v. *USA*) 136

Belilos v. *Switzerland* 130

Brogan et al. v. *UK* 109

Cyprus v. *Turkey* 84, 182, 194, 196, 199, 204

'*Death Penalty case*' 186

'*Death Penalty for Juveniles case*' 186, 215

'*Death Penalty: Racial Discrimination in its Application case*' 186

De Becker case. (*De Becker* v. *Belgium*) 143, 151

Fals Borda v. *Colombia* 81, 118–19

First Cyprus Case. (*Greece* v. *UK*) 16, 37, 42, 46, 65, 67–9, 76–7, 145, 149, 193

France, Norway, Denmark, Sweden, and the Netherlands v. *Turkey.* 16, 64, 76–7

Garcia Lanza et al. v. *Uruguay* 74, 158

The Greek case. (*Denmark, Norway, Sweden, and the Netherlands* v. *Greece*) 16, 18–20, 27, 32, 36, 43, 46–7, 97, 104–5, 149, 158, 194, 262

The Greek case 60–1, 62, 71–3, 76–7, 119, 181–2, 223–6, 255

Grille Motta v. *Uruguay* 21

Hiber Conteris v. *Uruguay* 81

Ireland v. *UK* 16, 28, 38, 60, 62, 76–7, 97, 107, 110, 113, 144–51, 179–82, 193–4

Landinelli et al. v. *Uruguay* 21, 70, 75, 101, 157–8

Lawless case. (*G. Lawless* v. *Ireland*) 2, 16–17, 27–9, 32, 38–9, 43, 44–7, 59, 60, 63–6, 71, 104–5, 107, 113, 119, 144–50, 193

Libya v. *Malta* 233

Macias case (v. *Nicaragua*) 97, 99

The Nicaraguan case. (*Nic* v. *US*) 217–18

Nicaragua-Miskitos case 38, 52–3, 61, 82, 159, 161, 165, 174, 176, 187–9, 197–9

North Sea Continental Shelf cases. (*FRG* v. *Den.*: *FRG* v. *Netherlands*) 140, 230, 232–3, 260

The Polish case 212, 223–6, 255, 261

---

[1] This table of cases includes the references of the decisions adopted by the UN HR Committee (referred to by the Committee as 'views') in individual communications under the Optional Protocol of the ICCPR, and decisions of the IACHR in individual cases.

Roach and Pinkerton case (v. *USA*) 135
Saldias de Lopez v. *Uruguay* 158
Salgar de Montejo v. *Colombia* 81, 120, 158–9
Suarez de Guerrero v. *Colombia* 97, 156
Tunisia v. *Libya case* 233
Velasquez–Rodriguez case. (*Velasquez* v. *Honduras*) 97
Weinberger v. *Uruguay* 120, 158, 184–5

American Convention on Human Rights: Reports of the Inter-American Commission on Human Rights

Report on the Situation of Political Prisoners and their Relatives in Cuba. OAS/Ser. L/V/II. 7, doc. 4 (17 May 1963).
Second Report on the Situation of Political Prisoners in Cuba. OAS/Ser. L/V/II. 23, doc. 3 (17 November 1970).
Report on the Situation of Human Rights in Chile. OAS/Ser. L/V/II. 34 doc. 21, corr. 1 (25 October 1974).
Third Report on the Situation of Human Rights in Chile. OAS/Ser. L/V/II. 40, doc. 10 (11 February 1977).
Report on the Situation of Human Rights in Paraguay. OAS/Ser. L/V/II. 43, doc. 13, corr. 1 (31 January 1978).
Report on the Situation of Human Rights in Panama. OAS/Ser. L/V/II. 44, doc. 38, rev. 1 (22 June 1978).
Report on the Situation of Human Rights in Nicaragua. OAS/Ser. L/V. 45, doc. 16, rev. 1 (17 November 1978).
Report on the Situation of Human Rights in El Salvador. OAS/Ser. L/V/II. 46, doc. 23, rev. 1 (17 November 1978).
Report on the Situation of Human Rights in Haiti. OAS/Ser. L/V/II. 46, doc. 46, rev. 1 (12 December 1979).
Report on the Situation of Human Rights in Argentina. OAS/Ser. L/V/II. 49, doc. 19, corr. 1 (11 April 1980).
Report on the Situation of Human Rights in Nicaragua. OAS/Ser. L/V/II. 53, doc. 25 (30 June 1981).
Report on the Situation of Human Rights in Colombia. OAS/Ser. L/V/II. 53, doc. 22 (30 June 1981).
Report on the Situation of Human Rights in Bolivia. OAS/Ser. L/V/II. 53, doc. 6 (1 July 1981).
Report on the Situation of Human Rights in Guatemala. OAS/Ser. L/V/II. 53, doc. 21, rev. 2 (13 October 1981).
Report on the Situation of Human Rights in Guatemala. OAS/Ser. L/V/II. 61, doc. 47, rev. 1 (5 October 1983).
Report on the Situation of Human Rights of a Segment of the Nicaraguan

Population of Miskito Origin. OAS/Ser. L/V/II. 62, doc. 10, rev. 3 (29 November 1983).

Report on the Situation of Human Rights in Chile. OAS/Ser. L/V/II. 66, doc. 17 (27 September 1985).

Report on the Situation of Human Rights in Suriname. OAS/Ser. L/V/II. 66, doc. 21, rev. 1 (2 October 1985).

Report on the Situation of Human Rights in Paraguay. OAS/Ser. L/V/II. 71, doc. 19, rev. 1 (28 September 1987).

Introduction

One of the most important problems in the international protection of human rights is that of identifying the standards governing these rights in situations of emergency. Public emergencies present a grave problem for States: that of overcoming the emergency and restoring order in the country while at the same time respecting the fundamental human rights of individuals. The derogation clause of human rights treaties establishes a legal regime regulating this crucial problem. This clause has been described as the 'cornerstone' of the entire system for protecting human rights, and as the most important provision of human rights treaties.[1]

Moreover, there are two additional reasons which make this topic highly relevant. First, in the last decades the gravest violations of fundamental human rights have occurred in the context of states of emergency. In these situations, States, using the emergency as an excuse, frequently deny the application of basic standards and take derogating measures which are excessive and in violation of international treaties on human rights. Therefore, in order to know the exact extent of the protection afforded by these treaties, a detailed examination of the treaty standards, undertaken in the light of the jurisprudence of the international monitoring bodies, is of fundamental importance.

Secondly, almost half of the States of the international community are not parties to the international treaties on human rights which establish a legal regime for emergencies, and therefore treaty standards are not applicable as such to these States. This fact, together with the notable absence of studies on this question, has created a dangerous uncertainty concerning the main criteria governing human rights in emergencies in terms of general international law. This uncertainty can be seen in the practice of the UN monitoring organs. At the same time, some international treaties on human rights have no derogation clauses (i.e. the African Charter, and some ILO Conventions); consequently the regime applicable in these cases also remains uncertain. For these reasons, a thorough analysis of the standards in general international law is of paramount importance.

In short, the main purpose of this study is twofold: to analyse the main

---

[1] See the remarks by Mr Prado Vallejo, a member of the UN HR Committee, in CCPR/C/SR. 351 (1982), p. 8, para. 32. See also the remarks of the Attorney-General of Ireland in the *Lawless case* (Counter-Memorial of the Government of Ireland), Ser. B: Pleadings, p. 224.

principles regulating human rights in emergencies as contained in the derogation clause of the treaties; and secondly, to examine the principles governing the same question in general international law. In respect of this latter question, attention will be paid to the hypothesis which contends that some of the principles of the derogation clause have become, or are in the process of becoming, customary international law.

THREE METHODOLOGICAL OBSERVATIONS

1. The treaties which are going to be examined in this study are: the European Convention on Human Rights and Fundamental Freedoms (1950), the International Covenant on Civil and Political Rights (1966), and the American Convention on Human Rights (1968). These treaties contain a derogation clause with specific standards for emergencies.[2]

These three treaties are considered the most important ones in so far as they establish a general and exhaustive regime concerning civil and political rights. This work is not going to deal with the question of the impact of emergencies on economic, social, and cultural rights, and this is for two reasons.[3] First, because most of the treaties dealing with this question do not contain a derogation clause for emergencies but simply a general limitation clause (e.g. article 4 of the 1966 International Covenant on Economic, Social, and Cultural Rights). Secondly, because the pro-grammatic character of these rights and the weak machinery of im-plementation established in these treaties has meant that there has been no relevant case-law in this area.[4]

2. Although municipal law and the doctrine employ a very wide terminology in order to refer to situations of emergency (terms like 'state of siege', 'situations of exception', 'suspension of guarantees', 'emergency powers', 'état d'urgence', and 'national emergencies' have been used), the term which will be utilized in this study is 'states of emergency', because of its more general use. In these human rights treaties, as opposed to the legislation of several States which establish different emergency situations with different legal regimes, there is only one

---

[2] See ICCPR, art. 4; ECHR, art. 15; and ACHR, art. 27.

[3] For a good analysis of the effects of emergencies on economic and social rights, see Int. Commiss. of Jurists, *States of Emergency: Their Impact on Human Rights* (Geneva, 1983), pp. 417–24.

[4] Although the European Social Charter has a derogation clause in article 30, for the above-mentioned reasons there has been no relevant case-law in this area. See D. J. Harris, *The European Social Charter*, (New York, 1984), p. 273. See the new attempts to imple-ment economic, social, and cultural rights under the ICESCR, in P. Alston and B. Simma, *First Session of the UN Committee on Economic, Social and Cultural Rights, AJIL* 81 (1987), 747–56; *Second Session, AJIL* 82 (1988), 603–15.

regime for all kinds of emergencies; therefore the expression 'states of emergency' will be used in order to refer to all kinds of emergencies.

3. In the study of this issue, special attention will be paid to the *travaux préparatoires* of the treaties and to the jurisprudence of the international bodies entrusted with the application of these standards in emergencies. In this respect, the fact that the ICCPR and the ACHR through the HR Committee and the Inter-American Court and Commission came into operation as recently as the 1970s makes the study of the jurisprudence of these bodies in the application of the derogation clause extremely interesting. Moreover, even though the case-law of the European Convention is better known due to its seniority, one has the impression that the pleadings and other relevant documents of the leading cases in emergencies have not received sufficient attention in the literature.

THE PLAN OF THE STUDY

The first part of the study will analyse the legal regime of the derogation clause of the three treaties. The legal regime of the derogation clause contains what could be called seven fundamental 'principles'.[5] Each of these principles will be studied in separate chapters. Thus, Chapter 1 deals with the *principle of exceptional threat*, in other words, with the question of the kind of emergency which the treaties foresee as justifying the declaration of emergency and derogations from human rights standards. Chapter 2 refers to the *principle of proclamation*, that is, to the requirement of officially proclaiming the state of emergency in the country for the valid operation of the derogation clause. Chapter 3 will study the *principle of notification*, that is, the obligation of the State to notify the exercise of the right of derogation, all the derogating measures taken, and the reasons for them to the other States that are parties to the treaty. The essential *principle of non-derogability of fundamental rights* will be studied in Chapter 4; this principle establishes that even in situations of emergency there are some rights which can never be derogated from. The different lists of rights under the three treaties, the other provisions of the treaties which are non-derogable by implication, the proposals to make some minimun guarantees against arbitrary detention and other proposals referring to the right of due process will also be analysed. Furthermore, Chapters 5, 6, and 7 will examine the important principles referring to the substantive conditions required in order to

---

[5] The word 'principles' is used here in a very wide and abstract way and not in the technical sense of 'principles of international law' or 'general principles of law'. Whether or not some, or all, of these principles could be considered 'general principles' in the technical sense is something that will be studied in the following chapters.

derogate from human rights provisions. These conditions are *proportionality*, *non-discrimination*, and *consistency* with other obligations under international law.

The second part of the study (Chapters 8, 9, and 10) will deal with the question of the principles governing human rights in emergencies in general international law. Two main lines of inquiry will be followed. The first will look at the main doctrines in general international law which justify non-compliance with international obligations. In the case of states of emergency, the doctrine which would justify non-compliance with human rights obligations arising from customary law seems to be the doctrine of State necessity. Therefore, this doctrine and the main principles regulating its application will be carefully examined. The second line of inquiry (Chapter 10) will analyse the contention that some of the principles of the derogation clause are emerging as principles of customary international law. Among the existing evidence available to prove the emergence of these principles as customary norms, the practice of the international monitoring bodies will be given special importance. Finally, in the light of all the evidence examined, an assessment will be made concerning the concrete principles which constitute emergent principles of general international law.

PART I

Human Rights Standards in States of Emergency in the Context of Multilateral Treaties: The Legal Regime of the Derogation Clause

Introduction: The Institution of the States of Emergency and the Origin of the Derogation Clause

States of emergency as a legal institution which justifies derogations from human rights standards is a well-known institution recognized in almost all systems of municipal law.[1] The historical origin of the institution goes back to Roman times, and is found in the nomination of a 'dictator' in exceptional circumstances of external attack or internal rebellion.[2] The derogation clause seems to recognize the same institution contained in municipal systems in the area of international law dealing with human rights.[3] Moreover, a recent study has pointed out that there is a striking convergence between the principles developed in public international law and the principles codified in constitutions or judicially evolved in different legal systems.[4]

THE ORIGIN OF THE DEROGATION CLAUSE IN THE THREE TREATIES

Even though the European Convention was signed in 1950, the origin of the derogation clause in the human rights treaties under consideration seems to come from a UK proposal to the UN Human Rights Commission in 1947 at the beginning of the drafting process of the Covenant. It is interesting to recall briefly the history of the derogation clause in the three treaties.

The first multilateral treaty dealing with a wide range of human rights was the European Convention, which was signed in Rome in 1950 and came into force in 1953, while the ICCPR was signed in 1966 and entered into force ten years later.[5] However, the drafting of the Covenant had

---

[1] See Ch. 10, pp. 242–3.

[2] For the institution of the 'dictator' see *inter alia* Guarino, *Storia del Diritto Romano* (Naples, 1981), pp. 108–9, 216–18. See also Bainville, *Les Dictateurs* (Paris, 1931), pp. 37 ff. See an interesting commentary on this institution by Rousseau, *The Social Contract*, in *Political Writings*, ed. Watkins, pp. 136 ff. For an account of the history of the institution of states of emergency see A. M. Singhvi, 'The Law of Emergency Powers: A Comparative Study', Ph.D. thesis (Cambridge, 1986), pp. 1–18.

[3] For the theory that the derogation clause is an adaptation to the particular field of human rights of the doctrine of necessity in international law, see Ch. 10.

[4] See Ch. 10, p. 377.

[5] The following abbreviations will be used for the three treaties: ICCPR for the Covenant, ECHR for the European Convention, and ACHR for the American Convention.

started much earlier, in fact in 1947. At that time, the UK forwarded to the Drafting Committee (1st session) a proposal for an International Bill of Rights which already contained a derogation clause whose wording was couched in similar terms to those of the final derogation clause.[6] The *travaux préparatoires* of the Covenant show that by June 1949 the UN Commission on Human Rights had adopted article 4, practically in its final version.[7] The European Convention then borrowed the derogation clause from the draft Covenant;[8] this is why the wording of both clauses is so similar.

In fact, in the first drafts of the European Convention, no derogation clause could be found at all. Thus, in the 'Teitgen Report' which was the first working paper of the Convention, there is just a general limitation clause (article 6).[9] In the meetings of the Committee of Experts held in Strasbourg from 2 to 8 of February 1950, the UK Government proposed for the first time the inclusion of a derogation clause similar to that of the draft Covenant (art. 4). However, one has to wait until the last draft of this Committee, which was submitted to the Committee of Ministers of the Council of Europe, to find a derogation clause in both alternative texts.[10] Finally, in the single text of the Convention submitted by the Conference of Senior Officials in June 1950, the derogation clause is found in its final version.[11]

Later, during the process of drafting the American Convention at the San José Conference in 1968, the derogation clause of these two treaties was the model taken by the drafters for the insertion of a derogation clause in the Convention.

A DEROGATION OR A LIMITATION CLAUSE?

Even though the derogation clause has been considered to be one of the most important articles of human rights treaties, the need for it was

---

[6] E/CN. 4/21, annexe B (art. 4). For the UK proposal, see *United Kingdom Draft of An International Bill of Human Rights* (London, June 1947). (I am very grateful to Sir Vincent Evans for sending me a copy of the original Draft). See also E/CN. 4/SR. 423, p. 4. For the *travaux préparatoires* of the Covenant see M. Bossuyt, *Guide to the* Travaux préparatoires *of the ICCPR* (The Hague, 1987). For the theory that the derogation clause was the result of the UK practice in the colonial context; see D. O'Donnell, 'Legitimidad de los Estados de Excepcion, a la luz de los instrumentos de Derechos Humanos', in E. Haba, *Tratado basico de Derechos Humanos* (San José, Costa Rica, 1986), ii. 679.

[7] E/CN. 4/335, 14 June 1949.

[8] For a short summary of the legislative history of the European Convention see vol. i of the preparatory work, ECHR, *Collected Edition of the* Travaux préparatoires *of the ECHR* (8 vols.; The Hague, 1975–1985).

[9] See *Collected Edition of the* Travaux préparatoires, i. 208.

[10] Ibid. iii. 312.

[11] Ibid. iv. 280.

actually contested by some delegations at the beginning of the *travaux préparatoires* of the Covenant. In fact, the first UK proposal was rejected by the UN Commission on Human Rights in 1947.[12] The main reason for the opposition to a derogation clause was that some delegations considered that a general limitation clause,[13] or several limitation clauses in specific articles, could do the work of the derogation clause.[14] However, the majority thought that there might be situations such as war, or other instances of extraordinary peril or crisis, in which far-reaching derogations to those principles foreseen in the limitation clauses would be necessary in order to ensure the continuing existence of the nation and the safety of the people. Moreover, the derogation clause would contain a very strict and detailed legal regime for permissible derogations in states of emergency in order to prevent abuses by States. This is why it was finally accepted.

However, the fact is that in the three human rights treaties under consideration, along with the derogation clause there are found several limitation clauses to specific rights.[15] Nevertheless, the functioning of these two different kinds of clauses now seems clear, and three main differences can be identified in their operation:

1. *As far as the situation justifying the operation of the clause is concerned*. The limitation clauses authorize restrictions on human rights on several grounds (public order, national security, public health and morals), in what can be called 'normal situations' or peacetime; in other words, for application in the daily life of States. Sometimes States need to restrict human rights on grounds of public order due to certain disturbances. On the other hand, the derogation clause operates in exceptional situations, that is, in public emergencies threatening the life of the nation.[16]

2. *As far as the extent of the rights affected is concerned*. The limitation clauses only affect specific rights, whereas the application of the derogation clause could affect all the rights contained in the treaty, except those made expressly non-derogable.

3. *As far as international accountability is concerned*. The operation of the limitation clauses does not require any special declaration by

---

[12] E/CN. 4/AC. 3/SR. 8, p. 11.

[13] The US delegation proposed a general limitation clause similar to that of the UDHR (art. 29). E/800 (art. 4. II) and E/CN. 4/AC. I/SR. 22, p. 3.

[14] A/2929, para. 36.

[15] For an analysis of the limitations clause, see *inter alia* R. Higgins, 'Derogations under Human Rights Treaties', *BYBIL* 48 (1976–7), 281–320; A. Kiss, 'Permissible Limitations on Rights', in L. Henkin (ed.), *The International Bill of Rights* (New York, 1981), pp. 290–310. See also for the principles governing limitations the 'Siracusa Principles' (repr. in *HR Quarterly*, 7 (1985) with commentaries; and also in *Int. CJ Review*, 36 (1986), 47–56).

[16] See the same position in Higgins, 'Derogations', pp. 281–2.

the State, whereas the operation of the derogation clause requires the notification to the other States parties of the proclamation of emergency, the derogated provisions, and the reasons therefor. This element of international accountability was an important factor for the choice of the derogation clause in order to prevent abuses by States in the exercise of emergency powers.[17]

[17] E/CN. 4/AC. 3/SR. 8, p. 10 (Mr Evans of the UK). See also E/CN. 4/SR. 195, p. 14, para. 70 (Mr Oribe of Uruguay). A/2929, para. 37 *in fine*.

I

The Principle of Exceptional Threat:
The Existence of the Emergency as Envisaged in the Three Main Multilateral Treaties

1. INTRODUCTION

One of the first and most important problems in the legal regulation of human rights in states of emergency is the precise definition of the kind of emergency which justifies the derogation of these rights.

In municipal law a great variety of grounds for declaring a state of emergency can be found; some of these grounds do not really represent a grave threat to the State.[1] Therefore the question arises whether these less grave emergencies are accepted in international law as legitimate justifications for derogations. One cannot automatically assume that the same kind of emergency is valid according to multilateral treaties on human rights.[2] For this reason, our first task should be to determine the precise definition of the concept of emergency as foreseen in the three main multilateral treaties. Then, in order to ascertain whether the international bodies have construed the concept of emergency in the same way as the drafters of the treaties, the jurisprudence of these bodies will be analysed. In the light of this analysis, the main characteristics of the concept of emergency will be summarized. Finally, reference will be made to the concrete kinds of circumstances which could justify prima facie the declaration of emergency.

2. THE DEFINITION OF EMERGENCY IN THE THREE TREATIES

Looking at the wording of the derogation clauses of these treaties as far as the definition of emergency is concerned, certain similarities and dif-

---

[1] See a list of grounds in E/CN. 4/826 5 Jan. 1962, 'Study of the Right of Everyone to be Free from Arbitrary Arrest, Detention and Exile', p. 257, para. 754. Another list in O'Donnell, 'States of Exception', *Int. CJ Review*, 21 (1978), 54; see below, pp. 30–1.

[2] The method of leaving the determination of the circumstances to municipal law was proposed by Uruguay and Chile in the preparatory work of the American Convention. However, this proposal was defeated at the San José Conference (1968). See below, p. 15.

ferences can be found. The ICCPR refers to a 'public emergency which threatens the life of the nation'. The ECHR refers to a 'war or other public emergency threatening the life of the nation', and the ACHR to a 'war, public danger, or other public emergency that threatens the independence or security of a State party'. The analysis of the concept 'foreseen' in these clauses will begin with the first two treaties because they are the oldest and their wording is similar.

(a)   The Concept in the Covenant and in the European Convention

The first text that can be found is the one proposed by the UN Commission on Human Rights to the ECOSOC, which contained the expression: 'in time of war or other public emergency'.[3] This expression was discussed in depth by the Commission in the following sessions. In the light of these discussions the principal choices made by the Commission were the following:

First, the suppression of any mention of war. In general, there was agreement on the suppression of any reference to 'war' because it was felt 'that the Covenant should not envisage, even by implication, the possibility of war, as the UN was established with the object of preventing war'.[4] However, war is obviously the greatest public emergency and therefore article 4 is fully applicable to that situation.[5] This construction is in line with the aim of the UK proposal of the derogation clause which tried 'to prevent States from arbitrarily derogating from their obligations in respect of human rights in time of war'.[6] The reason given was that under the general principles of international law in time of war, States were not strictly bound by conventional obligations unless the conventions contain provisions to the contrary.[7] It is interesting to note that when the European Convention was being drawn up, the UK proposed the adoption of a derogation clause similar to that of the draft Covenant; in fact, the Convention borrowed from the latter the wording which contained the expression 'in time of war or other public emergency'. Ironically, the UN Commission later suppressed the mention of war, whereas the European Convention approved the former wording.[8]

---

[3] UN HR Commission, Report to the ECOSOC on the 2nd session, 1947, ESC (VI) suppl. 1, annexe B 1, art. 4.

[4] A/2929, p. 67, para. 39. E/CN. 4/SR. 330, 1 July 1952. Several representatives opposed the inclusion of the word war.

[5] A/2929, pp. 66–7. E/CN. 4/SR. 330, Mr Valenzuela (Chile) and Mrs Roosevelt (USA).

[6] E/CN. 4/AC. 3/SR. 8, p. 10. See also Mr Wilson (UK), E/CN. 4/AC. 1/SR. 22.

[7] E/CN. 4/AC. 3/SR. 8, p. 10.

[8] E/CN. 4/335 (text of art. 4 adopted by the Commission on 14 June 1949); UK proposal of 6 Mar. 1950 to the European Convention, ECHR, *Collected Edition of the* Travaux préparatoires, iii. 280.

Secondly, the drafters preferred to adopt a broad term, such as 'public emergency', which in principle could embrace several situations, rather than to enumerate particular kinds of circumstances. In this respect, the proposal of a non-governmental organization which presented a more descriptive wording of the kind of emergency envisaged was defeated; this proposal stated: 'in the case of a state of emergency caused by an enemy invasion or a state of war or in the case of public commotion or disaster gravely upsetting the normal life in the territory of a State'.[9]

Thirdly, the provision of a further qualification of the term 'public emergency' in order to require that the situations justifying the declaration of emergency should be really *exceptional* and affect the whole nation. This would remove the risk of derogation in situations of lesser importance.[10] Several expressions were put forward as possible qualifications to the words 'public emergency': thus, 'directed against the interests of the people',[11] 'gravely threatening the vital interests of the people',[12] 'in case of exceptional danger',[13] and 'threatening the security and general welfare of the people'.[14] It was not easy to choose between these different expressions because concepts such as 'life of the nation' and 'interests of the people' are elusive, something which became obvious in the discussion which took place.[15] However, the proposal which finally won through was 'public emergency which threatens the life of the nation'.[16]

The term 'nation' was preferred to 'people' because the word 'people' might give rise to some doubt as to whether it denoted all the people or just a part of them.[17] This is because the term normally used to describe the state of emergency in the article was 'national emergency', which embraces all the people in the State and provides for the only justification for derogation. There was another reason for the adoption of this expression, namely, in order to follow closely the wording of article 15 of the European Convention which had been adopted in Rome in November 1950.[18]

This fact was to have an important effect as far as the final wording

---

[9] E/CN. 4/660, p. 10, para. 26 (the Co-ordinating Board of Jewish Organizations). See also the French proposal in E/CN. 4/365, p. 20.

[10] A/2929, p. 66, para. 39.

[11] E/CN. 4/319, p. 4 (proposed by the Soviet Union).

[12] E/CN. 4/365, p. 19 (Philippines); E/CN. 4/SR. 127, p. 12 (Lebanon).

[13] E/CN. 4/SR. 195, p, 9 (France).

[14] Ibid. P. 15 (Philippines).

[15] See E/CN. 4/SR. 330 (pp. 4–14) and 331. See also E/CN. 4/SR. 195–6.

[16] E/CN. 4/628, p. 2. This text was first proposed by the UK and its aim was to limit derogations to cases of grave emergencies threatening the life of the nation; the other texts seemed to the UK to be open to abuses.

[17] E/CN. 4/SR. 330, p. 14.

[18] E/CN. 4/528, Add. 1.

of the Covenant is concerned. Although it is true that the European Convention initially borrowed the derogation clause from the draft Covenant, it is equally true that in the final stage of the wording of the latter instrument, the drafters of the Covenant had in mind the then adopted text of the European Convention. The UK especially supported this position.

As far as the final wording of the European Convention is concerned, a very similar formula to that of the Covenant is found: 'in time of war or other public emergency which threatens the life of the nation'. The only difference is that it explicitly includes the case of war, which was suppressed in the Covenant at a late stage of the drafting for the reasons mentioned above.[19]

(b) The Wording of the American Convention[20]

Article 27(1) of the American Convention reads slightly differently from the other two treaties. The exact expression used to describe the emergency is 'in time of war, public danger, or other emergency that threatens the independence or security of a State Party'. The two main variations are these: (1) the explicit mention of the case of a *public danger*; (2) the kind of emergency envisaged is one which threatens *the independence or security of the State*.

The interpretation of the meaning of 'public danger' is clear in the light of an amendment made by El Salvador at the San José Conference; this amendment recognized that public calamity, such as flood and earthquake, was not necessarily a threat to internal or external security, but it occurs quite often and should justify the declaration of emergency.[21]

However, the most problematic variation is the substitution of the expression 'a threat to the life of the nation' for 'a threat to the independence or security of the State'. The crucial question touches upon the significance of this substitution. Prima facie it seems to be less restrictive than the kind of emergency envisaged in the other treaties.[22] In principle, the security of the State can be threatened by a small group of persons conducting unlawful activities. However, this does not necessarily amount to a threat to the life of the State. The attempt to make the meaning of this expression more precise has important practical consequences. In

---

[19] Council of Europe, *Problems Arising from the Coexistence of the UN Covenants on HR and the ECHR* (Strasbourg, 1970), doc. H(70) 7, p. 19.

[20] The American Convention was signed at San José (Costa Rica) on 22 Nov. 1969. It came into force in 1978. For the legislative history of the Convention, see T. Buergenthal and R. E. Norris (eds.), *The Inter-American System* (4 vols.; New York, 1984).

[21] Buergenthal and Norris (eds.), *The Inter-American System*, Minutes of the 14th session, 17 Nov. 1969, vol. i, booklet 12, p. 135.

[22] P. P. Camargo, 'The America Convention on Human Rights', *HRJ* 3 (1970), 356.

Latin American countries, many states of emergency have been declared for security reasons without any objective justification. This has produced gross violations of human rights.[23]

Unfortunately, the official records of the San José Conference do not give us any clear indication of the meaning of the expression. Nevertheless, what is clear is that the Conference established an international standard which requires that a situation be of a certain gravity in order to justify derogations from human rights; this is seen in the fact that the Conference rejected the Uruguayan and Chilean drafts which left the decision of the circumstances justifying the declaration of emergency entirely to the law of each state.[24] In addition, some of the documents presented to the Conference shed some light on the kind of emergency envisaged in the treaty.

The main working paper of the Conference was the IACHR draft. This draft, which contained a derogation clause in article 24, did not change during the Conference as far as paragraph 1 was concerned, the El Salvador amendment excepted.[25] However, two other documents are relevant, namely, Commissioner Martins' report and the 1968 IACHR Resolution on emergencies. In fact, when the IACHR submitted draft article 24 (which eventually became article 27) to the OAS Council, it suggested that the corresponding part of Commissioner Martins's report should be taken into consideration.[26] Martins's report was the result of a long study on the subject of human rights in emergencies and is considered to be an authoritative exposition of this question according to the IACHR. On the other hand, the 1968 Resolution is important because it was based on this report and its principles guided the drafting of the derogation clause; therefore, it is an important element in the interpretation of article 27.[27]

---

[23] For a brief commentary on the doctrine of national security in Latin America, see International Commission of Jurists, *States of Emergency: Their Impact on Human Rights* (Geneva, 1983), p. 416. For a wider study, see H. Montealegre, *La Seguridad del Estado y los Derechos Humanos* (Santiago de Chile, 1979).

[24] Draft Convention of Chile, in Buergenthal and Norris (eds.), *The Inter-American System*, p. 46, art. 22 (final 27); Draft Convention of Uruguay, in ibid., p. 70, art. 22 (final 27). See C. Grossman, 'Algunas consideraciones sobre el regimen de situaciones de emergencia bajo la Convencion Americana', in OAS, *Human Rights in the Americas* (Washington, 1984), p. 125.

[25] For the IACHR draft: 'Annotations on the Draft Inter-American Convention of H.R.' (document prepared by the Secretariat of the IACHR), OAS Ser. L/V/II. 19, doc. 53 (21 Mar. 69). See Buergenthal and Norris (ed.), *The Inter-American System*, vol. ii, booklet 13, pp. 25–82; for art. 24, see pp. 56–8.

[26] D. Martins, *The Protection of Human Rights in Connection with the Suspension of Guarantees or 'State of Siege'*, OAS Ser/L/V/II. 15, doc. 12 (11 Oct. 1966); repr. in OAS, *The Organization of American States and Human Rights: 1960–1967* (Washington, 1972), pp. 122–54.

[27] *Resolution on the Protection of H.R. in Connection with the Suspension of Guarantees or 'State of Siege'*, OAS Ser. L/V/II. 19, doc. 32 (16 May 1968). The history of the concern

In the light of these two documents, it seems that the concept of the emergency envisaged in the American Convention does not differ from that of the other two treaties. In fact, Martins' report includes in the concept only grave cases in which there is a threat to the integrity or existence of the three constitutive elements of the State, i.e. people, territory, and legal order.[28] On the other hand, the 1968 Resolution accepts the suspension of constitutional guarantees only 'when adopted in case of war or other serious public emergency threatening the life of the nation or the security of the State'.[29] The fact that the Resolution speaks about a serious threat and that it equates the security of the State with the life of the nation also confirms that the concept of emergency in the American Convention does not differ from that of the other two treaties. However, one should wait to see how this concept is constructed by the Inter-American organs in order to see if this impression is correct.

3. THE INTERPRETATION GIVEN TO THE CONCEPT OF EMERGENCY BY THE INTERNATIONAL BODIES ENTRUSTED WITH THE APPLICATION OF THE THREE TREATIES

(a) *The Jurisprudence Under the European Convention*

An interesting jurisprudence has been produced both by the Court and by the Commission on the derogation clause of the European Convention (art. 15).[30] The *Lawless case* and the *Greek case* are the two cases most relevant to this section because the central issue in these cases was the existence of the emergency envisaged in article 15(1).

In the *Lawless Case*,[31] the problem of the existence in Ireland in 1957

of the IACHR for the situations of human rights in states of emergency is a long one and covers almost a decade. At its 7th session in Oct. 1963, the first document prepared by the Secretariat entitled: *Preliminary Study of the States of Siege and the Protection of H.R. in American States* (OAS, Ser. L/V/II. 8, doc. 6) is found. In the same session Martins was elected Special Rapporteur. Commissioner Martins presented two reports, one in 1964 and a final one in 1967 (OAS, Ser. L/V/II. 15, doc. 12). For the history of this topic in the IACHR see OAS, Ser. L/V/II. 19, doc. 30, repr. in Buergenthal and Norris (eds.), *The Inter-American System*, vol. iv, booklet 23, p. 18.

[28] Martins, *The Protection of Human Rights*, p. 153.

[29] IACHR, 1968 Resolution on States of Emergency.

[30] The most important cases are: the *First Cyprus case* (1956–7); the *Lawless case* (1961); the *Greek case* (1969); the *Ireland* v. *UK case* (1978); the *Cyprus* v. *Turkey case* (three applications following the invasion of the island in 1974); and the *France . . .* v. *Turkey case* (1982).

[31] *Lawless case*, European Court of HR, Ser. A: Judgments (1 July 1961); Ser. B: Pleadings (1960–1).

of a public emergency threatening the life of the nation was discussed at some length. Mr Lawless was detained without trial between 13 July and 11 December 1957 in a military detention camp situated in the Republic. Lawless admitted that he had become a member of the IRA in January 1956, but said that he had left the organization in June 1956. Due to the terrorist activities of this organization, the legislature, from time to time, conferred upon the government special powers to deal with the situation; such powers occasionally included detention without trial.

On 8 July 1957, Ireland proclaimed a state of emergency and the special powers of detention without trial under Act N.2 1940 came into force. The proclamation of the emergency was due *inter alia* to an increase in IRA activities. This point was precisely one of the main issues in the case before the European Court of Human Rights. The proclamation of the emergency was challenged by Lawless before the Commission using the right of individual petition under article 25. The Court unanimously found that 'the Irish government were justified in declaring that there was a public emergency in the Republic of Ireland threatening the life of the nation' within the meaning of article 15(1).[32] This situation was

reasonably deduced by the Irish Government from a combination of several factors, namely, the existence in the territory of the Republic of Ireland of a secret army engaged in unconstitutional activities and using violence to attain its purpose; the fact that this army was also operating outside the territory of the State, thus seriously jeopardising the relations of the Republic with its neighbour; and thirdly 'the steady and alarming increase in terrorist activities from the autumn of 1956 and through the first half of 1957'.[33]

The unanimity of the Court contrasts with the different opinions within the Commission. The Court subscribed to the opinion of the majority. However, the view of the minority in the Commission was that the proclamation of a public emergency was not justified in Ireland in 1957.[34] Sir Humphrey Waldock, then President of the Commission, said in the hearings of 7 April that 'the differences between members of the Commission were primarily differences of degree and emphasis in appreciating the various elements in the case'.[35]

Although one should recognize the great difficulties in determining whether the facts of any particular situation fall within the concept of 'public emergency threatening the life of the nation', it seems that there were also differences in the way in which the various members of the

[32] Ibid., Ser. A: Judgment, para. 30.
[33] Ibid., para. 28.
[34] Dissenting opinions by Mr Eustathiades, Mr Susterhenn, Mr Dominedo, Mme Jansen-Pevtschin, and Mr Ermacora (Ser. B: Report of the Commission, pp. 90–120).
[35] Ibid., p. 358.

Commission construed the expression. The majority of the Commission held that the expression had 'a natural and ordinary meaning', and referred to 'a situation of exceptional and imminent danger or crisis affecting the general public, as distinct from particular groups, and constituting a threat to the organised life of the community which composes the State in question'.[36] The Court agreed with this construction.[37] The minority, however, emphasized that the emergency and the threat should be truly exceptional. One of its members, Mr Eustathiades, tried to show that according to the preparatory work 'the word "other" (in article 15(1)) shows that the threat to the existence of the nation must be of as exceptional a nature as war'.[38] Mr Susterhenn pointed out that in modern times 'war' means 'total war', so 'other public emergencies' should be close to 'total war'. Mr Dominedo pointed out that the emergency must be 'an extremely grave one and it must threaten the very life of the nation, in other words its very existence as such'.[39] In assessing the facts in Ireland in 1957, the minority thought that there was only a threat to public order or national security and by no means the kind of emergency envisaged in article 15(1) of the Convention.[40]

The majority rejected this construction of the concept of emergency as radical and extreme. They saw no grounds for interpreting the words 'or other public emergency threatening the life of the nation' any more strictly than is required by the natural and ordinary meaning of these words. According to a well-known rule of interpretation of the law of treaties, when the meaning is clear it is not necessary to resort to the preparatory work. Nevertheless, the majority also considered that the preparatory work of the Convention confirmed the natural and ordinary meaning of the words. Therefore, interpreting 'other public emergency' to the effect that it should amount to a 'war' is not required in the preparatory work and would amount, more or less, to a revision of the treaty.[41] The European Court in its finding on the merits agreed with this interpretation.

In conclusion, it seems as if the Court and the majority of the Commission in this case supported a workable and flexible standard of the concept of 'public emergency threatening the life of the nation', and not an extremely restrictive one.

The Greek case is also relevant on this point.[42] In 1967, the Military

---

[36] Ibid., p. 82, No. 90.
[37] Ser. A: para. 28.
[38] Ser. B: Report of the Commission, p. 92.
[39] Ibid., pp. 94 and 98.
[40] Ibid., p. 101. See the opinion of Mr Ermacora.
[41] Ibid., p. 81.
[42] *Greek case*, Report of the European Commission, *YBECHR* 12 (1969).

took power in Greece through a *coup d'état*. The legitimate government was replaced and the army established a 'Junta' to rule the country. By a Royal Decree, they declared martial law and suspended many articles of the constitution and placed opposition leaders in prison. Greece, according to article 15(3), notified the Secretary-General of the Council of Europe of the proclamation of the emergency and the derogation of certain rights.[43] Denmark, the Netherlands, Norway, and Sweden brought an application according to article 24 of the ECHR against these derogations. The Commission's report found that Greece had violated article 3 (torture and ill-treatment) and nine other articles of the Convention. The Committee of Ministers confirmed the Commission's report.

The Greek Government contended that the suspension of human rights was due to the existence of a public emergency threatening the life of the nation. The Commission, in order to ascertain the existence of the emergency according to article 15(1), divided the facts into two different periods of time, those that took place on 21 April 1967, the day of the *coup d'état*, and those that took place from that date onwards.

The Commission took a very strong stand, declaring that the burden of proof lay on the respondent government, even though they possessed a measure of discretion in assessing the situation. It seems as if the Commission applied a stronger standard of proof in this case than in the *Lawless case*. In the latter, the presumption was in favour of the government and the margin of appreciation quite broad. In the *Greek case*, the Commission defended the government's entitlement to rely on the right of derogation and acknowledged that it enjoyed a margin of appreciation. However, in practice, it adopted what can be called an 'objective test' by demanding strong proof of the actual emergency and by considerably reducing the margin of appreciation. This was one of the main reasons for the five dissenting opinions in this case.[44]

As far as the situation on 21 April 1967 was concerned, the Commission accepted that the government, in order to prove the existence of the emergency, could base its evidence on the situation that existed before that date. The government adduced three main factors which constituted a threat to the life of the nation, namely, the threat of a Communist take-over of the government by force, the state of public order, and the constitutional crisis. The Commission found it established beyond dispute that following the political crisis of July 1965, there had been a period of political instability and tension in Greece. In addition, there had been an expansion of the activities of the Communists and their

---

[43] *YBECHR* 10 (1967).

[44] *Greek case*, dissenting opinions of Mr Delahaye (pp. 76–82); Mr Eustathiades (pp. 82–6); Mr Susterhenn (pp. 87–92); Mr Ermacora (pp. 102–3); Mr Busuttil (pp. 113–18).

allies as well as some public disorder. Furthermore, these three factors were closely linked. However, the majority of the Commission concluded that 'the respondent government has not satisfied the Commission by the evidence it has adduced that there was on 21st April a public emergency threatening the life of the Greek nation'.[45] It is interesting to note that the Commission seemed to judge the situation facing the defeated constitutional government rather than the situation faced by the revolutionary one. The assessment that the situation on 21 April 1967 was not grave enough to amount to a public emergency threatening the life of the nation does not imply that there was no increased threat of Communist subversion and general disorder after the *coup d'état* by the right-wing military forces. The Commission did not consider this important factor.

The second period considered by the Commission was from 21 April onwards. The Greek Government added two new factors in order to justify the maintenance of the state of emergency. These were the bomb incidents, acts of sabotage, and the activities and formation of a number of illegal organizations. Nevertheless, the Commission did not find that these factors 'were beyond the control of the public authorities using normal measures, or that they were on a scale threatening the life of the Greek nation'.[46]

(b)   The Practice of the UN Human Rights Committee

Leaving aside the difficulties faced by the HR Committee in monitoring emergencies, due in part to the nature of its procedures and in part to the lack of collaboration from some governments, the Committee has expressed its views on the concept of emergency and in some particular cases has given its opinion on the existence of a true emergency both in reviewing States' reports and under the Optional Protocol.[47]

The declaration of a public emergency and the derogation of human rights are perfectly lawful under the Covenant provided they meet the conditions of article 4. The first condition, as the HR Committee pointed out in its 'general comment' on article 4, is precisely that the emergency which justifies derogating measures must be of an exceptional nature[48] in other words, not every emergency, even if legal under municipal law, is legitimate under the Covenant; it must attain a certain degree of gravity.

According to this position, members of the Committee, when reviewing

---

[45] Ibid., p. 76, para. 165.
[46] Ibid., p. 100, para. 207.
[47] For an analysis of the functions, competences, and difficulties that the HR Committee faces, see Ch. 2.
[48] CCPR/C/21/Add. 2, p. 2.

States' reports, have expressed the view that in some cases that degree of gravity had not been attained. Thus, the state of emergency declared in Chile was considered not to be justified by the circumstances. In reviewing the *1981 Chilean Report*, several members of the Committee noticed that the Government of Chile was continuing 'without any objective justification to apply measures intended for exceptional conditions of internal unrest'.[49] Similarly, when Chile presented its second report in 1985, the majority of the Committee found that at that time the emergency was not justified under article 4. In a rather strong statement, several members of the Committee said that 'what was called an emergency in Chile had nothing to do with what was intended by the same term in article 4'.[50] On this occasion, Chile adduced the existence of terrorist activities and the world economic situation, which was having an adverse effect on the country, as justifications for the state of emergency.[51] Chile did not send either the notice of derogation to the Secretary-General according to article 4(3), or the supplementary information required by the Committee since its first report in 1979. At the beginning of the meeting, the Committee expressed a strong protest against Chile's failure to comply with these requirements.[52]

Under the Optional Protocol mechanism, the Committee has established the sound policy that it is not enough to claim the existence of an emergency if the real extent of such an exceptional threat is not shown through the presentation of sufficient information and facts. This is why Uruguay has been found to be in violation of the Covenant in several cases. In the *Grille Motta case*, for instance, the Committee stated that article 4 'does not allow national measures derogating from any of its provisions except in strictly defined circumstances, and the Government has not made any submissions of fact or law to justify such derogations'.[53] In the *Landinelli case*, the Committee said that

Although the sovereign right of a State party to declare a state of emergency is not questioned . . . the HR Committee is of the opinion that a State, by merely invoking the existence of exceptional circumstances, cannot evade the obligations which it has undertaken by ratifying the Covenant.[54]

A second element that the Committee has underlined in its general comment on article 4 and in its practice is that the concept of 'public

---

[49] UNGAOR. A/34/40, Report of the HR Committee (1981), para. 78.
[50] UNGAOR. A/39/40, Report of the HR Committee (1984), p. 83, para. 449. See also CCPR/C/SR. 356, Report of Uruguay, para. 226.
[51] A/39/40, p. 84, para. 444.
[52] Ibid., p. 78, para. 437.
[53] Communication No. R. 2/11, Adoption of Views: 29 July 1980, in A/35/40, p. 136, para. 15.
[54] Communication. No. R 8/34, Adoption of Views: 8 Apr. 1981, in A/36/40, pp. 132–3.

emergency' has a temporary nature. As is well known, one of the greatest problems in situations of emergency is the permanent character that derogations assume in some countries; in other words, the maintenance of the derogating measures for a protracted period even though the emergency has ended. This *de facto* phenomenon has been called the 'institutionalization' of states of emergency.[55] Thus, for instance, in the revision of the Uruguayan report, members of the Committee stressed that 'the letter and the spirit of article 4 of the Covenant stipulated that when a country took measures to suppress human rights, those measures . . . must be *temporary*, they could not be institutionalised'.[56] Therefore, theoretically, once the exceptional circumstances have disappeared, the limitations clauses in force in normal times, and not the derogation clause, should be sufficient to deal with problems of public order. Colombia referred to this same temporary requirement of emergencies when it presented its report to the Committee; thus, it pointed out that the state of emergency at that time was different to the previous ones because it was 'legal, transitory, and limited in scope'.[57]

Among the various reasons for declaring states of emergency under the Covenant, States have mentioned the following: terrorist activities and economic crises (Chile), terrorism (Colombia, Uruguay, Peru . . .), natural disasters (Peru), external aggression (Nicaragua), the state of war in the Middle East and the refugee problem (Jordan), subversion (Uruguay), etc.

(c)   The Practice of the IACHR

The IACHR, as the principal organ of the OAS entrusted with the promotion of human rights in the region, has dealt with emergency situations on several occasions. The Commission has the function of promoting respect for human rights and monitoring the fulfilment of international obligations by the member States in this field. For States parties to the Convention, this treaty represents a detailed explanation of their legal obligation. In this respect, article 27 establishes clear standards applicable in states of emergency. For States which are not parties to the Convention, the main instrument applicable is the 1948 American Declaration on the Rights and Duties of Man.[58]

The Declaration is a more generic instrument, its style is similar to the UDHR, and it does not contain any specific derogation clause. It has only

---

[55] E/CN. 4/Sub. 2/1982/15, p. 31, paras. 129–45.
[56] See e.g. CCPR/C/SR. 356, p. 2 (Mr Dieye) (emphasis added). See also A/37/40 (1982), p. 58, para. 270.
[57] A/35/40 (1980), p. 59, para. 263.
[58] For the analysis of the functions of the IACHR, see Ch. 2.

a general limitation clause (article 28). The question of what principles and standards are applicable to States non parties to the Convention and the legal bases for that application will be examined in the final chapter of this study. In this section, and in the following chapters, the doctrine of the IACHR will be examined, first of all, in respect of those States parties to the Convention and therefore bound by the derogation clause as treaty law. Moreover, reference will be made to those cases in which the same principles of the derogation clause are applied by the IACHR to other States who are members of the OAS but not parties to the Convention; in this case, the principles of the derogation clause are applied outside treaty law.

(i) *The Doctrine under the American Convention.* The interpretation given by the Inter-American Commission to the concept of 'public emergency that threatens the independence and security of the State' does not differ from the interpretation given by the other international bodies under the European and the UN systems.

In its doctrine on states of emergency, in which the Commission summarizes its experience and the principles applicable to those situations, the Commission reaffirms the standards of article 27, namely, that derogations of human rights are only legitimate in exceptional circumstances and for a temporary period of time. A State of emergency was described as 'an institution essentially transitory in nature'.[59] In a recent advisory opinion, the Inter-American Court held the same position.[60] These exceptional circumstances, which can be tentatively described as political and social struggles, serious disturbances of public order, internal commotion and external attack, war and public disasters, should in any case represent a 'real threat to the public or to the security of the State'.

The IACHR has also noticed the abuses in the application of derogations: in practice in many instances these states of emergency have been enacted without the circumstances warranting it, as a simple means of increasing the margin of discretion of the exercise of public power'.[61] Accordingly, in the *Report on Bolivia* of 1981, and referring to the military *coup d'état*, the Commission found a violation of the principle of exceptional threat:

the Bolivian authorities exceed the limits of state action by disregarding the restrictions on the use of such measures stipulated in the American Convention both with respect to the gravity of the situation and the period of time such measures should remain in effect.[62]

---

[59] OAS, *The IACHR: Ten Years of Activities* (Washington, 1982), pp. 336–9.
[60] IA Court HR, advisory opinion, 'Habeas Corpus in Emergency Situations (articles 27(2), 25(1) and 7(6) of the ACHR)' (30 June 1987), para. 19.
[61] Ibid., p. 338.
[62] IACHR, *Report on Bolivia*, 1981, p. 23.

One of the problems often found in the IACHR reports is the lack of a thorough analysis of the application of the principles of the derogation clause; this is mainly due to the fact-finding nature of the Commission. This dearth of analysis makes a legal assessment of the reports problematic. However, a good legal analysis concerning the existence of the emergency as envisaged in article 27 is found in the IACHR *Report* on the situation of the Miskito population in Nicaragua in 1983.[63]

In one of its sections, the Commission tries to establish whether the compulsory relocation of Miskito Indians in January 1982 from the communities on the Coco River to the five camps in Tasba Pri was legally justified by the fact that there was an emergency in Nicaragua at that time. The right derogated from in this case was the right to residence and movement recognized in article 22 of the American Convention. After mentioning the jurisprudence of the European Commission and Court and the doctrine of the UNHR Committee, the IACHR considered, 'in interpreting the first part of paragraph 1 of article 27, that the emergency should be of a serious nature, created by an exceptional situation that truly represents a threat to the organised life of the State'.[64] In concluding its thorough analysis of the situation in that area, 'the Commission found that the security of the Nicaraguan State was truly threatened by the incursions of the groups of former members of the National Guard, which justified the declaration of a state of emergency and its maintenance'. Furthermore, 'the ongoing penetration of these armed groups into Nicaragua demonstrates that there was a real and imminent threat to the security of the State'.[65] Applying the criterion of the essentially temporary nature of these exceptional measures, the Commission established that 'once the danger that threatens the security of the State has been overcome, the special provisions should also be terminated'.[66]

A good example of the failure to analyse in depth the existence of a public emergency within the meaning of article 27(1) can be found in the 1981 *Report on Nicaragua*.[67] The reason given by Nicaragua in its notice of derogation on 23 January 1980 was that, as a consequence of the civil war, the country found itself in an abnormal situation. The end of Somoza's dictatorship provoked the imprisonment of 7,500 guards

---

[63] IACHR, *Report on the Situation of Human Rights of a Segment of the Nicaraguan Population of Miskito Origin*, OAS Ser. L/V/II. 62, doc. 10, rev. 3 (29 Nov. 1983), p. 113.
[64] Ibid., p. 115, para. 8.
[65] Ibid., p. 116, para. 12.
[66] Ibid., p. 117, para. 14.
[67] IACHR, *Report on Nicaragua*, 1981, OAS, Ser. L/V/II. 53, doc. 25. Farer has pointed out that the IACHR has consistently ignored 'the question whether the *initial* declaration of a state of exception was justified'. T. Farer, 'Elections, Democracy and Human Rights: Toward Union', *HR Quarterly*, 11 (1989) 508.

accused of committing various crimes against the Nicaraguan people. The notice of derogation finished by saying: 'when this type of situation exists, it is impossible to maintain a state of juridical normalcy, owing to the very nature of the events'.[68] Those prosecuted as Somocistas were deprived of the rights and judicial guarantees mentioned in the notice.

Although the Commission took a sympathetic stance on the difficulties faced by the Nicaraguan authorities in their attempt to reconstruct the nation and respect human rights after the civil war, it failed to analyse rigorously the existence in Nicaragua of the emergency envisaged in article 27 of the Convention. What the Commission does in the *Report* is simply to study the different rights as formulated in the Convention and ascertain the violations and inaccuracies found in Nicaraguan legislation and administrative practice. It seems to be an implicit assumption that there was an irregular situation in the country which justified certain limitations on the enjoyment of some human rights, but without an explicit and strict application of the doctrine of the states of emergency with its requisites set forth in article 27(1).

Similarly, in the 1981 *Report on Colombia*, even though the Commission sharply criticized the past use of the state of emergency by the Colombian authorities, it failed to give a clear opinion on the actual existence of a public emergency in Colombia at that time.[69] Colombia did not notify the Secretary-General of OAS of the emergency according to article 27(3). Yet Colombia had been living in a state of emergency since 1948, changing its temporary nature into an almost permanent system aimed at combating political and common violence in rural areas and, in recent years, in urban parts of the country as well. This systematic maintenance of the state of emergency gave way to a system of exception, the indefinite duration of which affects the institutional functioning of the Colombian state of law.[70]

Moreover, the establishment of the state of emergency was often unwarranted by the circumstances. This is why the Commission recommended that the Colombian government should apply the derogation foreseen in article 121 of its Constitution 'only in exceptionally serious cases'. Referring to the then present situation in Colombia, the Commission in a rather cryptic statement recommended that the country should 'lift the state of emergency as soon as circumstances allow; and comply with the provisions of article 27 of the American Convention'. This recommendation admits different interpretations on the actual existence of a public emergency and on the legal justification for its

---

[68] See T. Buergenthal, R. E. Norris, and D. Shelton (eds.), *Protecting Human Rights in the Americas: Selected Problems* (3rd edn., Kehl–Strasbourg, 1990), pp. 349–50.

[69] IACHR, *Report on Colombia*, 1981, OAS Ser. L/V/II. 53, doc. 22.

[70] Ibid., p. 40, para. 2.

declaration.[71] The first part of the sentence seems to accept that the circumstances justified the declaration, but the second part does not specify which actual principles of article 27 the Colombian Government had not complied with. It might simply refer to the formal requirement of notification, or perhaps, more plausibly, to the substantial requirements of the principles of exceptional threat and temporary nature of the emergency.

(ii) *The Principles Applicable to States Non-Parties to the Convention.* The same principles governing the assessment of the existence of a public emergency within the meaning of article 27, namely, that derogations are only legitimate in exceptional situations and for a limited period of time, were applied by the IACHR to States non-parties to the Convention. In several cases, the IACHR has pointed out that derogations from human rights standards were not justified because the disturbances did not amount to a public emergency threatening the independence of the State,[72] or because the state of emergency was extended for a long period of time, once the emergency had ended.[73]

In the 1985 *Report on Chile*, the Commission confirmed that international standards required that the suspension of rights must be due to extremely serious situations. It pointed to a discrepancy between this international standard and the 1980 Chilean Constitution. This was because transitory provision 24 'does not require that there be an actual disturbance of the internal peace but only that there be a *danger* of that occurring'. Nor does it requrie that 'a disturbance of the public order has occurred in practice as a result of acts of violence but only that the acts of violence that occurred were *intended* to cause that disturbance'.[74] This threshold falls short of the applicable international standards which require the actual and real existence of the emergency. The Commission also considered that the permanent state of emergency in Chile since 1973 contradicted the necessarily temporary nature of this institution.[75]

Two main conclusions can be drawn from the jurisprudence of the IACHR on the concept of public emergency as found in article 27 of the Convention.

[71] See Buergenthal, Norris, and Shelton (eds.), *Protecting Human Rights*, p. 234.
[72] IACHR, *Report on Chile*, 1974, OAS Ser. L/V/II. 34, doc. 21; IACHR, *Report on Paraguay*, 1987, pp. 15–22; IACHR, *Report on Argentina*, 1980, OAS Ser. L/V/II. 49, doc. 19, corr. 1, p. 22.
[73] IACHR, *Report on Paraguay*, 1978, OAS Ser. L/V/II. 43, doc. 13, p. 14, paras. 10 and 23; see also p. 15 and p. 88, para. 1.
[74] IACHR, *Report on Chile*, 1985, p. 44, para. 95 (emphasis added).
[75] Ibid., p. 45, para. 99. See also the 1983 *Report on Suriname* in which the Commission also found an inconsistency between the Suriname Constitution and the standards of the derogation clause as far as the principle of exceptional threat is concerned. See para. 80 of the Report.

1. The interpretation given to the expression 'public emergency which threatens the independence or security of the State', which provoked some anxiety because of its possible low threshold of application, has in fact been interpreted by the IACHR in a way very similar to the construction of the concept of 'public emergency threatening the life of the nation' within the European and UN systems. According to this interpretation derogations from human rights standards are only permissible in cases of extremely serious situations and for a limited period of time.

2. This principle of exceptional threat has also been applied by the IACHR in respect of States non-parties to the American Convention.

4. THE MAIN FEATURES OF THE EMERGENCY ENVISAGED IN THE TREATIES

The main characteristics of the concept of public emergency in the three treaties, as interpreted by the international bodies entrusted with its application,[76] can be summarized as follows:

(a) The Emergency Must be Actual or at Least Imminent

Under international law, so-called states of emergency of a 'preventive nature' are not lawful. In other words, it is not permissible for States to derogate from human rights in order to face possible exceptional situations which have not yet arisen. The emergency, therefore, must be present or at least imminent.

The possibility of declaring a 'public emergency' when it is 'imminent' but not yet present was recognized for the first time in the *Greek case*. The Commission quoted the definition of public emergency given by the Court in the *Lawless case* in its French version: 'une situation de crise ou de danger public exceptionnel et imminent'. The French version was given special weight because it was considered to be the authentic one.[77] However, the absence of convincing proof of the existence of an imminent danger was the main reason for the Commission's rejection of the claim of the Greek Government, in spite of the existence of social unrest provoked by the Communist Party, which wanted to take power; the Commission said:

---

[76] In the *Greek case*, the European Commission made a good summary of the characteristics of a public emergency, Report of the ECHR, *YBECHR* 12 (1969), 72, para. 153. See also T. Stein, Derogations from Guarantees Laid down in Human Rights Instruments, in I. Maier (ed.), *Protection of Human Rights in Europe*, Proceedings of the 5th International Colloquy about the ECHR, Frankfurt, 1980 (Heidelberg, 1982), pp. 123–33.

[77] *Greek case*, p. 71, para. 152.

The Commission has not found that the evidence adduced by the respondent Government shows that a displacement of the lawful Government by force of arms by the Communists and their allies was *imminent* on 21st April 1967.[78]

On the other hand, the European Court in the *Lawless case* found that the declaration of emergency made by the Irish government was lawful according to article 15(1), taking into account the 'imminent danger to the nation caused by the continuance of unlawful activities in Northern Ireland by the IRA and various associated groups, operating from the territory of the Republic'.[79]

(b)   Its Effects Must Involve the Whole Population[80]

In the preparatory work of the UNCCPR, the main concern of the UN Commission of Human Rights was to provide a qualification of the kind of emergency not open to abuse, so that a public emergency should be of such a magnitude as to threaten the life of the nation as a whole.[81] The question can arise whether it is possible according to the derogation clauses to declare a state of emergency only in one part of the territory. In this respect two different situations can be envisaged:

an emergency in one part of the territory but affecting the whole nation
an emergency in one part of the territory and affecting only that part of the nation.

The first situation presents no major legal problems in the present context. One can think of terrorist activities that take place mainly in one area of the country but which have an influence on the whole nation, or of an extremely serious natural disaster affecting the national economy. The finding of the European Commission and Court in the *Ireland* v. *UK* case would support this position, even if the issue was not discussed in detail because Ireland did not contest it.[82] Mme Questiaux, when interpreting this characteristic, said that the emergency must affect 'the whole nation and either the whole of the territory or certain parts thereof'.[83]

The second situation could be more controversial. One can think of a public emergency in a localized area of the country and affecting only the

---

[78] Ibid., p. 73 (emphasis added). See above, pp. 18–20.

[79] *Lawless case*, European Court of HR, Ser. A: Judgment, para. 29. Another good example of the application of this feature can be found in the IACHR *Report on Chile*, 1985, explained above, p. 26.

[80] *Lawless case*, para. 28 and the *Greek case*, para. 153.

[81] UNGAOR, 10th Session, Annexes, agenda item 28, pt. II, doc. A/2929, Report of the Secretary-General on the ICCPR Draft, para. 39.

[82] *Ireland* v. *UK*, European Court of HR, Ser. A: Judgment (1978), vol. 25; Ser. B: Pleadings, vol. 23. I.

[83] N. Questiaux, *Study of the Implications for Human Rights of Recent Developments concerning Situations Known as State of Siege or Emergency*, E/CN. 4/Sub. 2/1982/15, p. 15.

population living there. Even if nowadays, due to the close inter-
dependence of the different areas of a nation, it is hardly conceivable that
a grave emergency would not affect the whole nation, one may think, for
instance, of grave disturbances of public order taking place in a dependent territory of a State which do not affect the nation as a whole. The
numerous notices of derogation made by the UK under the ECHR from
1955 onwards are good examples of that situation.[84] No State contested
the lawfulness of this practice, even if it is hard to accept that, for
example, the emergency in British Guiana in 1955, 'owing to a dangerous
crisis in public order and in the economic life of the territory' affected the
whole of the UK population.[85] Nevertheless, in the ILA *Paris Report*
(1984), an emergency in a part of a territory and affecting only the
population established there is also accepted as a legitimate emergency
situation.[86]

(c)   The Threat Must be to the Very Existence of the Nation

This is understood as a threat to the organized life of the community
constituting the basis of the State. This threat could be to the physical
integrity of the population, to the territorial integrity, or to the functioning of the organs of the State. This test has been applied by the international bodies since the *Lawless case*.[87]

The last point of the test was controversial in the *Lawless case*.
Professor Ermacora construed the emergency required under article 15(1)
in a rather strict way, saying that when the organs of the State are
functioning normally, as was the case in Ireland, there is no grave threat
to the life of the nation, and therefore emergency measures are not
legitimate.[88] However, the majority in the Commission and in the Court
found that, because of the obstacles for the normal operation of the
tribunals in cases concerning IRA terrorists, preventive detention, with
due guarantees, was in fact a legitimate measure in order to overcome the
imminent danger.

(d)   The Declaration of Emergency Must be a Last Resort

The exceptional character of the derogations requires that those measures
only be taken when the normal measures to deal with minor infractions of

---

[84] R. Higgins, 'Derogations under Human Rights Treaties', *BYBIL* 48 (1976–7), 289–90
n. 5.

[85] *YBECHR* 1 (1955–7), 48.

[86] ILA Paris Report (1984), p. 58.

[87] See also: 'The Siracusa Principles on the Limitation and Derogation Provisions in the
ICCPR', 7 *HR Quarterly* (1985), principle no. 39.

[88] *Lawless case*, Ser. B: Pleadings, Report of the Commission, opinion of Mr Ermacora,
pp. 101–2.

public order have been exhausted and are not sufficient to deal with the threat. This is in line with the nature of the institution.

(e)   The Declaration of Emergency as a Temporary Measure

The institution of states of emergency is by its very nature temporary, and therefore derogations must end when the threat has disappeared. State practice shows that most constitutions authorize the declaration of a state of emergency for a well-defined period of time, which can only be extended according to constitutional procedures. This is why, especially under the Inter-American system, the so-called permanent states of emergency have been declared unlawful. Permanent states of emergency are those which are perpetuated, with or without proclamation, either as a result of *de facto* systematic extension or because the Constitution has not provided any time limit a priori.[89] The 1978 *Report on Paraguay* by the IACHR contains a good illustration of this kind of state of emergency.[90]

5.   CIRCUMSTANCES IN WHICH A PUBLIC EMERGENCY CAN BE DECLARED

The circumstances which can provoke the proclamation of a state of emergency and derogations of human rights are numerous. According to one report by the UN Commission on Human Rights which summarizes the circumstances in municipal law, the following are relevant:

International conflict, war, invasion, defence or security of the State or parts of the country; civil war, rebellion, insurrection, subversion, or harmful activities of counter-revolutionary elements; disturbances of peace, public order or safety; danger to the constitution and authorities created by it; natural or public calamity or disaster; danger to the economic life of the country or parts of it; maintenance of essential supplies and services for the community.[91]

To attempt a detailed study of these circumstances would be an extremely lengthy process. To illustrate the difficulties involved in such a study it is useful to look at the experience of the International Law Association (ILA). The ILA Committee, in its study on states of emergency in the first report to the Montreal Conference, tried to envisage such situations.[92]

---

[89] Questiaux, *Study*, p. 28.

[90] IACHR, *Report on the Situation of Human Rights in Paraguay* (1978), OAS Ser. L/V/II. 43, doc. 13, pp. 14–15 and p. 88, para. 1.

[91] UN Commission on HR, *Study of the Right of Everyone to be Free from Arbitrary Arrest, Detention and Exile*, E/CN. 4/826 (1962), p. 257. Another list can be found in D. O'Donnell, 'States of Exception', *Int. CJ Review* 21 (1978), 54. See also Martins, *The Protection of Human Rights*, p. 153.

[92] ILA Montreal Report (1982), p. 91, para. 15.

In its comments, added to the final text approved in the Paris Conference of 1984, the ILA said:

It is neither desirable nor possible to stipulate *in abstracto* what particular type or types of events will automatically constitute a public emergency within the meaning of the term; each case has to be judged on its own merits taking into account the overriding concern for the continuance of a democratic society.[93]

This recommendation of *ad hoc* definition seems sound, since it would be impossible to foresee *in abstracto* all possible types of emergency. Moreover, this has been the standard used by the drafters of the derogation clauses, who have preferred to put the emphasis on the gravity of the circumstances rather than to list all of them. During the preparatory work of the ICCPR, the proposal of the French Government and that of the Jewish World Organization, which contained a more descriptive definition of the circumstances, were rejected.[94] In any case, in the light of the circumstances mentioned in municipal systems and in the literature on derogations, a possible classification of situations justifying derogations can be made in principle:[95]

1. Political crises: war (international war, civil war, war of national liberation),[96] internal unrest, grave threats to public order or subversion.
2. Public or natural disasters.[97]
3. Economic crises.[98]

---

[93] ILA Paris Report (1984), p. 59, para. 1. The same attitude is in ST/TAO/HR/29, *Seminar on the Effective Realization of Civil and Political Rights at the National Level* (Kingston, 1967), p. 38 (hereafter 'Kingston Seminar').

[94] E/CN. 4/365, p. 20 and E/CN. 4/660, p. 10, para. 26.

[95] *Inter alia*: S. Marks, 'Principles and Norms of Human Rights Applicable in Emergency Situations', in K. Vasak (ed.), *International Dimensions of Human Rights* (2 vols.; Westport, Conn., 1982), i. 175–213. See also Questiaux, *Study*, and the 'Kingston Seminar'.

[96] For the relations between humanitarian law and human rights, see *Human Rights as the Basis of International Humanitarian Law*, Proceedings of the International Conference on Humanitarian Law (San Remo, 1970), esp. the reports by Robertson, Draper, and McBride. Also A. Calogeropoulos-Stratis, *Droit humanitaire et droits de l'homme* (Geneva, 1980); M. El Kouhene, *Les Garanties fondamentales de la personne en droit humanitaire et droits de l'homme* (The Hague, 1985). See also T. Meron, *Human Rights in Internal Strife* (Cambridge, 1987); Meron, 'Human Rights in Time of Peace and in Time of Armed Strife: Selected Problems', in T. Buergenthal (ed.), *Contemporary Issues in International Law: Essays in Honour of L. B. Sohn* (Kehl, 1984).

[97] See the *travaux préparatoires* of the American Convention, above, pp. 16–17, and those of the Covenant, E/CN. 4/365, p. 20 (French proposal) and E/CN. 4/SR. 196, p. 7 (remarks by Mr Valenzuela of Chile). See also the remarks of Mr Prado Vallejo, a member of the HR Committee, who seems to accept this circumstance in reviewing the Peruvian report. CCPR/C/SR. 430, p. 7, UN HR Committee, *Report on Peru*, 1983, A/38/40, p. 66.

[98] See E/CN. 4/826, p. 257. See also ILA Montreal Report (1982), p. 90. For a good summary of the emergency powers to assist in the maintenance of essential supplies and services in the UK legislation, consult: E. Bonner, *Emergency Powers in Peacetime* (London, 1985), pp. 211–70. See also M. Supperstone, *Brownlie's Law of Public Order and National Security* (London, 1981), pp. 216–17.

6. CONCLUSION

1. The kind of emergency which justifies derogations from human rights standards is defined by the treaties under consideration as a 'public emergency threatening the life of the nation'. The main aim of the drafters of the Covenant, which was the inspiration of the derogation clause of the ECHR and the ACHR, was to provide for a qualification for the public emergency which would not be open to abuses by States, rather than to list the concrete circumstances which would justify derogations. The qualification of the emergency that was approved ('threatening the life of the nation') means that the only emergency justifying derogations is an exceptional one which affects the whole nation. Therefore, less grave emergencies, even if accepted in municipal law, would not qualify for derogations under the treaties. The jurisprudence of the international bodies entrusted with the application of the treaties has clarified this concept of emergency.

2. Thus, in the *Lawless case* the European Court and Commission of Human Rights interpreted the concept as 'a situation of exceptional and imminent danger or crisis affecting the general public, as distinct from particular groups and constituting a threat to the organised life of the community which composes the State in question'. This construction of the concept constitutes the constant jurisprudence of the European organs and has been applied in all cases dealing with emergencies. When States have not given enough evidence to prove that an exceptional situation existed in the country, the European organs have not accepted the application of the right to derogate (i.e. in the *Greek case*).

3. The UN HR Committee has also considered that the public emergency justifying derogations must be one of an exceptional character and can only last as long as the life of the nation is actually threatened. In reviewing States' reports, members of the Committee have expressed their concern for the use of derogations when the circumstances were not of an exceptional nature, or when the derogations lasted longer than the emergency itself. Moreover, under the Optional Protocol, the Committee has not accepted the mere contention that there was a state of emergency, without supporting evidence; in the absence of that evidence, human rights standards cannot be derogated from.

4. The IACHR has construed the concept of emergency in a very similar way to that of the other bodies, even though the wording referring to the kind of emergency justifying derogations in the American Convention is slightly different ('threatening the independence or security of a State'); this wording could in theory have been interpreted as justifying derogations in less grave emergencies. Furthermore, the IACHR has applied this principle of exceptional threat outside treaty law, namely, to

those States which were not parties to the American Convention but were members of the OAS.

5. From the jurisprudence of these international bodies, the main characteristics of the emergency envisaged in the treaties can be summarized:

the emergency must be actual or at least imminent; therefore an emergency of 'a preventive nature' is not lawful;

the emergency should be of such a magnitude as to affect the whole of the nation, and not just a part of it;

the threat must be to the very existence of the nation, this being understood as a threat to the physical integrity of the population, to the territorial integrity, or to the functioning of the organs of the State;

the declaration of emergency must be used as a last resort once the normal measures used to deal with public order disturbances have been exhausted;

the declaration of emergency is a temporary measure which cannot last longer than the emergency itself; therefore, the so-called 'permanent states of emergency' are not lawful.

6. As an illustration of the kind of concrete circumstances which (if sufficiently serious) can in principle provoke the declaration of emergency and derogations from human rights standards, the following can be mentioned: (1) political crises, such as international or civil wars, internal unrest, grave threats to public order (these are the most common grounds for the use of the derogation clause); (2) public or natural disasters; (3) economic crises.

2

The Principle of Proclamation of The State of Emergency

1. INTRODUCTION

The derogation clause in the UN Covenant on Civil and Political Rights (ICCPR) establishes as a prerequisite for its valid operation the 'official proclamation' of the state of emergency (article 4(1)). In this chapter, the problems related to this requirement will be dealt with. First, there is the rationale behind this requirment of an 'official proclamation' under the Covenant, and its status under the other two main treaties on human rights which do not explicitly contain that requirement. Secondly, there is the issue of the organ competent to declare the emergency, i.e. the legislature and the executive, and the possible judicial control of the declaration by domestic courts. Although human rights treaties have not normally addressed these problems, several seminars organized by the UN and other international organizations have discussed them and some standards have been formulated. Finally, there is the problem of the international control of human rights in states of emergency by the organs created by the three main treaties. One of the first questions which arises when emergencies are in operation is precisely the competence of those organs to review such a sensitive matter which is largely related to the domestic jurisdiction of states. Due to the different nature of those organs, this problem will be studied in turn according to how they are dealt with under the European Convention, the UN Covenant, and the American Convention.

(a)   The Requirement of Official Proclamation in the Covenant

The act of proclamation (or declaration) of the state of emergency is basically an internal act of the State. It usually involves a decision taken by the political organs of the State (the executive and the legislature),[1] and it comes into operation only in cases of grave danger to the State. Municipal systems have established strict conditions, both substantive and procedural, for the declaration of emergency. This declaration is an important decision in the life of the State, and in some cases provokes not

[1] See pp. 38–40.

only derogations from human rights standards but also a certain alteration in the distribution of functions and powers among the different organs of the State.[2] In these situations, a formal declaration of emergency containing a clear account of all the exceptional measures taken provides an important element of publicity for those under the State's jurisdiction who would require to know the exact extent of the limitations of their rights and the alteration in the distribution of powers among the organs of the State; without this declaration, those under the jurisdiction of the State would not know the exact extent of their rights.[3] Although the act of proclamation of a state of emergency is prima facie an internal act of the State, it has been included in the derogation clause of the Covenant as a requirement. The reason for this incorporation appears in the *travaux préparatoires*.

It was, in fact, introduced by France at a late stage of the draft Covenant and was considered an essential requirement in order to prevent States from derogating arbitrarily from their obligations where such an action was not justified by events. Reference was made to the fact that in most countries a public emergency could be declared only under conditions defined by law, and that such guarantee would be lost unless a requirement of public proclamation was included. One of the aims of this requirement was to reduce the incidence of *de facto* states of emergency, obliging States to fulfil their obligations under municipal law.[4] Thus it seems that the main purpose of the drafters of the Covenant was to oblige States to comply with municipal systems, which usually contain clear regulations on the declaration of the emergency. Some of these regulations refer to *substantive* elements: for instance, that the state of emergency cannot be declared except in exceptional cases which involve a grave danger to the State; and some regulations refer to *procedural* requirements (normally the consent of the Parliament is needed if the emergency has been declared by the executive). Moreover, this obligation for States to 'officially proclaim' the state of emergency according to municipal law tries to suppress the so-called *de facto* states of emergency; in other words, those states of emergency which have not been officially proclaimed by the State, but in fact restrict *de facto* the human rights of the population. With this requirement, the Covenant tries to prevent States which have not officially proclaimed the emergency from relying on the right of derogation.

One of the problems that arises here is how far international monitoring organs should go in assessing this requirement of compliance with municipal law. In other words, should these organs declare null and void

---

[2] See Martins, *The Protection of Human Rights*, pp. 126–7.
[3] See *Cyprus case*, dissenting opinion of Mr Sperduti; see below, pp. 37–8.
[4] A/2929, p. 67, para. 41.

those declarations of emergency which have been proclaimed by the derogating State against its domestic law? One example is that of a government which seizes power illegally and declares a state of emergency against domestic law. This question arose in the UN Third Committee when discussing the requirement of proclamation. Mr Capotorti (Italy) wondered 'whether any officially performed act could be called illegal, because, even if the act was performed by a government which had seized power illegally, it was difficult for other states to pass on that Government's status'.[5] It is true that one could say that a revolutionary government which seizes power illegally can establish its own legality and declare the state of emergency according to the new legality; consequently it would be difficult to say that this declaration is against the (new) municipal law. However, it is undesirable for an international body in any case to enter into the question of the status of a government under its municipal law. The attitude taken by the European Commission in the *Greek case* is a good example of a correct approach to these cases. The Commission did not contest the fact that the proclamation of emergency by the revolutionary government was not in accordance with the Greek Constitution. In fact, as a result of the military coup, Parliament was dissolved and all the constitutional and legislative authority was vested in the revolutionary junta, altering the normal provisions of the Constitution in proclaiming the emergency.[6]

Even if governments in power do not change the law and declare the emergency against it, it is still undesirable for international bodies to enter into a long examination of whether or not the declaration was lawful under municipal law. In fact, neither the HR Committee nor any of the other international monitoring organs has so far found a declaration of emergency null and void according to the derogation clause just because the declaration was against municipal law. The most that has been done is to criticize the violation of the State's own legality in the declaration of the emergency. When the declaration, for instance, has been proclaimed in circumstances which were not justified according to municipal law (i.e. when there is no grave danger to the State), the application by international bodies of the principle of exceptional threat of the derogation clause would obtain the same result. A good example of this position being taken by an international monitoring organ can be found in the report on Chile by the Ad Hoc Working Group set up by the UN Commission on Human Rights in 1979; this Group found that the Chilean Government violated municipal law in declaring the state of emergency. The Government relied on Act No. 12. 927 which foresees the declaration of emergency on the grounds of 'public disaster'

[5] A/C. 3, 1261st meeting, p. 256. See also A/5655 (1963), para. 48. 50.
[6] *Greek case, YBECHR* 12 (1969), 695–7.

in anticipation of future situations. The situation of public disaster was added to the original Act No. 12. 927 in 1960 with the sole aim of covering disasters produced by natural phenomena and then only in the area affected by them. The Group found that the declaration of emergency, which in this case was in reality due to internal disturbances, and its extension for a further six months with express reference to its preventive nature 'can only be seen as a serious measure restricting human rights which has its sole basis in the will of the government since it lacks any legal foundation with the rules in force in Chile'.[7]

(b) *The Absence of this Requirement in the European Convention:*
Its Effects

The requirement of proclamation does not appear in the two other main treaties. Their *travaux préparatoires* do not give any indication of the reasons for omitting it. The absence of such a requirement in the European Convention was considered by the Committee of Experts to be 'a substantial difference' in relation to the Covenant.[8] Despite this, the European Commission in the *Cyprus case*, after refusing to decide if the lack of notification according to article 15(3) could attract the sanction of nullity, found that 'in any case, article 15 requires some formal and public act of derogation, such as a declaration of martial law or state of emergency, and that, where no such act has been proclaimed . . . art. 15 cannot apply'.[9]

At first sight, it seems as if the Commission considers that, even under the European Convention, a public proclamation, whatever its form, is an essential requisite in order to rely on article 15. However, the question is whether the emphasis of the European Commission on this requirement is the same as that of the drafters of the UN Covenant; in other words, if the Commission considers it to be essential under the Convention that the proclamation of emergency should be made according to all the conditions of municipal law, or if the emphasis of the Commission is on the publicity aspect of the requirement (to make everyone aware of the fact that there exists a situation of emergency and the need of derogation from normal standards). Unfortunately, the Commission does not adequately explore this point. However, two factors seem to favour the latter position: the dissenting opinion of Mr Sperduti, which put the finding in the context of publicity, and the same doctrine of the European bodies in other cases.

---

[7] A/34/583 (1979), p. 8. See also E/CN. 4/1362 (1980), p. 5, para. 14.
[8] Council of Europe, *Problems Arising from the Coexistence of the U.N. Covenants on Human Rights and the European Convention*, Report of the Committee of Experts (Strasbourg, 1970), doc. H(70) 7.
[9] *Cyprus case*, Report of the Commission (1976), *EHRR* 4 (1979), 556.

In his opinion, Mr Sperduti agreed with the majority of the Commission that some form of publicity in fact is necessary and in line with the Convention, but this does not mean that the publicity should immediately precede the measures. He pointed out that there are some situations in domestic and in international law (i.e. that applying in occupied territories) which necessitate the application of special rules as soon as the situation arises (i.e. the Fourth Geneva Convention); in these situations, one cannot deduce from article 15 of the ECHR the necessity to resort to further forms of publicity.[10]

In the *Lawless case* some aspects of this requirement of publicity were also discussed. The contention of the applicant was that the general proclamation was not enough and what was necessary was the proclamation of the concrete measures used and notified by the Irish Government to the Council of Europe to enable citizens to appreciate their position under the law. The European Court found that article 15 'does not oblige the State concerned to promulgate the notice of derogation within the framework of its municipal law'.[11]

On the other hand, the IACHR in the *Nicaragua–Miskitos case* stated that the proclamation of the emergency and of the measures necessary to deal with it by the Nicaraguan Government would have had a beneficial effect because it would have avoided 'an atmosphere of terror and confusion' in the relocation of Miskitos and the dramatic flight of 10,000 of them to Honduras.[12]

2. THE ORGAN COMPETENT TO PROCLAIM THE EMERGENCY: THE LEGISLATURE AND THE EXECUTIVE

Article 4 of the ICCPR and the other derogation clauses in the two main treaties are not explicit about which organ should be competent to proclaim the state of emergency. However, several seminars organized by the UN on the subject have reached some conclusions on what the legal

[10] *Cyprus case*, Mr Sperduti's dissenting opinion, joined by Mr Trechsel, p. 563.

[11] *Lawless case*, Ser. A: Judgment, p. 482, paras. 44–5. See the comments by J. Fawcett on this point in. *The Application of the European Convention on Human Rights* (2nd edn., Oxford, 1987), p. 313. The same issue was raised in the *Ireland v. UK case*, 1972, ECHR, *Collection of Decisions* (46 vols.; Strasbourg, 1960–74), xli. 68 and 63.

[12] OAS, IACHR, *Report on the Situation of Human Rights of a Segment of the Nicaraguan Population of Miskito Origin*, OAS/Ser. L/V/II. 62, doc. 10, rev. 3 (29 Nov. 1983), p 121. Although the American Convention does not explicitly refer to this principle, see the 1968 IACHR *Resolution on Emergencies*, OAS Ser. L/V/II/. 19, doc 30. See also the *Report on Paraguay*, 1987 in which the commission considered that the declaration violated the Constitution of Paraguay. IACHR, *Report on Paraguay*, 1987, pp. 15–22; see also the 1978 *Report on Paraguay*, p. 88.

regime guaranteeing human rights standards should be.[13] Most of these conclusions are based on the legislation of states.

The declaration of the emergency is a matter for the political organs of the State; that is to say, for the legislature and the executive. No democratic system would entrust this task to the military. Neither is the judiciary, due to its very nature, suitable to declare the emergency. The political organs of the State are in a better position to assess the facts, i.e. the real danger to the nation and the need for resort to a declaration of emergency. Most constitutions agree that, in principle, the legislature should declare the emergency through a special and expeditious procedure. In sudden and exceptional situations (*'ex necessitate'*), the executive could declare the emergency, but in this case the legislature should be summoned as soon as possible, if it is not already in session, in order to confirm the declaration. The legislature should have the power, if it considers that it is justified, to ratify the declaration, to modify its terms, or to revoke it. The legislature should also have an important role in controlling the emergency during the crisis.

The permanent control by the UK Parliament of the emergency in Northern Ireland has been important in guaranteeing an avoidance of abuses of human rights standards. At the same time, in the *Lawless case*, the fact that the Irish legislation made provision for parliamentary control was given some weight by the European Court in its assessment of the legitimacy of the measures.[14]

It seems that except in cases of absolute necessity, the legislature should be sitting during the emergency. In several countries, the legislature was not dissolved even in time of war. If the legislature is dissolved, it should be re-elected as soon as possible, provided that it is freely chosen and representative of the whole nation.

The emergency, according to its temporary nature, should be declared for a fixed period of time. Different enactments adopt different methods, but it has been suggested that the declaration should not exceed a maximum of six months.[15] Of course, if the emergency continues, the legislature could extend the declaration for a further period. However, if the circumstances which justified the declaration disappear before the end

---

[13] 'Kingston Seminar', ST/TAO/HR/29. See also Int. Commiss. of Jurists, *African Conference on the Rule of Law, Lagos, 1961* (Geneva, 1961).

[14] *Lawless case*, YBECHR 4 (1961), 438–88. See below, Ch. 5. See also the comments of the ILA on this point and on the importance of the principle of separation of powers and their reciprocal control in emergencies: ILA Paris Report (1984), p. 60. (See also Martins, *The Protection of Human Rights*, p. 128). An interesting criticism of the ILA approach from a Third World perspective is contained in A. Carty, 'Human Rights in a State of Exception: The ILA and the Third World,' in T. D. Campbell (ed.), *Human Rights from Rhetoric to Reality* (Oxford, 1986), pp. 60–79.

[15] See Int. Commiss. of Jurists, *States of Emergency*, p. 459.

of the period, there is no reason for the maintenance of the exceptional measures and in this case the declaration should end. All these procedures and safeguards should be clearly spelled out in the constitution or the legislation of the land. Thus, the legislation would provide clear guide-lines for governments in situations of emergency, and would prevent possible abuses.

3. JUDICIAL CONTROL

Another difficult issue is the judicial control of the declaration of emergency. On this point, one should distinguish between two different aspects: judicial control by domestic courts, and judicial, or quasi-judicial, control by international bodies in the application of human rights treaties. The former aspect presents particular problems, whereas the doctrine on international control of situations of emergency has consistently affirmed its justiciability, depending on the different terms of each treaty for its concrete application.

Judicial control by domestic courts. There is no agreement about the suitability of judicial control of the declaration of emergency. The justiciability of a declaration of emergency presents special problems, due to its political nature. From the point of view of international law, several different proposals have been advanced. There are those who affirm that, because of its political nature, there should be no control at all by the judiciary, whereas others maintain that a certain control would be salutary. Another proposal suggests that the question of judicial control should be resolved according to the legal traditions of each country, and, therefore, that international law should remain silent on this point.[16] The issue, due to the lack of precise standards in human rights treaties, was discussed at some length in the UN Kingston Seminar.[17] Even if the majority was against judicial control of the declaration, the question was controversial and different opinions were held.[18] However, in this seminar there was general agreement that all acts of application of emergency measures should fall under the jurisdiction of the courts. The courts should have full powers to quash, as null and void, all acts or measures which did not conform to the relevant lawful regulations. In that function, the courts should be guided by the principle of reason-

[16] Ibid; p. 435.

[17] 'Kingston Seminar.' Similar positions were held in: *Mexico Seminar on Amparo, Habeas Corpus and Other Similar Remedies (15–28 August 1961)*, ST/TAO/HR/12. And also: *African Conference on the Rule of Law, Lagos*, 1961.

[18] 'Kingston seminar', pp. 46–52. See the position of Mr Rowe of Jamaica. pp. 46–7. See also the position of the ILA in the Paris Report (1984), pp. 63–4. In favour of a judicial control of the declaration, see Dr Sorabje in ILA Montreal Report (1982), pp. 118–19.

ableness, determining whether a given measure or act was reasonably required or at least reasonably justified in the specific circumstances of each case. All ordinary remedies as well as special ones, such as *amparo*, habeas corpus, and so on should remain operative in order to check the unlawful restrictions of rights.[19] The ILA added to these functions of municipal courts the duty 'to ensure that there is no encroachment upon the non-derogable rights and that derogating measures from other rights are in compliance with the rule of proportionality'.[20]

A recent survey on the protection of human rights by municipal courts in situations of emergency in common law countries is not very encouraging.[21] It shows the vulnerability of the courts in times of crisis. The courts in general uphold far-reaching expansion of governmental powers with a corresponding contraction of individual rights. They have also been unwilling or unable to provide the same level of protection as they have accorded to human rights in normal times.

This is why several authors consider it more suitable to insist on the control of emergencies by the democratically elected legislature and the pressure of public opinion, rather than to insist on giving such control to the judiciary.[22] In some cases, international bodies have pointed out that 'substantial limitations imposed on the judiciary in reviewing the factual basis of the establishment of the states of emergency', and the excessive self-restraint on the part of the judiciary, have contributed to gross violations of human rights.[23]

On other occasions, national courts have affirmed the exclusive competence of the government in appreciating the existence of a public emergency and the measures necessary to deal with it.[24] The Chilean case is outstanding in this respect. The Supreme Court ruled that the

government is not required to state the reasons why it considers a particular person to be a danger to the security of the state and the courts cannot determine whether the decisions of the executive authority are reasonable and whether the measures adopted are strictly commensurate with the needs of the situation.[25]

---

[19] 'Kingston Seminar,' p. 51.
[20] ILA Paris Report (1984), p. 65.
[21] G. Alexander, 'The Illusory Protection of Human Rights by National Courts during Periods of Emergency,' *Human Rights Law Journal*, 5 (1984), 1–67. For a good analysis of the limited role of the judiciary in the emergency in Argentina, See A. Garro, 'The Role of the Argentine Judiciary in Controlling Governmental Action under the State of Siege,' *Human Rights Law Journal*, 4 (1983), 283–344.
[22] Alexander, 'The Illusory Protection of Human Rights', p. 65. See also: *African Conference on the Rule of Law, Lagos, 1961*, General Report by Dr T. O. Elias, pp. 47–9.
[23] IACHR, *Report on Chile*, 1985, OAS, Ser. L/V/II. 66, doc. 17, p. 45.
[24] Council of Europe, ECHR, *Collection of Decisions of National Courts Referring to the Convention*, art. 15, pp. 3–4, Greece, Council of State Decisions, 5 Dec. 1960 and 8 Feb. 1961.
[25] *Report of the* Ad Hoc *Group on Chile* (1979), E/CN. 4/1310, p. 16.

The limitations which the Chilean judiciary imposed on its power of supervision relate not only to the fact of detention and the reasons for it, but also its duration and the length of time the detainee is held incommunicado. The difficulties which municipal courts face in theory and practice when controlling declarations of emergency and derogating measures taken by the executive have increased the importance of, and confidence in, international control.

4. INTERNATIONAL CONTROL UNDER THE HUMAN RIGHTS TREATIES

The first question that must be addressed is that of the competence of international bodies entrusted with the application of human rights treaties to monitor the legality of such a complex matter as declarations of emergency. It is well known that the first defence of some States is their claim that those international bodies lack the competence necessary to analyse emergencies since they are a matter which belongs to the domestic jurisdiction of States. However, as Sir Humphrey Waldock has said, 'the effectiveness of any human rights convention is dependent on external supervision, in some form, of the exercise by governments of their powers in regard to public order, national security, public emergency, etc.'[26] On the other hand, the international control of emergencies by bodies established under human rights treaties depends on the specific powers conferred by each treaty upon these organs as interpreted by them. Due to the different nature of the organs established by the three main treaties, the problem of their competence will be studied under separate headings.

(a)   Under The European Convention

(i) *The Competence of the European Organs to Control Compliance with the Principles of the Derogation Clause in Exceptional Situations.* Since the *First Cyprus case*, the European Commission and Court have consistently declared themselves competent to examine emergency situations according to article 15 and in the light of their general duty 'to ensure the observance of the engagements undertaken by the High Contracting Parties' (art. 19).[27] This competence is not restricted only to what could be described as simple emergency situations, namely, those situations

---

[26] Sir H. Waldock, *Human Rights in Contemporary International Law and the Significance of the European Convention* (British Institute of International and Comparative Law, London, 1965), p. 16.

[27] *First Cyprus case*, YBECHR 1 (1956–8), 174–6.

which threaten the organization of legally and democratically constituted States, but is extended to extraordinary circumstances such as revolutions which overthrow a legally constituted government. This was precisely the contention made by the Greek Government at the admissibility stage of the case brought by Denmark, Norway, Sweden, and the Netherlands against Greece in 1967. On that occasion, the Greek Government said:

Generally speaking a revolution creates such a disturbance in the life of a State that it seems meaningless to try to assess the actions of a revolutionary government by the same criteria as would be applicable in normal circumstances or in the case of a simple public emergency threatening the life of the nation within the meaning of article 15. Any control exercised by the Commission would be equivalent, in the last resort, to an expression of approval or disapproval of the revolution itself. Clearly, this would no longer be 'control' in the proper sense of the term but would constitute an interference in a State's internal affairs.[28]

The Commission, rejecting the argument as unfounded, stated:

It is clear from Article 15 of the Convention, read together with article 19, 24 and 25, that the Commission is competent to examine the acts of government also in political situations of an extraordinary character such as after a revolution.[29]

It is now a well-established doctrine of the Commission that it is precisely in times of disturbance and danger which may well be caused by political tension that the most fundamental guarantees assume their greatest importance. In these situations, States tend to resort to arbitrary powers as the simplest and most expeditious way to quell a disturbance. The Commission, as a defender of the public interest, feels that to allow governments to resort so easily to the use of exceptional measures would greatly undermine the guarantees of the Convention.[30]

In assessing these situations the European organs have established a consistent jurisprudence on article 15. Three elements of this can be identified.[31] First, it is up to States to declare the emergency and to take the measures that it considers necessary to overcome the danger. This is a right of the State, an exceptional right in fact, which arises from its duty to protect the well-being of the nation against grave attacks. Secondly, the right of the State is not unlimited; the State is not the final judge of the matter. If the declaration of emergency and the derogating measures are challenged by other States Parties or by individuals, the European organs will examine whether all the conditions laid down in article 15 have been met. Thirdly, in assessing the situation and the measures

---

[28] *Greek case*, Decision on Admissibility, 31 May 1968, *YBECHR* 11 (1968), 716.
[29] Ibid., p. 724.
[30] *Lawless case*, European Court of HR, Ser. B, p. 395.
[31] See e.g. the construction of the European Court in *Ireland* v. *United Kingdom*, European Court of HR, 1978, Ser. A. p. 78.

required, States enjoy a margin of appreciation. In short, the right of States to declare a public emergency and to take measures derogating from their obligation in exceptional circumstances is combined with a judicial supervision of those political decisions by the European bodies.

(ii) *The Construction of the Irish Government: The Principle of Good Faith.* In the *Lawless case*, some differences appeared in the way in which the Irish Government and the Commission, and later the Court, construed article 15. In the proceedings before the Commission, the Irish Government maintained that it was for a government, and for a government alone, to determine when a state of emergency existed and what measures were required by the exigencies of the situation. If a government acts in good faith and its appreciation is reasonable, it should not be held in breach of its obligations under the Convention.[32]

The Commission's opinion did not follow this reasoning. The Commission analysed in great detail the situation and the facts in Ireland at that time in order to assess if there was a real public emergency threatening the life of the nation and if the measures were strictly required by the situation. Even though the Commission recognized that the good faith of the Irish Government was not open to question, it none the less examined the substance of the derogating measures according to the principle of proportionality.[33] The Commission, therefore, reaffirmed its competence to examine and express its own view on the government's judgment. Otherwise this would remain solely within the discretion of the government. This view is in line with general international law which states that

whenever a treaty or a customary rule of international law leaves any matter within the discretionary power of a Government, it is a normal function of international tribunals to appreciate whether in exercising its powers, the Government has acted within the proper limits of its powers under the treaty or the customary rule.[34]

In its last statement before the Court, the Irish Government slightly modified its argument and asked the Court to declare that the decision of a government arrived at in good faith should not be open to question if there is a prima-facie case which would reasonably warrant the government's arriving at the conclusions reached.[35] This last formulation takes into account not only the purely subjective test (good faith) which the Irish Government had held to before, but also some objective elements

---

[32] *Lawless case*, Ser. B: Counter-Memorial of the Irish Government, p. 77.

[33] Ibid., pp. 114–30. See the interesting opinion of Sir Humphrey Waldock on the construction of art. 15. See also the opinions of Mr Susterhenn (p. 152) and of Mr Ermacora (p. 156).

[34] Ibid., p. 334.

[35] Ibid., p. 343. See also the Counter-Memorial of the Irish Government, pp. 222 ff.

(namely the existence of a prima-facie case) which made its position closer to that of the Commission. However, at the end of the hearings, the Commission still insisted that article 15 laid down objective criteria; the margin of appreciation only comes into play after certain minimum objective conditions have been established. Good faith is an important element, but by no means the only one. The Court did not follow the reasoning of the Irish Government and came closer to the Commission's position.[36]

(iii) *The Concept of the Margin of Appreciation.* One of the major techniques used by the Strasbourg bodies when coming to decisions about how to treat these problems has been the concept of the margin of appreciation left to States.[37] This concept only appears in one provision of the Convention: article 1(2) of the First Protocol, under which a State has the right 'to enforce such laws as it deems necessary to control the use of property in accordance with the general interest'.

In principle, the judicial review of the Government's decision as to whether the facts of a particular situation amounted to a public emergency and which measures were strictly required could be made using a purely subjective criterion (the good faith of the Government), or an objective criterion (after examining all the objective conditions of art. 15). As has been seen above, this was at the centre of the discussion in the *Lawless case* between the Irish Government and the Commission. The concept of the margin of appreciation occupies a middle position and, as Professor Fawcett has pointed out, in using it, the Commission comes nearest perhaps to the objective criterion. In these situations, the Commission does not accept the judgment of the government as being conclusive, but rather places itself in the position of the citizens in a democracy who, because they have accorded to their legislature and executive the responsibility for their government, will accept derogations which appear, within the limits of reasonableness and good faith, to be necessary for government.[38]

Professor Fawcett's position seems to be based on the doctrine of the Commission as it was expressed by the then President Sir Humphrey Waldock in the *Lawless case*:

The concept of the margin of appreciation is that a Government's discharge of these responsibilities is essentially a delicate problem of appreciating complex

[36] *Lawless case*, Judgment of the Court, Ser. A, p. 22.

[37] For the concept in the European Convention, see C. Morrison, 'Margin of Appreciation in European Human Rights Law', *Human Rights Journal*, 6 (1973), 263–86. See also J. G. Merrills, *The Development of International Law by the European Court of Human Rights* (Manchester, 1988), pp. 136–59.

[38] J. Fawcett, *The Application of the European Convention of Human Rights* (2nd ed.). Oxford, 1987), pp. 310–12.

factors and of balancing conflicting considerations of the public interest, and that, once the Commission or the Court is satisfied that the Government's appreciation is at least on the margin of the powers conferred by article 15, then the interest which the public itself has in effective government and in the maintenance of order justifies and requires a decision in favor of the legality of the Government's appreciation.[39]

The concept of the margin of appreciation is a flexible one used by the Commission and the Court in the specific circumstances of each case. It is difficult to give a more precise definition to such an incommensurable concept adopted in the reasoning of judicial decisions. It was first mentioned by the Commission in the *Cyprus case* in 1956 when it left to the British Government 'a certain measure of discretion' in assessing the measures necessary in order to deal with the grave danger to public order in the island.[40] Since then, the theory of the margin of appreciation has been applied in the major cases on limitations and derogations under the European Convention.[41]

There has, nevertheless, been some criticism about the way in which it has been applied in particular cases. The major criticism, of course, centres on the legitimacy of applying this concept due to the fact that article 15 does not refer to it. The Commission held that 'the doctrine of the margin of appreciation derived from their (the Commission's) understanding of the responsibilities of governments, the difficulties of the situation and the inherent nature of the matters dealt with in article 15'.[42]

The dissenting opinions in the *Greek case* were mainly due to the fact that even if in theory the majority referred to the doctrine of the margin of appreciation, in practice they did not apply it, or, at least, the margin left to the Greek Government was much narrower than the one left to the Irish Government in the *Lawless case*.[43] It is interesting to note that the Greek Government asked for a wider margin of appreciation due to the fact that it was a revolutionary government; however, the Commission did not adopt this reasoning.

In his dissenting opinion, Mr Delahaye pointed out that the evidence before the Commission did not lead to any definite conclusion on the existence of an emergency according to article 15. In those circumstances, the margin of appreciation left to the government is a matter of particular

---

[39] *Lawless case*, European Court of HR, Ser. B: Verbatim report of the public hearings held by the Chamber on 7, 8, 10, and 11 Apr. 1961: Sir Humphrey Waldock, President of the Commission, p. 408.

[40] *First Cyprus case*, p. 174.

[41] For a survey of its application see Morrison, 'Margin of Appreciation', and Merrills, *The Development of International Law*.

[42] *Lawless case*, Ser. B, p. 397.

[43] *Greek case*. The majority referred to that doctrine in para. 154. Dissenting opinions: pp. 76–92.

importance. And, applying the test of the European Court in the *Lawless case*, he found that the existence of an imminent danger to the nation had reasonably been inferred by the Greek Government.[44] Mr Eustathiades, in the *Lawless case*, construed the expression 'public emergency threatening the life of the nation' in a restricted way, and consequently found that an emergency situation did not exist in Ireland at the time when Lawless was detained. In the *Greek case*, and after having taken into account the jurisprudence of the Court and the Commission in the *Lawless case* on the expressions 'public emergency' and 'margin of appreciation', Mr Eustathiades reached the conclusion that it was not possible to deduce that a situation of the type contemplated by the jurisprudence did not exist.[45]

On the other hand, Mr Susterhenn shared the view of the Irish Government in the *Lawless case*, and, applying the same subjective criteria of bona fides, reasonableness, and better position of the government, found that the Greek Government did not arbitrarily exceed its margin of appreciation when, after examining the facts, it came to the conclusion that there was a threat to the life of the nation.[46]

(iv) *The Problem of the Burden of Proof*. Another problem related to the judicial control of emergencies is that of deciding, as a matter of principle, where the burden of proof lies. Is the applicant the one who should prove that a public emergency does not exist or that a derogating measure is not strictly required by the exigencies of the situation, or is it the respondent government that has to prove and justify the emergency and the measures taken? The position of the Commission has been to ask the government which relies on the right of derogation to prove all the required facts.[47]

In conclusion, after examining the concepts of margin of appreciation, good faith, reasonableness, and the onus of proof, the boundaries of which are not clearly defined, one has the impression that they are no more than useful tools in the hands of those in charge of adjudicating difficult cases of a political nature. This is why they elude an absolute and consistent theoretical construction.

(b) Under the Covenant: The United Nations Human Rights Committee

The control of the Committee in situations of emergency is determined by the powers conferred upon it by the Covenant. According to the

---

[44] Ibid., dissenting opinion of Mr Delahaye at para. 178.
[45] Ibid., dissenting opinion of Mr Eustathiades at para. 184.
[46] Ibid., dissenting opinion of Mr Susterhenn, pp. 87–92.
[47] *Greek case*, p. 72. *The Lawless case*, opinions of Sir H. Waldock, p. 114, and Mr Susterhenn, pp. 144 ff. *Ireland. v. U.K.*, Ser. B, p. 75.

Covenant, there are three mechanisms of implementation: the States' reporting procedure under article 40, the inter-States' complaints under article 41, and individual communications under the Optional Protocol. While there has so far been no inter-state application, the Committee has acted under the other two procedures.

(i) *The Reporting Procedure.* The nature of the Committee is a special one, particularly as far as its functions under the reporting procedure are concerned. It is neither, strictly speaking, a judicial or a quasi-judicial body, as the European Court and Commission are, nor a fact-finding body (unlike the IACHR). It has the role of reviewing States reports, and, in the dialogue with States representatives, members of the Committee express their opinions about States' compliance with the Covenant. The Committee as a whole has decided not to make general comments on particular States. However, it has made useful comments on most of the articles of the Covenant. Unfortunately, the general comment on the derogation clause is nothing more than a restatement of that clause.[48] Therefore, the only way in which we can assess the interpretation of the Covenant and of the control in emergencies is through the reactions of the members in reviewing States' reports. These comments can be found in the Committee summary records and in the annual report of the Committee to the General Assembly.

The widespread failure to comply with the notification requirement according to article 4(3), the exclusively legalistic approach which is often unrealistic, and the unwillingness of some States fully to co-operate with the Committee are among the main setbacks of the reporting procedure. Moreover, the rule adopted by the Committee requiring States to submit reports every five years has provided an excuse for some States to avoid giving the supplementary information requested at the end of the previous report. The Chilean case is a good illustration of this. Chile's first report was presented in 1978; this report and the answers of its representatives did not satisfy the Committee which considered that the information provided was insufficient, and therefore invited the Government to submit another report and to furnish specific information concerning the restrictions on rights in force during that period of emergency.[49] The Chilean Government did not in fact submit any further information until the second periodic report five years later. In reviewing the second

[48] UNGAOR. A/36/40, Report of the HR Committee, 1981, Annexe VII, General Comment 5/13 (art. 4), p. 110. For a review of the role of the HR Committee, see ILA Seoul Report (1986), paras. 20–31. See also J. A. Walkate, 'The Human Rights Committee and Public Emergencies, *Yale Journal of World Public Order*, 9 (1982), 133–46.

[49] UNGAOR. A/34/40, Report of the HR Committee, 1981, pp. 19–25. See p. 23, para. 108.

report in 1984, the Chairman of the Committee protested that the representatives had not fulfilled that requirement which had been asked of them according to article 40(1)*b*.[50]

With all these shortcomings in the reporting mechanism, attempts have been made by some members to improve the procedure of monitoring situations of emergency, in which gross violations of human rights have often occurred. Perhaps the most serious attempt took place in 1982 at the 15th Session of the Committee. However, the proposals to improve the procedure encountered strong opposition from some members who construed the powers of the Committee in a very restrictive way.[51] Although the majority of the Committee were in favour of these proposals, the practice of reaching decisions by consensus prevented the adoption of these proposals. One may hope that in the future the HR Commitee will move in the direction defended by Ermacora and Opsahl in 1982 in order to strengthen the international protection of human rights in emergencies under the Covenant.

(ii) *The Practice of the Committee in Certain Specific Cases.* Nevertherless, despite all the shortcomings of the reporting procedure and the

---

[50] UNGAOR. A/39/40, Report of the HR Committee, 1984, p. 78, para. 437.

[51] See the interventions of Mr Opsahl and Mr Ermacora, and the formal proposal of Opsahl in CCPR/C/SR. 349, pp. 4–5. See a similar proposal by Mr Tarnopolsky in CCPR/C/SR. 463. See also; CCPR/C/SR. 404; 414; 575; for the discussions, see CCPR/C/SR. 334; 349; 351. For a wide construction of the powers of the Committee, see the position of Ermacora in CCPR/C/SR. 349, pp. 5–6. For the restrictive approach, see Mr Graefrath, CCPR/C/SR. 349, pp. 9–10. Graefrath's restrictive approach arises from his general conception of the powers of the Committee under the reporting procedure and ultimately from his socialist conception of the international protection of human rights. For Graefrath's general conception of the extremely limited powers of the Committee under the reporting procedure, see CCPR/C/SR. 231, pp. 2–5. See also CCPR/C/SR. 232. For an overview of the socialist conception of human rights, see A. Bloed and G. J. H. van Hoof, 'Some Aspects of the Socialist view of Human Rights', in A. Bloed and P. van Dijk (ed.), *Essays on Human Rights in the Helsinki Process* (The Hague. 1985), p. 46; A. Movchan. *Human Rights and International Relations* (Moscow, 1988). For a confirmation of the socialist general opposition to the development of effective international procedures until very recently, see P. Alston and B. Simma, 'First Session of the UN Committee on Economic, Social and Cultural Rights,' *AJIL* 81 (1987), 747–56 (see however the change of the atmosphere in the 2nd session after the resignation of the Soviet representative, *AJIL* 82 (1988), 603–15). For a good analysis of the traditional general opposition of members, from socialist countries to a wider interpretation of the powers of the Committee under the reporting procedure, see T. Meron, *Human Rights Law-Making in the United Nations: A Critique of Instruments and Process* (Oxford, 1986), pp. 123–6. See also D. Fischer, 'Reporting under the UNCCPR; The First Five Years of the Human Rights Committee', *AJIL* 76 (1982), 148; V. Dimitrejevic, *The Roles of the Human Rights Committee* (Europa Institut der Universität des Saarlandes, Saarbrucken, 1985). For the new attitude of the Soviet Union and the socialist States towards human rights issues, see Bloed, 'Recent Developments in Soviet Attitudes towards the International Protection of Human Rights', *Netherlands Quarterly of Human Rights*, 6 (1988), 80. See also: 'Soviet Union Accepts Compulsory Jurisdiction of the ICJ for Six Human Rights Conventions,' *AJIL* 83 (1989),

different opinions within the Committee on its competence to monitor emergencies, members of the Committee have openly commented (often, due to their legal expertise, in a highly significant way) on the States' compliance with the principles of the derogation clause. These comments have been made when States' representatives have been questioned about the presentation of the reports. However, two further problems make the members' task of monitoring emergencies more difficult: one is the lack of clarity on the legal status of countries in emergencies; and the other is the doubt over whether the obligation of sending in a periodic report can be derogated from in emergencies.

As far as the first problem is concerned, many States do actually fail to notify the declaration of emergency according to article 4(3). Other States do not proclaim the state of emergency even if the situation amounts to a *de facto* state of emergency due to the numerous derogations. Thus, when questioning States representatives, members of the Committee try to clarify the legal status in the country before making an assessemt of the situation.[52]

On the question of whether the reporting obligation can be derogated from due to the emergency, some conflicting opinions have been expressed. Thus, when reviewing the Lebanese report, Mr Bouziri stated that 'in view of its critical situation, Lebanon was not obliged to provide a report'.[53] However, when Nicaragua presented its report, Mr Prado Vallejo, a member of the Committee, pointed out that the fact that Nicaragua had submitted a very complete report on the situation of emergency was evidence that States consider that the reporting procedure is not suspended in those situations.[54] This position seens to be more in line with the actual practice of States under the Covenant. The contrary view would be damaging to the implementation of the Covenant and open to abuse by States wishing to avoid international accountability in emergency situations. This is not to deny that in very exceptional situations a State report could be delayed for special reasons which should be duly justified before the Committee.

On several occasions, members of the Committee have made clear statements not only about their general competence to monitor emergencies, but also about States' compliance with the principles of the

---

457; T. Franck, 'Soviet Initiatives: US Responses; New Opportunities for Reviving the UN System', *AJIL* 83 (1989), 531–43.

[52] See *inter alia* CCPR/C/SR. 443 (1983), *Report on Lebanon* (Remarks by Sir V. Evans), pp. 2–3; A/39/40 (1984), *Report on Sri Lanka*, p. 21, para. 104; CCPR/C/SR. 443. 446, *Report on Lebanon*, pp. 2 ff.; A/34/40, *Report on Syria* (1978), pp. 70 ff.; A/37/40 (1981), *Report on Jordan*, paras. 187 ff.; A/37/40 (1981), *Report on Iran*, p. 68, para. 306.

[53] CCPR/C/SR. 443, *Report on Lebanon*, p. 5. The Syrian representative took the same atitude in the presentation of his country's report.

[54] A/38/40, Report of the HR Committee, 1983, p. 50, para. 228.

derogation clause.[55] Moreover, the policy adopted by the Human Rights Committee under the Optional Protocol underlines the international control of states of emergency. In several cases, States accused of violations tried to defend themselves by claiming the existence of an emergency, but without giving any further details which would permit the Committee to assess the situation. In such cases, the Committee has held that 'a State by merely invoking the existence of exceptional circumstances, cannot evade the obligations which it has undertaken by ratifying the Covenant'. In one of these cases, despite the fact that the Uruguayan Government sent a notice of derogation according to article 4(3), the Committee felt unable to accept that the requirements of article 4(1) had been fully met. The State concerned is duty-bound to give a sufficiently detailed account of the relevant facts. Without full and comprehensive information, the Committee cannot conclude that there exist valid reasons for derogation from normal standards.[56]

The Committee has established important guide-lines in the monitoring of emergencies under the Optional Protocol, namely, that there can be no justification for the violations of non-derogable rights; that rights which are not mentioned in the notice of derogation cannot be derogated from; that the failure of States to provide precise responses to allegations will justify the finding that a reliable attested violation has occurred, and, finally, that a State has the duty to investigate allegations under the Optional Protocol in good faith and to report the result of its investigations.[57]

(c) Under the American Convention: The Control Exercised by the Inter-American Commission on Human Rights

The Inter-American Commission on Human Rights (IACHR) was set up by the Council of the OAS in 1960 as a consequence of Resolution VII of

---

[55] See Ch. 1 on the principle of exceptional threat (pp. 20–2); see also Ch. 5 on the principle of proportionality (pp. 152–9). See also the remarks of Mr Dieye in reviewing the Uruguayan report in A/38/40 (1983), p. 50; CCPR/C/SR. 356, paras. 2–26.

[56] See *inter alia*, A/36/40, Communication No. R 8/34, Uruguay, pp. 132–3; A/37/40, Communication No. R 11/46, Colombia, pp. 172–3.

[57] See the summary in ILA Seoul Report (1986), p. 12. For the role of the HR Committee in general under the Optional Protocol, see P. R. Ghandhi, 'The Human Rights Committee and the Right of Individual Communication, 57 *BYBIL* (1986), 201–51. See also Zayas *et al.*, 'Application of the International Covenant on Civil and Political Rights under the Optional Protocol by the Human Rights Committee', *German YBIL* 28 (1985), 1–64. A survey of decisions given by the HR Committee up till July 1989 can be found in M. Nowak, *HRLJ* 1 (1980) 136–70; *HRLJ* 2 (1981), 168–72; *HRLJ* 3 (1982) 207–20; *HRLJ* 5 (1984), 199–219; *HRLJ* 7 (1986), 287–307; *HRLJ* 11 (1990), 139–56. See also P. R. Ghandhi, 'The Human Rights Committee and Derogations in Public Emergencies', *German YBIL*, 32 (1989), 321–61.

the 5th Meeting of the Ministers of Foreign Affairs in 1959, and it was entrusted with the duty of furthering respect for human rights in the Continent.[58] In 1965, in the OAS Conference held in Rio de Janeiro, the Commission received new powers and functions, namely, to examine individual communications, to address governments, to obtain the pertinent information and to make recommendations to them, and finally, to submit an annual report to the OAS General Assembly.

In 1967, as a consequence of the so-called 'Protocol of Buenos Aires', the OAS Charter was amended and the Commission became one of the principal organs of the OAS, through which the Organization actually fulfils its purpose. Article 112 of the Protocol establishes that its principal functions are 'to promote the observance and protection of Human Rights, and to serve as a consultative organ of the organization in those matters'.[59] In 1979, when the American Convention on Human Rights entered into force, a 'new' IACHR started functioning. This Commission inherited the powers of the previous one, as far as non-parties to the Convention were concerned, and it had new powers in respect of States parties to it.[60] A new Statute and new regulations were established thereafter.[61]

Since the beginning of its work, the Commission has produced several reports on violations of human rights by states in the context of emergencies. The Commission has mainly acted as a fact-finding body and has produced thorough reports on the legal and factual situations of human rights in the Americas. The great flexibility of its *modus operandi* has enabled it to use all kinds of techniques in order to assess these situations: it has accepted information from a variety of sources, it has held hearings, and most effectively of all, it has made on-site visits. These *in loco* visits have made the Commission the most experienced international body from the point of view of organizing visits and protecting witnesses.[62]

The Commission, unlike the UN Human Rights Committee, is able to act ex officio and produce the reports which it deems necessary in the performance of its duties. In the *Bolivia Report* in 1980 and in the *Nicaragua–Miskitos Report* in 1983, the Commission applied the stan-

---

[58] OAS Resolution VII, pt. II, Santiago de Chile, 1959, Actas y Documentos, OAS (OEA Ser. F/III. 5), p. 308.

[59] OAS, *The Organization of American States and Human Rights, 1960–1967* (Washington, 1972), pp. 2 ff. See also pp. 601–9 for the First Statute of the Commission and pp. 611 ff. for the First Regulations.

[60] ACHR, arts. 34–51.

[61] OAS, *Handbook of Existing Rules Pertaining to Human Rights in the Inter-American System, 1985*, OEA Ser. L/V/II. 65, pp. 103–42.

[62] Ibid., arts. 55–9 of the Commission Regulations. See also E. Vargas Carreno, 'Las Observaciones *in loco* practicadas por la Comision Americana de Derechos Humanos,' in OAS, *Human Rights in the Americas* (Washington, 1984), pp. 290–305.

dards of article 27 even though the Governments did not file a notice of derogation. On the other hand, it is interesting to note that the Commission has not been divided when interpreting its competence to review emergencies. Even though the Commission is very reluctant to see itself as a judicial body, it has not hesitated in making value judgments about States' compliance with the terms of human rights instruments. As far as states of emergency are concerned, the Commission has since 1980 improved the quality of its legal analysis.[63] However, with the exception of a few cases like the *Nicaragua–Miskitos case*, there has been no thorough legal analysis of the compliance of the measures of derogation with the criteria laid down in article 27.[64]

In 1968, the IACHR approved an important resolution concerning the protection of rights in connection with the suspension of constitutional guarantees. In the resolution, the Commission stated:

That the imposition of the state of siege or the suspension of instituted guarantees can be made the object, without prejudice to national sovereignty or the principle of non-intervention, of international control by means of signing and ratification of a convention establishing the reciprocal undertaking . . .

The undertaking involved conforming to the substantial requirements later recognized in article 27, and informing the other Contracting Parties of the derogations. Moreover, the resolution entrusted the Commission with the observance of compliance on the part of the States Parties with those substantive and formal requirements. When non-compliance takes place the IACHR has the power to recommend that the State concerned adhere immediately to its undertaking, and in the event of non-observance, to report to the Inter-American Conference or the Meeting of Consultation of Ministers of Foreign Affairs.[65]

In its practice, the Commission has always declared itself to be competent to assess States' compliance with the requirements of article 27 in emergencies. Moreover, the Commission has applied several of these standards to States not parties to the Convention, relying on its theory that article 27, together with article 4 of the ICCPR and article 15 of the European Convention, embody 'the most accepted doctrine in inter-

---

[63] See also C. Grossman, 'Algunas consideraciones sobre el regimen de situationes de excepcion bajo la Convencion Americana de Derechos Humanos,' in OAS, *Human Rights in the Americas*, p. 123.

[64] The 1987 *Report on Paraguay* and the 1985–6 *Report on Nicaragua* are also good examples of the present tendency of the IACHR to improve the legal analysis of derogations. In the *Report on Bolivia* of 1981 there was a timid attempt to make more explicit a profound analysis of the emergency according to art. 27. For the analysis of the *Nicaragua–Miskitos case*, see below, Ch. 5, p. 164, and Ch. 6, pp. 187–8.

[65] IACHR, *Resolution on the Protection on Human Rights in Connection with the Suspension of Constitutional Guarantees or 'State of Siege'*, OEA/Ser. L/V/II. 19, doc. 32 (16 May 1968). This resolution was based on Commissioner Martins's report.

national law' on emergencies.[66] There is no evidence so far that proves that States have rejected the competence of the Commission to pronounce on these matters. Even the Chilean Government's strong criticism of the 1985 IACHR *Report* was based on grounds other than the competence of the Commission to assess the lawfulness of the state of emergency.[67]

The IACHR has found many states of emergency to be in violation of article 27:

(*a*) due to the absence of an exceptional threat to the organized life of the State.[68]

(*b*) due to the institutionalization of the emergency, which is against its inherently temporary character.[69]

(*c*) due to non-compliance with the principle of proportionality; all derogating measures should be strictly required by the exigencies of the situation.[70]

(*d*) due to non-compliance with the fundamental principle of non-derogability of certain rights.[71]

(*e*) due to non-compliance with the formal principle of notification.[72]

In the case of Bolivia, the Commission found that the Government had violated the substantive and the formal requirements of article 27.[73]

As a consequence of these judgments on the unlawfulness of the emergencies, the Commission has explicitly recommended the lifting of

[66] IACHR, *Report on Chile*, 1974, OEA, Ser. L/V/II, 34 Doc 21 p. 212. Report on Paraguay 1978. OEA Ser/L/V/II. 43, doc. 13, p. 14. IACHR, Annual Report, 1974, OEA, Ser. P/AG, doc. 520/75, p. 36.

[67] IACHR, *Report on Chile*, 1985, pp. 291–305.

[68] Ibid., p. 44, para. 94; IACHR, *Report on Chile*, 1974, pp. 2–4; IACHR, *Report on Bolivia*, 1981, p. 23, para. 6.

[69] IACHR, *Report on Colombia*, 1981, p. 40; IACHR, *Report on Paraguay*, 1978, p. 14, paras. 10–13; IACHR, *Report on Bolivia*, 1981, p. 23, para. 6; IACHR, *Report on Chile*, 1985, p. 44, para. 99; IACHR, *Report on Nicaragua*, 1981, pp. 113 ff. on freedom of thought and expression, and pp. 126 ff. on political rights.

[70] IACHR, *Annual Report*, 1986, Nicaragua (many measures considered out of proportion), pp. 165 ff.; IACHR, *Report on Colombia*, 1981, p. 219, with respect to the right of personal liberty.

[71] IACHR, *Report on Guatemala*, 1983, pp. 18–19; IACHR, *Report on Argentina*, 1980, p. 26; IACHR, *Report on Suriname*, 1985, p. 17.

[72] IACHR, *Report on Bolivia*, 1981, p. 23, para. 6; IACHR, *Report on Guatemala*, 1983, p. 33, para. 6; IACHR, *Report on Nicaragua–Miskitos*, 1983, p. 121, paras. 32 ff.

[73] OAS, 11th Session, 1981, Official Records and Documents, vol. 2 (pt. 2), First Commission: Legal and Political Affairs, Introduction to the Bolivian Report by Vice-President Monroy Cabra, p. 46.

the state of emergency, or alternatively, if an emergency actually exists, compliance with the provisions of article 27.[74]

5. CONCLUSION

1. The requirement of 'official proclamation' of the emergency appears only in the UN Covenant. The reason for introducing this requirement into the derogation clause was to reduce the incidence of the *de facto* states of emergency by obliging States to declare the emergency following the procedures of municipal law. In fact, most municipal systems require the 'official proclamation' of the emergency according to strict conditions. These guarantees would have been lost if this requirement had not been included in the derogation clause. However, this requirement has not been construed as requiring international monitoring bodies thoroughly to analyse the compliance of the derogating State with municipal law, and to declare null and void any declaration made in violation of municipal law; this would put international monitoring bodies in a difficult position when trying to pass judgment on the legality of a State's own acts according to municipal law.

2. The requirement of 'official proclamation' has not given rise to a large case-law under the treaties. Although the European and American Conventions do not explicitly contain this requirement, their monitoring organs have made reference to it in some cases; however, it seems that the main concern of the monitoring organs when insisting on the proclamation of the emergency has been to make the emergency public (publicity) rather than to insist on the State's compliance with domestic law procedures. In any case, these bodies have on some occasions criticized the violation by Governments of their own legislation when proclaiming the emergency.

3. Although there is no indication in the treaties of which organ of the State is competent to proclaim the emergency, several seminars organized by the UN have established guide-lines based on States' practice. The proclamation of the emergency is a matter for the political organs of the State, namely, the legislature and the executive. Even if in some exceptional cases the executive declares the emergency, the legislature should have an important role not only in approving the declaration but also in controlling the application of the exceptional measures during the crisis.

4. There has been a wide range of views on the question of the

[74] IACHR, *Report on Paraguay*, 1978, p. 88; IACHR, *Report on Guatemala*, 1983, p. 19; IACHR, *Report on Suriname*, 1985, p. 67; IACHR, *Report on Bolivia*, 1981, pp. 23 and 221; IACHR, *Report on Argentina*, 1980, p. 265, para. 4.

suitability of judicial control over the declaration of emergency. Due to the political nature of the proclamation, the majority of the doctrine think that the judiciary should confine itself to checking whether all the constitutional and legal procedures have been complied with and should not pass judgment on the substance of the declaration (i.e. the need for it, etc.). However, the judiciary should play an important role in monitoring the legality of the application of the emergency measures, and should be able to nullify those measures which are unlawfully taken; the principles of proportionality and reasonableness in the circumstances should also assist the judiciary in the determination of the legality of the measures.

5. The theoretical and practical difficulties that municipal courts face when controlling the declaration of emergency and derogating measures have increased the importance of international control. Although states of emergency are undoubtedly a complex matter to monitor (because they are so close to the States' domestic jurisdiction), and although the first defence of some States on these occasions is to deny the competence of international bodies to control them, these bodies have in general terms affirmed their competence to assess States' compliance with the standards of the derogation clause. However, the fact that these bodies established under the three treaties are different in nature and also have different competences is an important factor in the assessment of their capacity effectively to monitor States' compliance with the principles of the derogation clause. Thus, the European and the Inter-American Courts are judicial bodies, the European Commission is a quasi-judicial body, the IACHR acts mainly as a fact-finding body, and the UN HR Committee has limited powers in the reviewing of States' reports.

6. The European organs have consistently interpreted their task acording to the view that the right of States to declare a public emergency and to take measures derogating from their obligations under the Convention is recognized but also supervised by them. In assessing the existence of a public emergency and the need for derogating measures, States enjoy a margin of appreciation. After a thorough examination of concepts like good faith, margin of appreciation, reasonableness, and onus of proof in the case-law of the European organs, the conclusion is that these concepts elude an absolute and consistent theoretical construction, and that they are no more than useful tools in the hands of those in charge of adjudicating difficult cases of a political nature.

7. The effective monitoring of the HR Committee is hampered by its limited powers under the reporting procedure. Moreover, there have been different opinions within the Committee about the course of action open to it in order to improve the monitoring of states of emergency. In this respect, the debate which took place in 1982 is revealing; a restric-

tive interpretation of the powers of the Committee seems to have won through for some time. However, in spite of all the shortcomings of the reporting procedure and the different opinions within the Committee about its powers to monitor emergencies, members of the Committee have openly made comments, which are often of a high quality due to the legal expertise of most of the members, concerning States' compliance with the principles of the derogation clause; this happened when States' representatives were questioned about the presentation of reports. Under the Optional Protocol, the Committee has established important guidelines for dealing with derogations.

8. Although the Inter-American Court has produced several advisory opinions relating to emergencies, the organ which under the OAS system has so far most effectively monitored states of emergency has been the IACHR. Despite the fact that its fact-finding nature has meant that the legal quality of its analysis of the emergencies contained in the reports has not been very high, it has strongly affirmed its competence to assess States' compliance with the principles of the derogation clause. Moreover, in many of the IACHR reports, there is a clear condemnation of fundamental violations of these principles. Finally, when defining States' obligations in emergencies, the IACHR has also applied the main principles of the derogation clause as the relevant standards in respect of States non-parties to the Convention but members of the OAS.

3

The Principle
of Notification

1. THE RATIONALE OF THE PRINCIPLE OF NOTIFICATION

In the three main treaties, the derogation clause contains a similar provision according to which States which avail themselves of the right of
derogation shall inform the other States parties to the treaty of the
measures taken, the reasons therefor, and the date at which the derogation terminates.[1] The reason for including this provision in the derogation
clause seems to be quite clear. The derogations from human rights obligations in situations of emergency is a very serious measure and a matter of
concern for the other States parties.[2] These States, according to the
travaux préparatoires of the Covenant, have the right to be notified in
order to know exactly what the situation of the derogating State is in
respect of the treaty, and accordingly to be able to exercise their own
rights.[3] One of the most important rights under these treaties is that of
presenting an inter-State complaint against the derogating State if the
applicant believes that a violation of the standards of the treaty has taken
place. In fact, under the European system, the inter-State complaints
procedure has been the most successful in ensuring respect for the
Convention in emergencies. This procedure would not be possible if
States parties to the treaty were not fully informed of the measures
of derogation through notification. In a system as integrated as the
European one, the aim of which is the creation of a system of collective
guarantees of human rights, a kind of European 'public order' within
which all States have accepted the procedure of inter-State complaints,
the notification requirement plays a major role. Due to the importance of
this requirement, there was some discussion in the preparation of the UN
Covenant about the suitability of extending the scope of the legal interest

---

[1] ICCPR, art 4(3); ECHR, art. 15(3); ACHR art. 27(3).

[2] UNGAOR, 10th Session, Annexes, agenda item 28, pt. II, Draft Covenant, Report of
the Secretary-General, A/2929, p. 69, para. 47.

[3] Ibid. It is interesting to note the Soviet representative's opposition to the notification
requirement in the early stages of the Covenant's drafting, E/CN. 4/SR. 127 (1949), p. 14.

in challenging derogations to all members of the UN. India suggested that the Secretary-General should communicate all notices of derogations to the General Assembly.[4] However, it was felt that it was dangerous to allow States not parties to the Covenant to express criticism about the way in which States parties fulfil their obligations under the Covenant.[5]

In other words, since the main aim of the derogation clause is to avoid abuses in emergencies, States undertake the commitment to adopt the decision involving derogations 'in public'.[6] This fact has another advantage. Due to the necessity of declaring the state of emergency and notifying the parties to the treaty, derogating States should think very carefully about which measures are 'strictly necessary' to overcome the emergency. At the same time, the notices of derogation, which are usually sent by the Secretary-General to the bodies entrusted under the treaties with monitoring their implementation, keep these bodies informed in order that they may exercise their own jurisdiction under the different procedures. Otherwise these bodies, when receiving inter-State complaints or individual communications, or when exercising jurisdiction under the reporting procedures, will not be aware of the exact situation of the derogating State in respect of the treaty. There is no doubt that an emergency situation could affect their jurisdiction.[7]

The notification requirement stresses the international accountability introduced by the derogation clause in situations of emergency. Thus the declaration of a state of emergency, when derogations from provisions of the treaty are involved, is no longer exclusively a matter of domestic concern. Mr Oribe (Uruguay) pointed out in the *travaux préparatoires* of the Covenant that the derogation clause set forth 'a new principle in international law—that of the responsibility of States towards the members of the community of nations for any measures derogating from human rights and fundamental freedoms'.[8]

The notification requirement is so significant that a failure to notify the other States parties could result in the loss of the right of derogation. The European Commission and Court have always refused to accept as a matter of principle that a failure to meet the notification requirement can never attract the sanction of nullity of the derogation.[9]

---

[4] The Indian proposal: E/CN. 4/SR. 330 (1952), pp. 3–7.
[5] A/2929, p. 69, para. 47.
[6] See the intervention of Rene Cassin, E/CN. 4/SR. 127 (1949), p. 7.
[7] See *inter alia* the statement of the delegate of the European Commission with respect to the Counter-Memorial of the Irish Government in the *Lawless case*, European Court of HR, Ser. B: Pleadings, pp. 335–6.
[8] E/CN. 4/SR. 195 (1950), p. 10.
[9] See below, pp. 64 ff.

2. THE CONTENTS OF THE NOTICE OF
DEROGATION OR THE FORMAL REQUIREMENTS
OF THE PRINCIPLE OF NOTIFICATION

It was pointed out in the preparatory work of the Covenant that precisely because the emergency powers have been abused in the past, 'a mere notification would not be enough'.[10] And even though paragraph 3 of the derogation clause does not foresee a specific form for the notification, certain elements are required.

(a)   The Element of Time

The requirement that the information should be sent in time is expressly contained in the Covenant and in the American Convention ('immediately'), and has been considered to be a necessary element also under the European Convention by the jurisprudence of the European organs; accordingly, the information should be sent 'without any avoidable delay'.[11] This requirement does not mean that the notice of derogation should be prior to the derogation; it simply means that it must be sent within a reasonable time. What constitutes a reasonable time has been the object of some clarification by the international bodies.

The European organs found in the *Lawless case* that the Irish notice of derogation, which was sent twelve days after the entry into force of the suspension of guarantees, was made in reasonable time.[12] They also found that the three UK notices of derogation (20 August 1971, 23 January 1973, and 16 August 1973) fulfilled the requirements of article 15(3).[13] The first and the last notices were sent eleven and eight days after the derogations had entered into force, respectively.[14] The second notice was sent forty-three days after the approval of the Terrorist Order by the House of Commons (11 December 1972).[15]

An interesting problem occurred in the *Greek case*. Greece sent two notices of derogation with incomplete information about the emergency, within one month of the declaration of the emergency. However, the reasons for derogation were sent more than four months later. There-

---

[10] A/2929, p. 69, para. 47.
[11] *Lawless case*, Ser. B: Report of the Commission, para. 80. See also the *Greek case*, Report of the Commission, para. 78.
[12] *Lawless case*, Ser. A: Judgment, p. 61, para. 47; Ser. B: Pleadings, Report of the Commission, p. 73, para. 80.
[13] *Ireland* v. *UK*, European Court of HR, Ser. A: p. 84, para. 223.
[14] Ibid., paras. 80, 88. See also *YBECHR* 16 (1973), 26–8.
[15] *Ireland* v. *UK*, Ser. A, p. 84, para. 85. In the notice of derogation of 23 Jan. 1973 there is no indication of the date on which the derogating measures came into force. See *YBECHR* 16 (1973), 24–5.

fore, the Commission found that Greece failed to comply with this time requirement of article 15(3). This finding suggests that all the necessary information, and not just part of it, must be sent in time.[16]

In reviewing the *Colombia Report*, members of the UN HR Committee have also pointed out that the notice of derogation had been sent very late, and therefore did not comply with the time requirement. In fact, the state of emergency was proclaimed in Colombia some months after the entry into force of the Covenant (in 1976), whereas the notice of derogation was sent on 18 July 1980, that is, four years later.[17]

The IACHR also found that Guatemala, by sending its notice of derogation nine months after the proclamation of the emergency, did not comply with the notification requirement.[18] In the *Nicaragua–Miskitos case*,[19] it was established that the Nicaraguan Government, which had carried out the compulsory relocation of the Miskito Indians in January 1982, did not declare the state of emergency until 15 March 1982. The notice of derogation was sent on 22 March. In its report, the IACHR found that 'the formal requirements set forth in paragraph 3 of article 27 were not observed by the Government at the time of the relocation'.[20]

(b)   The Contents of the Notice of Derogation

The Derogated Provisions. Obviously the first elements which have to be indicated in the notice of derogation are the provisions which the State has derogated from. In the *Greek case*, the applicants contended that the Greek Government had failed to comply with the notification require-ment because *inter alia* it did not indicate the provisions derogated from in the Convention. The Commission found that even if the Greek Gov-ernment did not directly mention those provisions it nevertheless did so indirectly when it communicated the suspended articles of the Greek Constitution.[21]

This issue arose again in the *Ireland* v. *UK case* when the Irish Govern-ment in its application held that the UK could not derogate from article 14 of the European Convention (the discrimination provision) because there had been no mention at all of this article in the notice of deroga-tion. This issue is an important one. The Commission held in the *Greek case* that paragraph 3 did not oblige the Government to indicate expressly

---

[16] *Greek case*, Report of the Commission, paras. 80–1; pp. 42–3.
[17] See the remarks of Mr Tarnopolski, p. 370, para 35 and Mr Movchan, p. 374, para. 18, in the *Year Book HR Committee* (1979–80), vol. 1.
[18] IACHR, *Report on Guatemala*, 1983, p. 33, para. 6.
[19] IACHR, *Nicaragua–Miskitos case*, 1983.
[20] Ibid, p. 114.
[21] *Greek case*, Report of the Commission, p. 40, para. 74 (applicants); p. 42, para. 80(3) (Commission's opinion).

the articles of the Convention which had been derogated from, because this was done indirectly by the Government when it communicated the full texts of the articles suspended in its Constitution. However, in the *Ireland* v. *UK case*, the problem was whether the lack of any direct or indirect indication of an article of the Convention in the notice of derogation would have deprived the Government of the possibility of suspending that right.[22] Unfortunately, the issue was not resolved in this case. The Commission, when referring to article 15(3), left the question open until the analysis of the substantive issue relating to article 14 had been carried out. Then, after a thorough analysis, the Commission found no evidence that the measures taken by the UK Government in Northern Ireland (detention and internment) had been applied with discrimination contrary to article 14. Therefore, the Commission did not need to answer the first question.[23]

At the same time, in at least two cases in which the HR Committee has not received notices of derogation, it has asked the UN Secretary-General to address the governments concerned in order to obtain the information about the articles derogated from.[24] Furthermore, the UN HR Committee, under the Optional Protocol, has consistently held that a State cannot lawfully rely on the right of derogation from one article of the Covenant if the notice of derogation, or the information under that procedure, does not state that it is going to do so.[25]

Emergency Legislation and Administrative Decrees. The indication of the provisions derogated from is not all that is required. At least under the European Convention, States must keep the Secretary-General 'fully informed' of all the measures taken.[26] This is why the European Commission held that the Greek Government had not complied with article 15(3), since it had failed to provide information on the texts of a number of legislative measures, in particular of the new Greek Constitution of 1968 and other administrative measures.[27] In the light of this finding of the Commission, States usually enclose with the notice of derogation the relevant legislation and administrative decrees relating to the emergency; this information about the legal and administrative measures has become part of the notification requirement.

The Reasons for Derogation. States must also indicate the reasons for

---

[22] *Ireland* v. *UK*, European Court of HR, Ser. B, Report of the Commission, p. 118.
[23] Ibid., p. 241. For the finding on discrimination, see below, Ch. 6.
[24] UN HR Committee, Cases of Uruguay and Colombia (see below, pp. 74–6, 80).
[25] See doctrine of the HR Committee, below, pp. 80–1.
[26] In this respect, it could be argued that the wording of the principle of notification in the European Convention establishes a higher standard than the one in the other two treaties.
[27] *Greek case*, Report of the Commission, pp. 42–3, para. 81.

derogation. The aim of this requirement is to enable the other States parties to understand the real situation in the country so that they can assess the need for derogations. This is sometimes done through a historical explanation of the causes of the emergency.[28] Moreover, States under the European system should explain the reasons for all the measures taken. The lack of a clear statement of the reasons for derogation and the insufficiency of the information provided by the Irish Government were also of relevance in the *Lawless case*. In this case, the applicant contended that the notice of derogation sent by the Irish Government to the Secretary-General in 1956 was not a valid compliance with article 15(3) on two grounds. First, the reasons given by the Irish Government for derogation did not show that there was a public emergency threatening the life of the nation according to paragraph 1 of article 15. The notice only stated that the aim of the measures was merely 'to prevent the commission of offences against public peace and order' and to prevent the maintaining of military or armed forces other than those authorized by the Constitution. Secondly, the information about the measures and the reasons therefor was insufficient.[29] In its reply before the Commission, the Irish Government pointed out that paragraph 3 of article 15 did not demand the provision of any specific form of notice of derogation; it only required that the Secretary-General be informed of the measures taken and of the reasons therefor.[30]

The Commission found unanimously that even if paragraph 3 was drafted in general terms, certain particulars were required. It refers to the necessity of notifying the derogating measures 'without any avoidable delay' and of furnishing 'sufficient information' about them in order to enable the other States parties and the Commission to appreciate the nature and extent of the derogations. The Commission reaffirmed the primary importance of the notification requirement in the system of collective guarantees of the Convention. Taking into account these elements, the Commission found that the Irish notice was sent 'without delay' and that it provided sufficient information about the nature of the measures. Furthermore, the Irish Government attached to the notice some of the legislative acts affecting the rights suspended. However, the Commission thought that the notice of derogation was open to criticism because it did not indicate with 'sufficient clearness' the reasons which led the respondent Government to derogate from its obligations under the Convention. The Commission drew attention to the need for fuller information concerning the reasons.[31]

---

[28] Ibid. See the Greek notices of derogation, *YBECHR* 10 (1967), 26–44 and *YBECHR* 11 (1968), 10–35.

[29] *Lawless case*, Ser. B: Report of the Commission, p. 71, para. 77.

[30] Ibid., p. 72.   [31] Ibid., pp. 73–4, para. 80.

In the *France . . . v. Turkey case*, the applicant Governments contended that Turkey had not complied with its obligations under article 15(3), in so far as the reasons for derogations were stated in very general terms and no complete information was given about the various methods of derogation.[32]

The Communication of the Date of Termination. Finally, States are also obliged according to the notification requirement to make a further communication of the date on which they have terminated the derogations and from which the provisions of the treaty are again fully in force. Therefore, as was stated in the Third Committee of the General Assembly, there are two separate notifications, the first at the beginning of the derogation and the second at the end. The UN Third Committee in its eighteenth session in 1963 amended an early wording of the paragraph just to make the two separate obligations clearer.[33]

Even though the practice of States has been scarce and not always consistent, it seems that States should also notify any extension of the emergency.[34] This has been the position of the IACHR in the case of El Salvador.[35]

3. THE EFFECT OF THE FAILURE TO COMPLY WITH THE NOTIFICATION REQUIREMENT

One of the crucial questions as far as the notification requirement is concerned is the effect of a failure to comply with it on the right of derogation and on the measures taken. On the one hand, the notification requirement seems at first sight to be a formal requirement which becomes necessary after the right of derogation has been exercised and which simply has the aim of informing the other States parties. Therefore, the failure to notify would not seem to be very serious and would not put in question the validity of the exercise of the right of derogation. On the other hand, if one takes into account the importance given to it by the drafters of the treaties, it seems that, in certain circumstances, the failure to notify could affect the validity of the right itself.

[32] *France, Norway, Denmark, Sweden, and the Netherlands* v. *Turkey.*, Application No. 9940 (1982), *EHRR* 6 (1984), 245–6.
[33] UN Third Committee, 18th Session (1963), A/5655, para. 54 (see also paras. 44, 45, 46, and 55).
[34] See the practice on this point of Turkey under the ECHR and of Peru under the ICCPR. For the Turkish notices of derogation see the *YBECHR* since 1962. For the Peruvian notices see *Human Rights: Status of International Instruments (up to 1 Sept. 1987)* (UN Publications, New York, 1987), pp. 58–85.
[35] IACHR, *Annual Report*, 1984–5, p. 141. For the notice of derogation sent by El Salvador see T. Buergenthal, R. E. Norris, and D. Shelton (ed.), *Protecting Human Rights in the Americas* (3rd edn., Kehl-Strasbourg, 1990), p. 349.

This question arose in the *Lawless case* and provoked an interesting debate concerning the interpretation of the notification requirement between the Irish Government and the European Commission on Human Rights. The Irish Government introduced the issue at an early stage of the proceedings when the applicant contended that the Irish notice of derogation was not valid according to article 15(3). The Government replied that, even if the notice was defective, the right of derogation was not conditional on giving such information to the Secretary-General. Later in the proceedings, the Government construed paragraph 3 as imposing a wholly independent liability on the derogating Government; accordingly, the failure of the Government to comply with this requirement could never attract the sanction of nullity of the right of derogation.[36]

The Commission took a different position, and in the final paragraph of its report said:

The Commission is of the opinion that in the circumstances of the present case there is no question of the measure taken by the respondent Government under paragraph 1 of Article 15 being invalidated merely by reason of the inadequacy of the reasons given in the letter of 20 July 1957 for the bringing into force of the Act of 1940. In stating this opinion, however, the Commission is not to be understood as having expressed the view that in no circumstances whatever may a failure to comply with the provisions of paragraph 3 of article 15 attract the sanction of nullity of the derogation or some other sanction.[37]

The Irish Government did not agree with this position and in the hearings before the European Court insisted that the Court should rule as a matter of principle that a failure to comply with this requirement could never attract the sanction of nullity.[38] The Commission, on the other hand, maintained and further clarified its position during the hearings.

The Commission had two strong arguments for acting in this way. First, it recognized that the obligation to inform the Secretary-General was an essential link in the machinery provided in the Convention for ensuring the observance of the engagements undertaken by the States. Without such information, the other States Parties to the Convention would not know their position under article 24 (inter-State complaints). Moreover, without such information the Commission itself would be unaware of facts which may radically affect the extent of its own jurisdiction to entertain the application with respect to the acts of the States in question.

---

[36] *Lawless case*, Ser. B: Pleadings, Counter-Memorial of the Government of Ireland, pp. 227 ff.

[37] *Lawless case*, Report of the Commission, pp. 72–3, para. 80.

[38] See a similar construction of this requirement by Mr Sperduti in his dissenting opinion in the *Cyprus case*. Report of the European Commission on HR (1976) (reproduced in *EHRR* 4 (1982), 562).

Secondly, the Commission found it impossible to foresee all possible circumstances in which a government may fail to comply with the notification requirement. There could be cases of bad faith, in which a government might deliberately withhold information from the Secretary-General in order not to attract attention to controversial measures.[39] The Commission finally added:

Whether in some circumstances failure to inform the Secretary-General may attract some form of sanction as, for example, the drawing of inferences adverse to the Government or even the sanction of the nullity of the rights to derogate are questions upon which the Commission has felt unwilling to pronounce in advance of being called upon to do so in a concrete case. Neither the Cyprus case nor the Lawless case called for the decision of these questions, and all that the Commission has sought to do in these cases is to keep its own position open for future cases.[40]

Therefore, the Commission found it unnecessary for the Court to rule on this point as suggested by the Irish Government.

The European Court agreed with the position of the Commission and did not rule according to the wishes of the Irish Government that, as a matter of principle, the failure to comply with paragraph 3 of article 15 can never attract the sanction of nullity of the right of derogation.[41]

Even if the question of the consequences of a failure to notify was discussed in the *Lawless case* as a matter of principle and without referring to any particular situations, it seems that two different situations should be distinguished: one is the *total failure* to meet the notification requirement because of the absence of any notification whatsoever, and the other is a *partial failure* to meet that requirement because the notice of derogation is incomplete, does not provide sufficient information, or has not been sent on time.

(a) Total Failure to Meet the Notification Requirement: The Question of the Application ex officio of the Derogation Clause in Cases in which no Notification has been Sent

This situation is obviously more serious and there is no doubt that, in principle, a State which does not notify could lose the right of derogation. The question is whether the State which does not notify automatically loses its right of derogation, or whether there can be certain situations in which it does not lose it and therefore the international body could apply

---

[39] *Lawless case*, Ser. B: Pleadings, pp. 335–6.
[40] Ibid., p. 336. See the interesting developments of these arguments by the President of the Commission, Sir Humphrey Waldock: ibid., Hearing of 8 August 1961, pp. 384–90.
[41] *Lawless case*, Ser. A: Judgment, para. 47.

the derogation clause ex officio. Here the question does not concern a possible delay in sending the notice of derogation, but rather a total failure of notification. The problem is far from being merely theoretical. Should the international bodies apply the derogation clause ex officio when a situation of emergency exists in a given State but no notification has been sent? The problem is not easy to solve. In principle, if the derogation clause is construed in human rights treaties as a sovereign right of the State, one cannot see how, if a State does not rely on it, the international organ could apply the derogation clause *sua sponte*. It is for the holder of the right, in this case the State, to exercise it if it wishes to do so. If the international organ were to apply the right of derogation ex officio, it would put itself in the position of the State in exercising a right which only belongs to the State. On the other hand, if a State fails to notify a state of emergency for whatever reason, even for a sound reason, and there exists a *de facto* emergency situation, it would be unrealistic to accept that the standards applicable by the international body should be those normally applied in peacetime. This is especially true when some of the international organs applying human rights standards act not through an adversarial proceeding but rather under the reporting procedure (the UN HR Committee) or as fact-finding bodies (i.e. the IACHR). Of course, the risk is that an ex officio application of the derogation clause without notification would weaken this requirement. This interesting problem has provoked different reactions from the European Commission, the IACHR, and the HR Committee in cases in which no notification was sent.

(i) *The Cyprus Case.* In the *Cyprus case*,[42] the European Commission had to face for the first time a *de facto* state of emergency without any notification. The main problem discussed in the *Lawless* and in the *Greek case*, namely, whether a failure to comply with the notification requirement would attract the sanction of nullity of the derogation, seemed to push the Commission into taking a decision on it in this case in which no notification at all was sent. Surprisingly enough, the Commission managed to avoid having to rule directly on the issue. The relevant facts of the case were these.

Following the occupation of the northern part of the island by Turkey in 1974, Cyprus filed an application for several violations of human rights committed by the Turkish armed forces. Cyprus recognized that the situation was a public emergency due to a military conflict. However, it contended that the Commission was prevented from applying the derogation clause ex officio because Turkey had neither relied on the derogation

---

[42] *Cyprus case*, Report of the Commission (1976), *EHRR* 4 (1982), 482 ff.

clause nor had it notified the other States parties according to article 15(3) of the European Convention. The Cyprus Government said:

The derogation provided in Article 15 was 'a right of the State concerned': Article 15(3) spoke of the High Contracting Party 'availing itself of this right of derogation'. If the State concerned did not exercise the right of derogation no other person could invoke it, and neither the Commission nor the Court could apply it ex officio. Turkey had not invoked any right of derogation in the present case, although she had done so in the past on other occasions.[43]

Although Turkey recognized that article 15 should be examined by the Commission, it refused to recognize the competence of the Cyprus Government ('the Greek-Cypriot Administration') to bring that application, because of the violation of the Zürich and London Agreements and of the Treaties of Nicosia of 1960. It therefore refused to submit observations on the merits.

The Commission's position was difficult. On the one hand, there was clearly a *de facto* state of emergency recognized by both parties. On the other hand, Turkey did not formally rely on the right of derogation and did not give notification of it. Therefore the question was: in these circumstances, could the derogation clause be applied ex officio? The Commission did not want to rule that because of the lack of notification the right of derogation could not be applied. Its finding was more subtle. After recalling the doctrine of the Commission since the *First Cyprus case*, it found that 'in the present case the Commission still does not consider itself called upon generally to determine the above question'. And it added:

It finds however, that, in any case, Article 15 requires some formal and public act of derogation, such as a declaration of martial law or state of emergency, and that, where no such act has been proclaimed by the High Contracting Party concerned, although it was not in the circumstances prevented from doing so, Article 15 cannot apply.[44]

It seems, therefore, that the Commission, without solving the question of the effect of a lack of notification on the right of derogation, considered that a formal proclamation is an essential requirement according to article 15 of the Convention. Moreover, it seems to be so essential that, without it, the right of derogation cannot be exercised.[45]

The possibility of applying the derogation clause ex officio seems to have been explored by the Commission. In his separate opinion, Mr Ermacora discussed this question. For him, the notification requirement

[43] Ibid., p. 533, para. 511.
[44] *Cyprus case*, Report of the Commission, p. 556, para. 527.
[45] See Ch. 2, pp. 37–8.

is an essential condition for the derogation; consequently the Commission cannot apply the derogation clause ex officio, because only the State (which is the holder of the right) can put the right of derogation into practice.[46]

The decision of the Commission upheld the importance of the notification requirement, and in relation to it, the necessity of a formal promulgation, according to the terms of the Convention.

(ii) *The Bolivian Case*.[47] In the *Bolivian case*, however, the IACHR acted differently. The Commission knew through press reports that a *coup d'état* had taken place in Bolivia on 17 July 1980. The military junta which took power suspended constitutional guarantees but did not send any notice of derogation to the Secretary-General of the OAS as required by article 27(3). Nevertheless, the IACHR asked the Government in an official letter for information about those legal provisions enacted which could affect the observance of human rights and which had suspended the obligations that Bolivia undertook according to the American Convention. Surprisingly enough, the Government in answering the letter pointed out that 'no legal provision has been enacted that is repressive or prejudicial to the observance of human rights'. However, the Commission, in its report, applied to Bolivia the standards of the derogation clause despite the absence of notification and the Government's continuing denial of any suspension of human rights.[48]

(iii) *The Attitude of the UN HR Committee*. The UN HR Committee has generally been quite cautious about the consequences of a failure to notify and about the possibility of applying the derogation clause ex officio.

1. Under the *reporting procedure* it has consistently pointed out that, if a State has not declared and notified the state of emergency, it cannot take into account the difficulties of a possible emergency situation in order to apply the derogation clause.[49] For example, in the dialogue following the *Sri Lankan Report* in 1983, one of the members of the Committee states quite clearly that if there is no notification of the state of emergency, as was the case then, the State should be held accountable under the normal standards of the Covenant.[50]

---

[46] *Cyprus case*, separate opinion of Mr Ermacora, pp. 565 ff. See, however, the interesting dissenting opinions of Prof. Dr Daver (pp. 579 ff.) and Mr Sperduti (pp. 561 ff.).

[47] IACHR, *Report on Bolivia*, 1981. (See also a good exposition of this case in Buergenthal, Norris, and Shelton (eds.), *Protecting Human Rights*, pp. 352–4.)

[48] IACHR, *Report on Bolivia*, 1981, p. 23.

[49] UN HR Committee, Review Reports on Rwanda, Sudan, Syria.

[50] Ibid., Report of Sri Lanka, CCPR/C/SR. 473 (1983), Mr Opsahl. See also A/37/40, Report of Rwanda (1981), p. 54, para. 247.

2. Under the *Optional Protocol*, even if the Committee has, as a matter of principle, left open the question of the nullity of the right of derogation because of the absence of notification, it has held that a State cannot rely on this right if it does not provide enough information through either the notice of derogation or the proceedings under the Optional Protocol.[51]

To conclude, it seems that, as a general rule, States will lose the right of derogation if they do not rely on it and do not comply with the notification requirement. International bodies have so far rejected the application of the derogation clause ex officio, especially when there is an adversarial procedure. This position is very much in line with the construction of the derogation as a sovereign right which belongs only to the State. However, one cannot foresee all possible situations in which States might find themselves; it could be that, due to particular circumstances, a failure to notify derogations would not automatically provoke the nullity of the derogating measures. Consequently, this possibility should be left open.

On the other hand, international bodies, when exercising fact-finding functions (i.e. when they produce a report on the situation of a given country) may take into account the *de facto* state of emergency in order to apply the derogation standards, even if the State in question did not notify it to the other States parties. This has been the position of the IACHR in the *Bolivian case*. Nevertheless, these bodies should be aware of the possibility of weakening the notification requirement if they do not insist on that obligation and if they apply the derogation clause ex officio.

(b)   Partial Failure to Meet the Notification Requirement

The question of the consequences of a partial failure to comply with these formal requirements upon the right of derogation has not been resolved by the international bodies. So far, they have left the question open. At the same time, they have refused to declare as a matter of principle that a failure to meet the formal requirements of the notification procedure can never attract the sanction of nullity. The best view seems to be that the duty to furnish timely and complete information is an autonomous obligation of the derogation clause, and that failure to provide this information cannot normally deprive the derogating State of its right of derogation. However, as the European Commission pointed out, in some cases in which there might be bad faith and the intention of hiding information

---

[51] *Landinelli case*, UN HR Committee, Communication No. R. 8/34, A/36/40 (1981), p. 132, para. 8(3). See below, pp. 75–6.

from the rest of the States or from the international organs, the finding could be different.

In the *Lawless case*, the Court and the Commission found that, even if the Irish notice of derogation did not clearly state the reasons for derogation, the notice provided sufficient information according to article 15(3).[52] In the *Greek case*, the European Commission found that Greece did not meet the formal requirements of paragraph 3 on the following two grounds: first, the reasons for derogation were sent four months after the declaration of the emergency; secondly, the Greek Government failed to send both the emergency legislation, including the new Greek Constitution of 1968, and the administrative decrees concerning measures of derogation.[53]

However, it did not decide on the effect that this failure would have on the right of derogation, because the Commission also found that the situation did not constitute a public emergency threatening the life of the nation and, therefore, the right of derogation could not apply.

In his dissenting opinion, Mr Delahaye pointed out that the Commission had interpreted article 15(3) in a very strict way. In his opinion, what is required is a liberal interpretation of the notification requirement. If this does not happen, all derogating states will be found to be violating it.[54] It should be recalled that Greece in fact sent two timely notices of derogation with some information.[55] It is interesting to note that the UK representative in the drafting of the Covenant also expressed his concern about a strict interpretation of this requirement.[56]

Professor Fawcett expressed an interesting dissenting opinion in this case precisely to avoid the problem of a derogating State which has sent one or more notices of derogation and is later found to be in violation of the notification requirement.[57] Professor Fawcett did not agree with the conclusion of the Commission that the Greek Government was in breach of article 15(3), and he raised the problem of the depositary powers of the Secretary-General. According to his view, in order for article 15(3) to be effective, it must also impose obligations on the Secretary-General of the Council of Europe. He said:

(If the notification) did not meet the requirements of full information under Article 15, paragraph 3, then it was the duty of the Secretary-General ... to request fuller information from the respondent Government. In the absence of

---

[52] *Lawless case*, Ser. A: Judgment, p. 62, para. 48; Ser. B: Report of the Commission, p. 73.

[53] *Greek case*, Report of the Commission, *YBECHR* 12 (1969), P. 81, paras. 42–3.

[54] Ibid., Dissenting opinion of Mr Delahaye, pp. 43–4, paras. 82–5.

[55] *Greek case*, Report of the Commission, p. 42, para. 80.

[56] E/CN. 4/SR. 331, p. 8, Mr Hoare (UK). See also A/2929, para. 47.

[57] *Greek case*, Dissenting opinion of Mr Fawcett, p. 44, para. 86.

any such request, the respondent Government was entitled to assume that it had complied with Article 15, paragraph 3, and that it was not necessary to send further information.[58]

Professor Fawcett's construction of paragraph 3 of the derogation clause is sound. The exercise of these powers seems to be more feasible under the European Convention.[59] However, it would be equally desirable under the ICCPR; thus, the UN Secretary-General should ask derogating States for more information when the notice of derogation is not sufficient.[60] In fact, he has exercised these powers in relation to notifications on derogations on at least two occasions.[61] The IACHR Special Rapporteur on States of Emergency held the same position as far as the American Convention was concerned.[62]

4. THE DISTINCTION BETWEEN THE OBLIGATION OF NOTIFYING AND THE OBLIGATION OF PROVIDING INFORMATION TO INTERNATIONAL BODIES UNDER OTHER PROCEDURES

Another important problem which has to some extent reduced the effectivenes of the principle of notification has been what seems to be a lack of a clear distinction in some international bodies, particularly within the HR Committee, between two different obligations of States: one is that of notifying the other States parties, and the other that of giving full information on the emergency to, for instance, the HR Committee, under other relevant mechanisms (in the case of the Covenant through the

---

[58] Ibid.

[59] In support of this power of the Secretary-General of the Council of Europe, art. 57 of the ECHR has been adduced. See also 'Declaration of the Secretary-General of the Council of Europe on Article 57 of the European Convention on Human Rights, made before the Juridical Commission of the Consultative Parliamentary Assembly at Oslo, 29 Aug 1964, Council of Europe, *Collected Texts of the ECHR* (Strasbourg, 1979), p. 91 (quoted in N. Questiaux, *Study of the Implications for Human Rights of Recent Developments concerning Situations Known as State of Siege or Emergency*, E/CN. 4/Sub. 2/1982/15, p. 15 n. 11).

[60] See art. 77(*d*) of the Vienna Convention on the Law of Treaties. See also International Law Commission, A/CN. 4/L. 116/Add. 9, p. 4, Commentary adopted at the 894th meeting on art. 29(1) (*d*), draft 1966 (Vienna Convention on the Law of Treaties). See also S. Rosenne, 'The Depositary of International Treaties', *AJIL* 61 (1967), 923–45. These articles of the Vienna Convention on the depositary powers of the Secretary-General are based on the practice of the UN Secretary-General as a depositary of multilateral treaties; see 'Summary of the Practice of the Secretary-General as Depositary of Multilateral Agreements', ST/LEG/7.

[61] See the notice of derogation sent by the Colombian Government in *Human Rights: Status of International Instruments*, pp. 61–2. See also in the case of Uruguay, A/34/40, p. 16, para. 65.

[62] Martins, *The Protection of Human Rights*, p. 148.

reporting procedure and the Optional Protocol). This important distinction, which could have serious consequences, was perceptively noted by the European Commission of Human Rights in the *Greek case*. The Commission said,

in any event, information given to the Commission or a Sub-Commission in proceedings under Article 24 or 25 cannot rank as, or replace, information required under Article 15, paragraph (3), since information communicated under this provision is to be brought to the knowledge of all High Contracting parties and of the Convention organs while that given to the Commission or Sub-Commission is limited to that organ and the parties before it.[63]

As a matter of principle and as far as the UN Covenant is concerned, it is important to distinguish between these two different obligations, which also have different aims and arise from different sources.

1. The first obligation is aimed primarily at informing the other States parties of the exact situation of the derogating State in respect to the Covenant in order to enable them to exercise their rights, and also to inform the HR Committee whose jurisdiction could be affected. The second is aimed exclusively at informing the HR Committee, and eventually the applicant under the Optional Protocol, in order to enable it to discharge its functions according to these procedures.

2. The first obligation arises from article 4(3), the second from a completely different source: from article 40, in the case of the reporting procedure, and from article 4(2) of the Optional Protocol, in the case of an individual communication.

Because these are two different obligations with different aims and sources, a State which only fulfils the latter should not be considered to have fulfilled the former. Furthermore, these two obligations should not be confused by the Committee. Even if the HR Committee can get information through the notice of derogation in order to discharge its functions, the main aim of the principle of notification is to keep the other States parties fully informed. This should be emphasized by the HR Committee when addressing derogating States on article 4. The Committee should not accept that the information given to the Committee under other procedures could compensate for an incomplete, insufficient, or late notice of derogation. In these cases, it must also require that supplementary information be sent to the other States parties according to article 4(3). If this information is not sent to the other States parties, the implementation of the Covenant and the respect for human rights standards in emergencies would suffer.

It is true that the HR Committee, in its 'general comment' on article

---

[63] *Greek case*, Report of the Commission, p. 42, para. 80(6).

4, after insisting on compliance with the substantive provisions of the derogation clause, has pointed to the importance of the notification requirement. At the same time, it mentioned in the same paragraph the necessity of providing information on derogations under the reporting procedure. The HR Committee said:

> The Committee also considers that it is equally important for States parties, in times of public emergency, to inform the other States parties of the nature and extent of the derogations they have made and of the reasons therefor and, further, to fulfil their reporting obligations under article 40 of the Covenant by indicating the nature and extent of each right derogated from together with the relevant documentation.[64]

Of course, this is a perfectly healthy policy and very much in accordance with the Covenant, as long as it is made clear that both obligations must be complied with. This is what the Committee seems to have failed to do, provoking some confusion in the practice of States. Three examples will illustrate this assertion.

The first is the case of *Uruguay*. On 30 July 1979, the UN Secretary-General received a notice of derogation from the Government of Uruguay.[65] The HR Committee decided, after receiving what was clearly an insufficient notice of derogation, to request the UN Secretary-General to address the Uruguayan Government and to request it do submit its report, which had been due in 1977, as soon as possible, and to include in it detailed information about the rights derogated from, the extent of the derogation, and the justification for the derogation in respect of each right derogated from.[66] The Committee did not ask for more information to be sent to the other States parties in order to keep them fully informed of the situation according to the notification requirement, but only for the purpose of the knowledge of the Committee under the reporting procedure.[67]

A second illustration of the confusion of the two obligations mentioned above can be found in the original case of the *oral presentation* of the derogation to the HR Committee by the representative of *Ecuador*. In 1982, Ecuador declared the extension of the state of emergency from 20 to 25 October owing to serious disorders brought about by the suppression of subsidies. At that time, the Committee was in session and the Ecuadorian representative asked for the Committee's permission to make

---

[64] A/36/40 (1981), p. 110.

[65] See the Uruguayan notice in *Human Rights: Status of International Instruments*, pp. 84–5.

[66] A/34/40 (1979), p. 16, para. 65.

[67] See, however, the remarks of some members on the failure of Uruguay to meet the requirements of art. 4(3), in A/37/40 (1982), p. 59.

an oral presentation of the state of emergency with a due explanation of all the measures taken and the reasons therefor.[68] In his submission, the representative said that his Government had sent the UN Secretary-General a communication dated 25 October 1982. It is interesting to note that in the official records of notification, the Ecuadorian notice of derogation appeared with the date of 12 May 1983, that is, almost seven months later.[69]

Even though the oral presentation of the derogation to the Committee, when in session, was undoubtedly an interesting development (especially as it occurred between periodic reports) and was welcomed by the Committee, it should never in fact become a substitute for the obligation of immediate notification to the other States parties with all the formal requirements of paragraph 3 of article 4.[70]

The third example is the *Landinelli case*,[71] this time under the Optional Protocol procedure. The HR Committee has received a very large number of communications from Uruguayan citizens under the Optional Protocol. In many of these complaints, the issue of derogation was under discussion. In the *Landinelli case*, in which the Government relied on the notice of derogation, the Committee felt unable to accept that a public emergency threatening the life of the nation existed in Uruguay at that time, due to the fact that factual details concerning the emergency and the nature and scope of the derogations had not been provided by the Government. The Committee was of the opinion that a State cannot evade the obligations which it has undertaken by ratifying the Covenant merely by invoking in abstract terms the existence of exceptional circumstances. And in any case, the Committee made it clear that

Although the substantive right to take derogatory measures may not depend on a formal notification being made pursuant to article 4(3) of the Covenant, the State party concerned is duty-bound to give a sufficiently detailed account of the relevant facts when it invokes article 4(1) of the Covenant in proceedings under the Optional Protocol.[72]

It seems that what the Committee wanted to emphasize in this case was the need for full and comprehensive information in order to assess whether a public emergency actually exists. If the respondent Government does not give that information either through the notice of deroga-

---

[68] CCPR/C/SR. 406 (1982), p. 203.

[69] *Human Rights: Status of International Instruments*, p. 62.

[70] CCPR/C/SR. 406, p. 3, para. 10. In fact, Mr Tomuschat considered that Ecuador had complied with this requirement, although the notice was sent 7 months after the proclamation and contained very little information about the emergency and the measures taken.

[71] *Landinelli case*, UN HR Committee, A/36/40, pp. 130 ff.

[72] Ibid., p. 132, para. 8 (3).

tion foreseen in article 4(3), or through the procedure under the Optional Protocol, article 4(2), the HR Committee 'cannot conclude that valid reasons exist to legitimise a departure from the normal legal regime prescribed by the Covenant'.[73] The Committee did not want to pronounce as a matter of principle on the effect of the failure fully to comply with the notification requirement, and therefore left the question open; however, even though it emphasized the importance of the obligation of informing the Committee according to the Optional Protocol (article 4(2)), it failed to do the same with the State obligation of fully notifying the other States parties.

5. THE PRACTICE OF NOTIFICATION UNDER THE THREE TREATIES

(a)   The European Convention

If it is compared with the other two treaties, the practice of notification under the European Convention has been very consistent. In other words, derogating States have consistently fulfilled the notification requirement since the beginning of the Convention.[74] By way of illustration, Turkey has sent forty-eight notices to the Secretary-General of the Council of Europe up to 1986. This practice has allowed the other States parties to be informed of the extent of derogations and to exercise their rights when they consider that a violation of the derogation clause has taken place. It is interesting to realize that most of the relevant cases on derogation under the European Convention have been initiated through the inter-State complaint procedure (article 24).[75] The great advantage of the European system is that the inter-State complaint procedure automatically comes into operation on the signing of the Convention (article 24). Since the very early days of the Convention, States have challenged notices of derogations; thus, in 1956 Greece filed a complaint against the UK because of the UK derogations in Cyprus, and it contended that the UK notice was 'irregular in form'.[76]

The practice of notification in the European system has been the following:

1. The derogating state addresses to the Secretary-General a *note verbale* summarily indicating the grounds invoked for the emergency, a

---

[73] Ibid.
[74] The notices of derogation under the European Convention can be found in the *YBECHR*.
[75] *First Cyprus case*; *Greek case*; *Ireland v. UK*; *Cyprus case*; *Turkey case*.
[76] *First Cyprus case*, *YBECHR* 2 (1958–9), 174.

list of provisions of the Convention which are to be suspended, and
the expected period of derogation and its geographical extension. The
emergency clauses of municipal law referred to in the note are often
appended;

2. The Secretary-General acknowledges receipt;

3. He then notifies the derogation to the other States parties by trans-
mitting to them a copy of the *note verbale*. The other documents appended
could be sent by request;

4. Finally, the Secretary-General transmits a copy of the *note verbale*,
for information, to the Presidents of the Commission, the Court, and the
Parliamentary Assembly.[77]

Since the beginning, the European Commission and Court have always
affirmed their competence to examine whether notices of derogation meet
the requirements of article 15(3). Moreover, in the *Ireland* v. *UK Case*,
the Court decided to examine *proprio motu* the fulfilment of the requirement, without any allegation of irregularities from the applicant.[78] The
two major issues in the case-law under the European Convention have
been the question of what the exact formal requirements of the notice
are, and what effect a failure to meet those requirements could have upon
the right of derogation. The doctrine on notification applied by the
European Court and Commission in the most relevant cases, namely, the
Lawless case, the *Greek case*, the *Cyprus case*, the *Ireland* v. *UK case*,
and the *Turkey case* has already been studied in the sections above.

(b)   The ICCPR

The first impression that one gets when coming across the practice of
notification under the derogation clause in the Covenant is that there is
a widespread failure to notify the other States parties of the measures of
derogation taken, according to article 4(3). Many states do not notify at
all; other States' notices of derogation are too general, too brief, and do
not give a clear indication of what articles of the Covenant have been
suspended. Sometimes, the notices are also too legalistic, and do not
specify the reasons either for the derogation or for the concrete measures
taken.[79] Some others do not arrive within a reasonable period of time,

---

[77] Questiaux, *Study of the Implications for Human Rights*, E/CN. 4/Sub. 2/1982/15, p. 13.

[78] *Ireland* v. *UK*, European Court of HR, Ser. A: Judgment, vol. 25, p. 84, para. 232.

[79] All notices of derogation under the ICCPR can be found in *Human Rights: Status of
International Instruments*. For notices which are *too brief*: Ecuador, 20 Mar. 1984; Sri
Lanka, 21 May 1984. In contrast, see the long explanation by Nicaragua: 13 May 1987. For
notices which are *too general*: Uruguay, 30 July 1979. With no indication of the derogated
provisions: Bolivia, 29 May 1986; Colombia, 18 July 1980. No indication of the reasons:
Ecuador, 12 May 1983; El Salvador, 14 Nov. 1983.

and often there is no indication as to when the derogation terminates.[80] In addition, if the derogation is extended after the fixed period mentioned in the notice, some States do not give notification of it. This widespread failure happens in spite of the clearly stated terms found in article 4(3), the great importance given to this requirement by the drafters of the Covenant, and the occasional insistence of the HR Committee on the need to comply.

Leaving aside some surprising statements, such as the one made by the Chilean representatives[81] before the HR Committee in which they claimed that they were unaware of the obligation of notifying according to article 4(3), and cases of bad faith, the main reasons for this widespread failure to comply with the notification requirement seem to be the following:

1. The lack of clarity on the part of many Governments about their legal situation. In the presentation of reports to the HR Committee, it appears that States are sometimes not clear at all about the true legal situation of the country as far as derogations from the Covenant are concerned. This is due in part to the complexity of the emergency legislation with its accumulation of decrees, laws, and administrative orders, which makes it very difficult, even for the State concerned, to know the exact extent of derogations, or, in other words, the exact extent of the enjoyment of human rights in the country.

2. The fear of international criticism. Some states seem to assume that derogations necessarily imply human rights violations, and they fear international criticism for using the derogation clause and therefore do not want to acknowledge officially that a public emergency exists in the country. However, the derogation clause in human rights treaties has been established for the precise purpose of lawfully suspending some human rights in emergency situations.

3. The explicit refusal of some States to rely on the right of derogation even if a *de facto* state of emergency exists in the country. One of the most common situations in certain countries is a *de facto* state of emergency due to an international war, or internal strife. In these States, the emergency is often not declared, and is certainly not communicated to the other States parties according to article 4(3).

In line with the nature of the right of derogation it is true that, even if a *de facto* state of emergency actually exists, States are free to decide whether to rely on it or not. Several states have given good reasons to the

---

[80] Chile, 7 Sep. 1976 (almost 6 months later); Colombia, 18 July 1980 (four years later); Ecuador, 12 May 1983 (7 months later). In one single note, the derogation and the termination: Ecuador, 12 May 1983.

[81] CCPR/C/SR. 531 (1984), p. 5, para. 15.

HR Committee for not using the right of derogation. Thus, *Cyprus* declared that even though they recognized the existence of a *de facto* state of emergency in the country after the occupation of the northern part of the island by Turkish troops, they did not proclaim it and they did not therefore derogate from normal standards, because this could have had adverse effects on the population.[82] Similarly, *Suriname* did not declare a state of emergency despite the fact that an emergency situation had existed in the country for two months, because they could cope with it using normal peacetime powers.[83]

An important point to note is that the obligation of notifying the other States parties arises only when a State takes measures derogating from the provisions of the Covenant. This means that it is perfectly possible to have proclaimed a state of emergency in the country for internal purposes, without derogating from any provision of the Covenant; in this case, the obligation of notifying the state of emergency does not arise.[84] A similar situation can be found when a State declares that the derogations have terminated not because the emergency has finished but because it can cope with it without derogations. This was the case in the *UK* derogations as far as Northern Ireland is concerned.[85]

4. As far as some technical failures in the notices of derogation are concerned, some States representatives have acknowledged their lack of expertise in the functioning of the derogation clause. For example, the Nicaraguan representatives stated in 1983 before the Committee that they had included in their notice of derogation all the non-derogable rights of the Covenant as derogated provisions due to their lack of expertise, but that in fact many of those rights were really still in force.[86]

5. Finally, another possible explanation for this widespread failure to comply with the principle of notification under the Covenant can be found in the lack of operation of the inter-State complaints procedure. So far very few States have accepted the procedure found in article 41. In the European system, the inter-State complaints procedure has been the one which has best ensured respect for human rights in emergencies. In other words, States parties have exercised their rights under the Convention when violations in emergencies have taken place in other States. In those complaints, States have several times pointed out the violation of the notification requirement by the derogating State, thus making it easier for

---

[82] A/34/40 (1979), p. 91, para. 383.

[83] A/35/40 (1980), p. 59, para. 297.

[84] See the clear statement by Sir Vincent Evans in the review of the Chilean report. CCPR/C/SR. 520 (1984), p. 6, para. 27.

[85] A/40/40 (1985), p. 100, para. 529. For the UK notices of derogation: *Human Rights: Status of International Instruments*, p. 84.

[86] A/38/40 (1983), p. 55, para. 244.

the European bodies to examine the fulfilment of that requirement and to affirm the right of the States parties to the Convention to be fully informed on derogations.

The Practice under the ICCPR. Another problem that members of the Committee have found in reviewing States' reports has been the total unwillingness to answer questions on derogations by some representatives;[87] this fact has prevented the Committee from making an objective assessment of the situation and consequently from insisting on compliance with the notification requirement. However, on some occasions when there has been a clear emergency situation, the Committee has asked the governments concerned officially to declare the state of emergency and to notify it to the other States parties. It is difficult to understand, for instance, why States like Syria, El Salvador (before 1983), Iraq, Sri Lanka, Jordan, Iran, Lebanon, Afghanistan, and Colombia (before 1980), had not declared and notified the emergency situation to the other States parties. As a result of the Committee's suggestion of officially proclaiming and notifying the *de facto* states of emergency, some States have fulfilled the latter obligation immediately after the review of the reports.[88]

Under the Optional Protocol. The HR Committee has so far never found that a government could not rely on the right of derogation merely because its notice of derogation was incomplete, insufficient, or actually non-existent. It has left that question open in principle, and in practice it has combined the information provided by the notice with the information provided by the proceedings. Accordingly, it seems that a defective notice of derogation can always be compensated for by information provided under the Optional Protocol procedure. If this is so, the importance of the notification requirement will suffer a set-back. Nevertheless, the HR Committee has adopted some interesting policies relating to derogation and the requirement of notification under the Optional Protocol procedures.

1. If a State does not rely on article 4 through a notice of derogation or in the proceedings under the Optional Protocol, the Committee cannot apply it ex officio. This construction is logical and in line with the nature of the right of derogation as a sovereign right of the State. If a State does

----

[87] The case of the representatives of Iran in 1982, A/37/40, p. 68, para. 306; of Iraq in 1980, A/35/40, pp. 27 ff.; and of Afghanistan in 1985, A/40/40, pp. 117 ff.

[88] Colombia (report reviewed on July 1980; notification sent on 18 July 1980). El Salvador (report reviewed on 27 Oct. 1983; notification, 14 Nov. 1983; see also CCPRC/SR. 468, p. 5, para. 24). Sri Lanka (report reviewed on 3 Nov. 1983; notification, 21 May 1984; see also A/39/40, p. 25, para. 123). Jordan: A/37/40 (1981), para. 201.

not rely on it (even if a *de facto* state of emergency exists) it cannot be applied.[89]

2. As was mentioned above, the Committee pointed out in the *Landinelli case* that even if a State relies on the right of derogation, if it does not give enough information about the emergency situation and the need for derogation from normal standards, either in the notice or in the procedures under the Optional Protocol, the derogation clause cannot be applied.[90]

3. At the same time, if a State fails to mention in the notice of derogation or in the following proceedings that it has derogated from a particular provision of the Covenant, it cannot rely on it by merely referring to the general situation of emergency. In the *Salgar de Montejo case*, the Colombian Government declared that temporary measures had been adopted which limited the application of article 19(2) and article 21 of the Covenant. However, the issue under discussion in that case was related to article 14(5). The Committee found, therefore, that Mrs Salgar de Montejo was denied the right to a review of her conviction by a higher tribunal (article 14(5)), because that right was not lawfully derogated from by the Colombian Government.[91]

(c) The ACHR

The first thing that strikes those who study the practice of derogations under the American Convention as compared to the practice under the other two treaties is that no official publication of the notices of derogation can be found in it. The European System publishes the notices in the *Year Book of the European Commission on Human Rights*, and, under the Covenant the notices are periodically published in the Multilateral treaties deposited with the UN Secretary-General. The notices of derogation cannot be found in the recent works on the American Convention, not even in the very complete four-volume work produced by Buerghental. Indirect, though incomplete, information is found in Despouy's study on derogations.[92]

---

[89] *Hiber Conteris case*, UN HR Committee, Communication No. 139/1983, A/40/40 (1985), p. 201, para. 7 (5).

[90] See above, pp. 75–6.

[91] *Salgar de Montejo case*, UN HR Committee, Communication No. 64/1979, A/37/40 (1982), p. 172, paras. 10 and 11. Also *Fals Borda case*, UN HR Committee, Communication No. 46/1979, A/37/40 (1982), paras. 13 and 14. In this case the Committee found a violation of article 9(3)(4), which had not been officially derogated from.

[92] UN Sub-Commission on Prevention of Discrimination and Protection of Minorities, *First Annual Report and List of States which, since 1 January 1985, have Proclaimed, Extended or Terminated a State of Emergency, Presented by Mr. Leandro* Despouy, *Special Rapporteur Appointed Pursuant to Economic and Social Council Resolution 1985/37*, E/CN. 4/Sub. 2/1987/19 (updated yearly).

At the same time, there are very few references to the principle of notification in the IACHR reports either in relation to particular States or in its annual report, let alone an analysis of the fulfilment of the formal requirements of article 27(3) by the notices of derogation sent by States. By way of example, there was a special section in the IACHR *Annual Report 1980–1* on states of emergency in ten countries; among these countries there were six States parties to the American Convention. However, there was no reference at all to the obligation of notifying, although only two out of these six States parties had partially complied with the notification procedure. Moreover, there is a set of recommendations at the end of the report on states of emergency, but, surprisingly enough, there is no mention of the need of notifying. The Commission only raised the question in the *Bolivian case*, but even there it provided no detailed study of this question.[93]

Although the IACHR has emphasized in its reports other substantive requirements of the derogation clause, but not the notification requirement, there have been three cases in which the Commission has called attention to the failure of States to comply with that formal requirement— in the case of El Salvador for not notifying the extensions of the emergencies,[94] and in the cases of Bolivia and Colombia for not sending the notice of derogation.[95] In the case of Colombia, the recommendations of the IACHR produced some results: after the Commission's report, Colombia lifted the state of emergency and gave notification of this to the other States parties through the Secretary-General of the OAS.[96] Colombia also notified a new state of emergency in 1984.[97]

In two other cases, the IACHR has pointed out that certain States have not complied with the notification requirement of article 27(3), because they did not send the notice of derogation in time. These were the *Guatemala case* and the *Nicaragua–Miskitos case*.[98]

In conclusion, it seems that compliance with the principle of notification within the Inter-American system and the possibility of effectively monitoring states of emergencies would depend, to some extent, on the vigour with which the IACHR insists on the importance of complying with this requirement. The pattern over recent years is that several Latin-

---

[93] IACHR, *Annual Report, 1980–1*. The six states parties were Bolivia, Colombia, El Salvador, Grenada, Haiti, and Nicaragua. Only El Salvador and Nicaragua partially complied with the requirement. See also IACHR, *Report on Bolivia*, 1981, pp. 22 and 116.

[94] See above, p. 64.

[95] For Bolivia, see IACHR, *Report on Bolivia*, 1981, pp. 22 (para. 3.5) and 116; see also p. 24. For Colombia, see IACHR, *Report on Colombia*, 1981, OEA/Ser. L/V/II. 53, doc. 2, p. 221.

[96] IACHR, *Annual Report, 1981–2*, p. 110.

[97] See UN Sub-Commission . . . , *First Annual Report . . . by Mr. Leandro Despouy*, p. 22.

[98] See above, p. 61.

American States have complied with the notification requirement under the UNCCPR, but not under the ACHR; this is a worrying state of affairs.[99] Having said that, it is also true that, even under the ACHR, more Latin-American States have recently notified states of emergency.[100]

6. CONCLUSION

1. The three main treaties under consideration contain the requirement of notification. According to this requirement, the States availing themselves of the right of derogation shall inform the other States parties of the provisions derogated from and the reasons therefore. The rationale behind this requirement is that States have the right to be notified in order to be able to exercise their rights under the treaty (i.e. to challenge derogations through the inter-State complaints procedure).

2. In the light of the jurisprudence of the international bodies, the notices of derogation have to fulfil some formal requirements. First of all, the notice must be sent in reasonable time (twelve days after the proclamation was found to be acceptable, but not four months). Secondly, the notice must give information about the provisions of the treaty which have been derogated from; this could be done *directly*, that is, by explicitly mentioning the articles of the treaty which have been suspended, or *indirectly*, by mentioning the Constitutional provisions suspended and the measures taken. The notices of derogation usually include copies of the emergency legislation and administrative decrees affecting human rights guarantees. Thirdly, the reasons for derogations should be included in the notice, in order to enable the other States to understand the real situation in the country so that they can assess the need for derogation. Finally, States are also obliged to send a further communication concerning the date on which they have terminated the derogations and from which the provisions of the treaty are again fully in force.

3. One of the main questions that arises, as far as this principle of notification is concerned, concerns the effect that a failure to comply with this principle could have on the right of derogation and of the measures taken. On the one hand, it seems that as a formal requirement, the failure to comply with it cannot invalidate the right of derogation; on the other, if one takes into account that the drafters of the treaties considered

---

[99] Bolivia in 1985 and 1986; Ecuador in 1986; Colombia in 1980; El Salvador in 1983, 1984, and 1985; Panama in 1987. Paraguay notified the state of emergency to the UN Secretary-General even though it was not party to the Covenant. It did not notify the OAS Secretary-General.

[100] Argentina in 1985; Nicaragua in 1985 and 1986; Peru from 1980 onwards; Suriname in 1986.

it as an essential requirement because of its very important aim, it could attract in some cases the sanction of nullity. However, two situations could be distinguished here: one is the total failure to notify States parties of the derogations; the other, a partial failure, that is, when the notice of derogation contains insufficient information or was not sent on time. In this latter case, it seems that international bodies would not deny the application of the right of derogation, although they would point out that there has been a violation of a treaty obligation (the principle of notification). Nevertheless, international bodies have not agreed to rule as a matter of principle that a failure to comply with these formal requirements in no case whatsoever could attract the sanction of nullity of the right of derogation.

The former case (when there is no notification at all) presents more problems, especially when the lack of notification is accompanied by the State's non-reliance on the right of derogation during the eventual proceedings (i.e. the *Cyprus case* under the European Convention). In these cases, international bodies, when assessing the situation, are faced with the problem of whether to apply the principles of the derogation clause ex officio. International bodies have not resolved the problem of whether a lack of notification attracts the sanction of nullity of the right of derogation; however, when States in the proceedings have not relied on the right, they have not applied the principles of the derogation clause ex officio (i.e. the European Commission in the *Cyprus case*, the HR Committee under the Optional Protocol). This is very much in line with the nature of the derogation clause, which is conceived as a right of States. Therefore if States do not rely on it, international bodies could not apply it *sua sponte*. However, in cases in which fact-finding bodies (i.e. the IACHR) have had to assess the situation of human rights in a given country which finds itself in an emergency, they have applied the principles of the derogation clause in spite of the absence of notification. This position, which is understandable, could weaken the principle of notification.

4. In practice, the notices of derogation do not only serve to inform States parties to the treaties, but also to provide information for the monitoring organs; this is important because the situation of emergency and the derogations could affect the extent of the jurisdiction of these bodies. Some of these bodies, especially the UN HR Committee, have not always made a clear distinction between the two different obligations which States have: one is to notify the other States, and the other to provide those bodies with information under the different procedures (i.e. the reporting procedure of article 40, or under the Optional Protocol). The HR Committee has put greater emphasis on this second obligation, with the risk of weakening the obligation of notifying because

they have not sufficiently insisted on the obligation of States to send full information to the other parties. This could have damaging consequences for the international protection of human rights, because States would not be aware of the exact situation in the derogating State and therefore would be prevented from exercising their rights to ensure the fulfilment of treaty obligations through, for instance, the inter-State complaint procedure.

5. The practice of notifications under the three treaties presents different features. Under the *European Convention*, the practice of notification has been very consistent in the sense that derogating States have fulfilled this obligation since the beginning of the Convention. These notifications have allowed the other States parties to be informed of the situation and to exercise their rights under the Convention. Thus, due to the fact that the inter-State complaint procedure automatically comes into operation on the signing of the Convention, States have effectively challenged derogations when they were prima-facie not justified by events or disproportionate to the emergency. In fact, most of the leading cases on derogations were brought to the European organs under this procedure. The European organs have always affirmed their competence to check whether the notices of derogation met the requirements of article 15(3), and on some occasions have found violations in this respect due to the lack of sufficient information or because the notices had not been sent in time.

Under the ICCPR, there has been a widespread failure to notify States parties of derogations. Thus, many States have not notified at all; other States have sent notices which were too general, too brief, or without giving any indication of the articles of the Covenant derogated from, or of the reasons therefor. Moreover, some other notices were not sent in within a reasonable period of time. This widespread failure has occurred in spite of the clear terms of article 4(3), the great importance given to this requirement by the drafters of the Covenant, and the Committee's occasional insistence on the need to comply with it. Among the reasons which might explain this failure are the following: the relatively brief experience of the implementation of the Covenant; the lack of clarity on the part of some governments in emergencies about the true legal situation as far as derogations from the Covenant are concerned; the fear of international criticism; the explicit refusal of some States to declare and notify actual states of emergency; and the lack of expertise of some governments concerning the functioning of the derogation clause. Moreover, two more elements concerning the HR Committee could also help to explain this situation: the limited powers of the Committee under the reporting procedure, and the absence of a stronger emphasis on the fulfilment of this requirement of notification; this latter failure is perhaps due to the above-mentioned lack of a clear distinction between the two different obligations of States, that of notifying other States and that of providing information to the Committee under other procedures. Finally, the fact that not many States have accepted the inter-State complaints procedure (article 41) (in contrast to what has happened within the European system) could also explain a certain lack of interest both in States and in the Committee in fulfilling this requirement.

Under the *American Convention*, the situation is no better. First of all, because notices of derogations, unlike under the other treaties, are not even published by the IACHR, and secondly, because the IACHR has not sufficiently insisted on the importance of this requirement. However, in a few cases, the Commission has made indirect reference to the principles, and, in two cases, it has pointed out that the notices of derogation had not been sent in on time. Many of the reasons adduced above in order to explain the failure of notification under the Covenant are equally applicable here, especially the lack of operation of the inter-State complaint procedure.

4

The Principle of Non-Derogability of Fundamental Rights

I. INTRODUCTION

One of the most important principles in the regulation of human rights in states of emergency contained in the derogation clause of the three human rights treaties is the principle of non-derogability.[1] According to this vital principle, even in situations of public emergency threatening the life of the nation, there are some rights which can never be suspended. This principle establishes a clear limitation on the right of States to take measures derogating from human rights standards when they face an emergency. In the drafting of the UN Covenant, as in the other treaties, there was unanimity on the necessity of including that principle in the derogation clause.[2] As early as the second session of the Drafting Committee (1948), the importance of this principle, which was based on a proposal of the World Jewish Congress, was emphasized by several members.[3] However, it is interesting to note that in the UK Draft of an International Bill of Rights submitted as a working paper for the Commission of Human Rights, in which the origin of the derogation clause in all three human rights treaties can be found, there is no mention of this principle.[4] These rights which cannot be derogated from are technically called 'non-derogable rights'.

Even though there was total agreement on this principle, the problem of establishing which rights should be made non-derogable was far from an easy task for the drafters not only of the Covenant, but of the two other main treaties as well. One of the striking features of the derogation clause of these treaties is that they contain a different list of non-derogable rights. Article 15(2) of the European Convention has four

---

[1] The UK representative in the drafting of the Covenant at the time, Miss Bowie, described the principle as 'an essential one', E/CN. 4/SR. 127 (27 June 1949), p. 7 Mr René Cassin and other representatives made the same point, E/CN. 4/SR. 195, paras. 69 and 90. See also Mr Oribe (Uruguay) and Mr Malik (Lebanon), paras. 71 and 87 respectively.

[2] A/2929, para. 45. See para. 2 of art. 4 of the ICCPR.

[3] E/CN. 4/AC. 1/SR. 34 (20 May 1948), Commission of HR, Drafting Committee, 2nd Session, pp. 4–5, Mr Heyward (Australia), Mr Malik (Lebanon), and Mr Santa Cruz (Chile). For the draft of the World Jewish Congress see E/C. 2/194, pp. 2 ff.

[4] UK Draft of an International Bill of Rights (London, June 1947); see E/CN. 4/21, Annexe B: art. 4 (UK), Drafting Committee, 1st Session.

rights; article 4(2) of the UN Covenant seven; and article 27(2) of the American Convention eleven—and the last entrenches also the judicial guarantees essential for the protection of these non-derogable rights. The difficulty of agreeing on a concrete list of non-derogable rights is well illustrated in the *travaux préparatoires* of the treaties.

(a)   *The* travaux préparatoires *of the Three Treaties*

The UN Covenant. The *travaux préparatoires* of the UN Covenant show that it was difficult to reach an agreement on which rights should be made non-derogable.[5] Different proposals with different lists of rights were presented to the Human Rights Commission at its Fifth (1949), Sixth (1950), and Eighth (1951) Sessions. At the Fifth Session, the United States, France, and the UK presented their views.[6] Whereas the USA proposed to include a positive enumeration of the rights which can be subjected to limitations (this was in line with the general US approach of having simply a general limitation clause and not a derogation), France and the UK proposed to enumerate those rights which can never be suspended. The UK proposal contained only four non-derogable rights: the right to life, freedom from torture and ill-treatment, freedom from slavery, and the principle of non-retroactivity of penal laws. The French proposal was much wider; it contained the present articles 6, 7, 8, 9, 11, 12, 13, 14, 15, 16, and 18 of the Covenant. Due to lack of time for detailed discussion of these different lists (the US proposal was soon put aside), France proposed to leave blank the enumeration of the non-derogable rights in the first draft at the end of the Fifth Session, while approving the principle of the existence of such a list.[7] This was accepted.

The detailed discussion of the rights to be included in the list of non-derogable rights was held at the Sixth Session of the HR Commission (1950). The UK proposal for a reduced list was not adopted.[8] It is interesting to note that the same proposal was made at the same time by the UK representatives in the drafting of the European Convention, where it was in fact approved and became the final text of the provision on non-derogable rights (art. 15, para. 2). The UN HR Commission preferred to take the French proposal as the discussion text, and it decided to vote on each specific article. According to this procedure, articles 6, 7, 8, 11, 15, 16, and 18 were adopted, whereas the important

---

[5] See A/2929, para. 45.
[6] See the USA proposal in E/CN. 4/170, Add. 1: only limitations on current arts. 9, 12, 14, 18, 21, and 22. The UK proposal is in E/CN. 4/188 (16 May 1949). The French proposal is in E/CN. 4/324 (13 June 1949).
[7] See E/CN. 4/SR. 127.
[8] E/CN. 4/365 and E/CN. 4/SR. 195.

articles included in the French proposal, i.e. articles 9 (right of personal liberty and security), 14 (right to due process of law), and 12 (liberty of movement) were withdrawn by the French delegation before putting them to a vote.[9] This final decision on the list of the non-derogable rights, which was later to become the final paragraph 2 of the derogation clause in the Covenant, did not satisfy all the members of the Commission. Mr Malik (Lebanon) reserved the right of his delegation to propose on the second reading of the Covenant that other rights should be included in the list. He mentioned especially article 26 (non-discrimination) as a non-derogable right.[10] This article was not included because it was pointed out that 'enemy aliens in time of war cannot be treated equally as the citizens of the State'.[11]

René Cassin explained the withdrawal of article 9 (right of personal liberty) in the French proposal, saying that nevertheless they believed that 'there should be no total derogation from article 9 and that States should make every effort to avoid arbitrary arrests'.[12] On the other hand, Mrs Roosevelt considered that among all the guarantees against arbitrary detention, only article 9(5), the writ of habeas corpus, should be made non-derogable.[13] The opposition of Mrs Roosevelt to the inclusion of the whole of article 9 was put on the footing that no Government would in time of war be in a position to guarantee those provisions referring to arrest, bail, and compensation for prisoners-of-war or enemy aliens. Since these matters were covered by the 1949 Geneva Conventions, recently approved by an international conference, there was no need to rewrite those conventions.[14] The successor of René Cassin in the French Delegation, Mr Leroy Beaulieu, took a more radical approach and opposed the inclusion of the whole of article 9, even the writ of habeas corpus, on the same grounds as Mrs Roosevelt, but he pointed out that the suspension of that writ can never be justified by minor events like earthquakes. Taking into account this new position in the French delegation, he withdrew the proposal of including article 9 in the list of non-derogable rights.[15] Another article which was later the object of some debate was the one referring to freedom of religion (current art. 18). This article was included in the list because it contains a limitation clause (para. 3) which allowed the necessary restrictions on emergencies[16].

During the Eighth Session (1952) of the UN HR Commission, Mr

---

[9] E/CN. 4/365 and E/CN. 4/SR. 195 and 196.
[10] E/CN. 4/SR. 195, p. 23.
[11] E/CN. 4/SR. 196, pp. 4–5. See below, Ch. 6.
[12] E/CN. 4/SR. 195, para. 153.
[13] Ibid., para. 154.
[14] E/CN. 4/SR. 196, para. 20.
[15] Ibid., paras. 21 and 23.
[16] E/CN. 4/SR. 330, pp. 8 ff.

Azkoul (Lebanon) proposed to add to the list the then article 6, paragraphs 1 and 2 (prohibition of arbitrary arrest), and article 8, paragraph
2(a) (the rights of detainess to be presumed innocent, and to be informed
promptly of the nature and cause of the accusation against them).[17]
However, the Lebanese delegation had to withdraw its proposal in order
to accelerate the work of the Commission, but they put on record that
'the list had not been adequately studied and should be supplemented'.[18]
It is very interesting to note that the Israeli Government proposed at the
Seventh Session to include among the non-derogable rights the right to a
fair trial by an independent and impartial tribunal with all the judicial
guarantees established in current article 14. The Israeli Government did
not consider it necessary to derogate from any of those guarantees, even
in states of emergency.[19]

In the UN Third Committee in 1963, although there was some debate
on the suitability of the list of non-derogable rights, no alteration was
made. Among those rights discussed were the right to marry and the
freedom of thought, conscience, and religion. There were proposals to
include the right to marry (current art. 23) in the list, considering that it is
a strictly private matter which should not be subject to interference by the
State even in emergencies. However, several delegations pointed out
that, according to the legislation of many States parties, the act of marry-
ing a foreigner implies the right to acquire citizenship; therefore, in time
of war, a State may feel obliged to bar marriages between its nationals
and enemy aliens.[20] At the same time, the proposal of the Mexican
delegation to exclude paragraph 3 on the right to freedom of thought,
conscience, and religion did not succeed.[21]

After this concise review of the legislative history of the non-derogable
rights provisions of the Covenant, some conclusions emerge:

1. From the beginning of the drafting process, there was a consensus
on the need for providing within the derogation clause a list of rights
which cannot be suspended by the State even in public emergency. This
was considered an essential principle of the derogation clause.

2. The drawing up of the list presented real difficulties due to the fact
that several proposals with different lists of rights were put forward.

3. It is not clear, according to the *travaux préparatoires*, what criteria
were adopted in the elaboration of the list. The final list was not com-

---

[17] Ibid., p. 14.
[18] E/CN. 4/SR. 331, pp. 6–7.
[19] E/CN. 4/515, Add. 6, Commission of HR, 7th Session (1951), 'Observations of
Governments of Member States on the Draft International Covenant on Human Rights . . .
as Drafted at the 6th Session'.
[20] UN Third Committee, 18th Session (1963), A/5655, para. 51.
[21] Ibid., para. 52.

pletely satisfactory for some delegations, especially because some important rights were left out (i.e. arts. 9 and 14). It seems, as one of the
delegations pointed out, that there was no careful study of each right and
the implications of its derogation for the protection of human rights in
public emergencies. The fact that the derogation clause covers the situation of war, which is in principle the gravest emergency threatening the
life of the nation, seemed to have pushed the drafters of the Covenant to
adopt a cautious and restrictive approach to the list of non-derogable
rights. The case of war seems always to have been in the minds of the
drafters when assessing the possibility of accepting any other right. This
legislative context could be important in order to prevent sweeping
derogations by States parties in situations (such as a public emergency
due to internal disturbances) which fall short of war.

The European Covention. Unfortunately, the *travaux préparatoires* of the
European Convention of Human Rights do not give any indication of the
existence of debates on the elaboration of the list of non-derogable rights
(art. 15, para. 2). The UK presented to the Committee of Experts in 1950
the same proposal as the one presented to the UN Commission of Human
Rights in the drafting of the Covenant. It contained, as was said above, a
reduced list of four non-derogable rights. This list was accepted without
modifications.[22]

The American Convention. The list of non-derogable rights in the
American Convention of Human Rights is the largest one of the three
treaties under consideration. It contains eleven rights and, in addition,
the entrenchment of the judicial guarantees essential for the protection of
those rights.

It is interesting to recall how the drafters of the Convention agreed
upon that extensive list. In the first draft of the Convention, prepared by
the Inter-American Council of Jurists in September 1959, there is the
same list of four rights as in the European Convention.[23] This draft was
criticized by a group of experts meeting in Uruguay one month later.
They decided to include in the list of non-derogable rights fourteen
rights, among them those related to guarantees against arbitrary detention, habeas corpus, and due process of law.[24] The IACHR Special

---

[22] Second Meeting of the Committee of Experts held in Strasbourg, 6–10 Mar. 1950,
ECHR, *Collected Edition of the Travaux préparatoires*, iii. 280. It should be added that
according to art. 4(1) of the 7th Protocol the principle 'non bis in idem' is also entrenched
from derogation. At the same time, according to arts. 2 and 3 of the sixth Protocol the death
penalty (art. 2, para. 1 of the Convention) can only be imposed in case of war, and not in
other public emergencies. See also the similar system established by the Second Protocol to
the ICCPR (arts. 2 and 6), aiming at the abolition of the Death Penalty (15 Dec. 1989).

[23] Inter-American Council of Jurists, Draft American Convention. See Buergenthal and
Norris (eds.), *The Inter-American System*, vol. ii.

[24] Martins, *The Protection of Human Rights*, pp. 145–7.

Rapporteur on Emergencies took a similar approach and included seven-
teen rights in his report to the Commission.[25] On the other hand, in 1968
the IACHR adopted an important resolution on human rights in states of
emergency which contained a reduced list of non-derogable rights: the
right to life, the right to liberty and personal security, the right to
protection against arbitrary detention, the right to due process of law,
and the right of freedom of thought, conscience, and religion.[26]

The IACHR presented to the San José Conference its own Draft
Convention on Human Rights in 1969.[27] In this draft, the Commission
adopted a middle approach and proposed to protect nine rights from
suspension: the four of the European Convention, plus three contained in
the 1968 Resolution (the right to protection against arbitrary arrest, the
right to due process of law, and the right of freedom of thought, conscience, and religion), and, in order to co-ordinate this list with the UN
Covenant, the right to recognition of juridical personality and the right
not to be deprived of liberty for debt. The latter right was also recognized
in a preliminary draft of the American Convention.

This Draft Convention of the IACHR was the main working document
at the San José Conference (1969). There were two other drafts presented
by Chile and Uruguay, which contained practically the same reduced
list of non-derogable rights as the European Convention, although the
Uruguayan draft included the right to juridical personality.[28] According
to the Minutes of the Conference, which unfortunately do not give a
full account of the reasons why some rights were included and others
rejected, there were two types of objections to the list proposed by the
IACHR Draft.[29]

First of all, there was a general objection: some States wanted to
suppress the whole provision due to the fact that it conflicted with their
constitutional clauses. Mexico was especially interested in this, and
wanted no restrictions whatsoever on its sovereignty in dealing with
emergencies. In its written observations to the Conference, the Mexican
Government pointed out that 'the Mexican Constitution establishes the
possibility of suspending all those rights that may be an obstacle to
dealing rapidly and effectively with an emergency situation, with the only
restrictions being that the suspension be for a limited time, that it be

---

[25] For the list of these 17 rights, see ibid., p. 154.
[26] IACHR, *Resolution on the Protection of Human Rights in Connection with the Suspension of Guarantees or 'States of Siege'* (1968).
[27] IACHR, *Draft of the American Convention*; see Buergenthal and Norris (eds.), *The
Inter-American System*, vol. ii, booklet 13. See also 'Annotations to the Draft by the
IACHR', ibid., pp. 25 ff.
[28] Buergenthal and Norris (eds.), *The Inter-American System*, vol. ii, booklet 16(1), p. 46
(art. 22 of the Chilean Draft); p. 70 (art. 22 of the Uruguayan Draft).
[29] For the Minutes of the Conference. see ibid.. booklet 12.

by means of general provisions, and that it not be limited to a given individual'.[30] The Mexican proposal to delete the whole provision on non-derogable rights did not succeed because the provision on non-derogable rights was considered an 'essential' one; therefore Mexico included a formal reservation to this provision in the Final Act of the Conference. According to this reservation stemming from its Constitution, Mexico could suspend 'any right that may be an obstacle in order to cope, rapidly and easily, with situations of extreme emergency'.[31]

The second type of objection to the list proposed by the IACHR Draft dealt more specifically with particular rights. The US delegation first of all proposed to identify by article, number, and title the rights which may not be suspended, rather than to include a vague reference to them in the form of general concepts as was the case with the IACHR Draft. At the same time, it introduced an alternative text which protected from derogation only the right to juridical personality, and the rights to life, freedom from torture, freedom from *ex post facto* laws, and freedom of religion.[32] Due to the debate on these different lists of rights, a working group was set up in the Conference to study the final list of non-derogable rights.[33] Unfortunately, there are no records of the work of this group and therefore it is not possible to know the reasons for the changes introduced in the final list. The changes in respect of the IACHR Draft were substantial: the important rights which protect human beings from arbitrary arrest and the rights of due process disappeared, as did the right to freedom of thought and the prohibition of imprisonment for debt. On the other hand, new rights were included in the list: freedom from slavery, the right of the family, the right to a nationality, and the important right of participation in government.[34] However, the US delegation was not happy with this list and insisted on the desirability of including in it the fundamental articles 6 and 7 which dealt with protection from arbitrary arrest and with the right to a fair trial ('the protection of an individual arrested without being informed of the cause or charges, and his right to have a hearing to respond to the accusations that are brought against him').[35]

---

[30] 'Observations of States Submitted to the Conference with the Draft and Annotations of the Draft by the IACHR', ibid., booklet 13, p. 138 (Mexico). See also pp. 116–17 (Ecuador). For the discussions at the Conference and the proposals to delete para. 2, see the comments by the Mexican and Costa Rican representatives, ibid., booklet 12, pp. 134–5.

[31] Reservation of Mexico to art. 27(2), statements included in the Final Act, San José Conference (1969), in Buergenthal and Norris (eds.), *The Inter-American System*, vol. i, booklet 3, p. 27. Mexico did not confirm this resevation when it ratified the American Convention.

[32] Buergenthal and Norris (eds.), *The Inter-American System*, vol. ii, booklet, 13, p. 164.

[33] Ibid., booklet 12, p. 137.

[34] Ibid., booklet 12, p. 137.

[35] Ibid., pp. 136–7.

This US proposal was not accepted. However, in the Second Plenary Session of the Conference, the US delegation again introduced a new amendment in order to include as a non-derogable right at least 'the judicial guarantees essential for the protection of such rights'.[36] This important amendment was approved and included in the final text of article 27(2). Thus, as the IACHR has pointed out, the American Convention can be proud of being 'the first international human rights instrument to include among the rights that may not be suspended essential judicial guarantees for the protection of non-derogable rights'.[37]

(b)   The Rationale behind the List of Non-Derogable Rights

One of the problems that one faces in the study of non-derogable rights is the ascertainment of the rationale at work behind the choice of different lists of non-derogable rights in the three treaties. Unfortunately, the records of the *travaux préparatoires* give little information on this point. It seems that as Professor Hartman has pointed out,[38] there are two different criteria which could be used in adopting a list of non-derogable rights, and which were implicitly present in the drafting of the treaties. First, to include those rights which are absolutely fundamental and indispensable for the protection of the human being. Secondly, to include those rights the derogation of which by the State in public emergencies would never be justified because they have no direct bearing on the emergency.

It is interesting to note that neither of these criteria seems to have been consistently held to by any of the treaties. Perhaps the clearest case of a treaty adhering to one of these criteria is the European Convention, which contains a short list of non-derogable rights which could be deemed to be absolutely fundamental for the protection of the human being in emergencies. However, one could say that there are some other rights which seem to be no less fundamental and which have not been included, for example, some minimum guarantees for persons under administrative detention and guarantees of due process of law. It is true that the fact that a right is not listed as non-derogable does not mean that it could automatically be derogated from, because the principle of proportionality should also be strictly applied. However, the use of the right of derogation by States, especially when there is no strong monitor organ to control it, could lead in practice to a complete derogation of those rights which

---

[36] Ibid., p. 254.
[37] IA Court HR, advisory opinion, 'Habeas Corpus in Emergency Situations (articles 27(2), 25(1) and 7(6) of the ACHR)' (30 June 1987), para. 36.
[38] J. F. Hartman, 'Working Paper for the Committee of Experts on the Article 4 Derogation Provision', *HR Quarterly*, 7 (1985), 113–14.

are also fundamental but which have not been explicitly entrenched in the treaty. This has happened in recent years when gross violations of human rights have taken place in states of emergency, and this is the reason why some international organizations have proposed to enlarge the list of non-derogable rights in order to ensure the effective protection of the most fundamental rights in emergencies.

The UN Covenant and the American Convention seem to have applied both criteria, but not consistently. It is true that both included the same 'core' of fundamental rights as the European Convention, but even though their list is larger, the same criticism can be made: there is an absence of other equally fundamental rights. At the same time, some of the rights included are prima-facie not so fundamental, but were incorporated because they do not directly relate to emergencies and, therefore, States would have no reason to derogate from them.[39] Moreover, the American Convention includes two other non-derogable rights which may have some importance in emergencies. i.e. the right to nationality (art. 20) and the right to participate in government (art. 23), and it has also entrenched the judicial guarantees essential for the protection of non-derogable rights.

In short, the list of non-derogable rights contained in the three treaties can be criticized for two reasons: first, because it does not contain certain rights which are very important and very much at risk in public emergencies, and second, because it contains (at least in the case of the ICCPR and the ACHR) other rights which are not so fundamental and at risk in emergencies. It is true that the fact that a particular treaty contains a large list of rights, some of which are not particularly at risk in emergencies, is not *per se* a harmful thing; however, a short list of non-derogable rights would have been preferable, because it would have strengthened the concept of non-derogability and would have been more likely to have been put into practice by States. This is an important, albeit psychological, factor. A shorter list would have focused attention on those fundamental rights which should be especially protected in emergencies. This short list, therefore, should have included only those rights which must be protected because they are so fundamental *and* in danger of being violated in emergencies.

However, the purpose of this chapter is not the examination of a hypothetical list of new non-derogable rights. As Meron has pointed out, 'the prospects for a consensus reaching beyond these few rights (the core of the 4 rights common to the three treaties), are not immediate . . . The international community as a whole has neither established a uniform list of non-derogable rights nor ranked non-derogable rights ahead of

---

[39] For the analysis of these rights, see below, pp. 97 ff.

derogable rights.'[40] The purpose of this chapter is twofold: first, to examine the protection achieved in the three treaties through the list of non-derogable rights and other provisions of the treaties which are considered to be non-derogable by implication. Special attention will be paid to the consequences that the enlarged list of non-derogable rights of the ICCPR and the ACHR may have for emergencies. Second, there are two areas in which the urgency to formulate some minimum guarantees has been stressed by international monitoring organs and legal experts, namely, guarantees against arbitrary detention and guarantees of due process of law. The importance of these minimum guarantees is due to the fact that gross violations of the most fundamental rights (the right to life and freedom from torture) have been possible in part due to the absence of these guarantees. This is why they should be considered non-derogable, at least in principle. Therefore, due to their great importance, these guarantees need to be examined.

2. THE FOUR COMMON NON-DEROGABLE RIGHTS

The three lists of non-derogable rights in the three treaties under consideration contain four rights which are considered non-derogable in all of the treaties. These four rights are: the right to life, the right to be free from torture and other inhuman or degrading treatment or punishment, the right to be free from slavery or servitude, and the principle of non-retroactivity of penal laws. These rights are so fundamental that they are considered to be not only customary international law but also norms of *ius cogens*.[41] These four non-derogable rights constitute what has been called the 'irreducible core' of human rights. Even though no derogation whatsoever from these rights is permitted in emergencies, the international monitoring bodies have consistently denounced gross violations of them. A detailed study of the case-law of these four rights under the three treaties is outside the scope of this book. However, some of the most common violations of these rights can be mentioned. Among the gravest violations to the right to life are executions carried out without due process guarantees, deaths resulting from torture or ill-treatment in prison, enforced disappearances, and deaths resulting from excessive use of force by law enforcement officials.[42] Other violations are the adminis-

---

[40] T. Meron, 'On a Hierarchy of International Human Rights', *AJIL* 80 (1986), 11 and 17.
[41] See *inter alia* Questiaux, *Study of the Implications*, p. 19. See also the US *Restatement*, pp. 174–5.
[42] For the issues relating to the right to life, see B. A. Ramcharan, *The Right to Life in International Law* (The Hague, 1985). For an analysis of the findings of the three UN

trative practice of torture and ill-treatment, prolonged incommunicado detention, and the retroactive application of penal laws.

3. THE OTHER NON-DEROGABLE RIGHTS UNDER THE ICCPR AND THE ACHR

In addition to the four common non-derogable rights in the three treaties under consideration, the UN Covenant and the American Covention contain a further number of non-derogable rights. Among these rights, some bear a relation to public emergencies, but many have no direct implication for the security of the State. These rights have been included in the two treaties because it is not necessary to derogate from them in order to overcome the emergency, rather than because they are among the most fundamental rights for the protection of the human being in emergencies.

(a)   Rights Unrelated to Public Emergencies

The first right entrenched against derogation by both the UN Covenant and the American Convention, and which seems to have no direct relation with emergencies, is *the right to be recognized as a person before the law*.[43] No corresponding provision can be found in the European Convention. However, the Committee of Experts of the Council of Europe has pointed out that 'it is most unlikely that derogations from this provision would ever be made'.[44] As far as the content of the provision is concerned, it does not seem to add any new obligations with respect to the European Convention. When the Council of Europe was preparing the Fourth Protocol, which was intended to incorporate new rights into the Convention, the Committee of Experts rightly considered it

Special Rapporteurs on summary executions, torture, and disappearances, see D. Weissbrodt, 'The Three "Theme" Special Rapporteurs of the UN Commission on Human Rights', 80 *AJIL* (1986), 685–95. For the issues related to treatment of detainees see N. Rodley, *The Treatment of Prisoners under International Law* (Oxford, 1987). Among the leading cases, the following should be mentioned: as far as torture and ill-treatment is concerned, the *Ireland* v. *UK* and the *Greek case* under the ECHR. On excessive use of force by law enforcement officials, see the *Suarez de Guerrero case* and the *Baboeran* et al. *case* on arbitrary killings (views expressed by the UN HR Committee under the Optional Protocol). On disappearances: the *Velasquez–Rodriguez case* by the IA Court HR. Moreover, the reports of the IACHR have contained important findings on gross violations of these fundamental rights, e.g. the 1985 *Report on Chile*, the 1980 *Report on Argentina*, etc.

[43] ICCPR, art. 16; ACHR, art. 3.

[44] Council of Europe, *Problems Arising from the Coexistence of the UN Covenants on Human Rights, and the European Convention*, Report of the Committee of Experts (Strasbourg, Sept. 1970), p. 21, para. 75(ii).

unnecessary to introduce this provision, on the grounds that 'its content could be deduced from other articles of the Convention, especially articles 4, 6 and 14'.[45]

In article 11 the UN Covenant also prohibits derogation of *the right to be free from imprisonment merely on the ground of inability to fulfil a contractual obligation*. This provision is also contained in article 7(7) of the American Convention and in article 1 of the Fourth Protocol of the European Convention, but in these treaties both articles are derogable. However, it is very unlikely that derogation from this provision would ever be made in a state of emergency.[46]

The American Convention further entrenched from derogation some other *rights relating to the family*, namely, the right to marry and to build a family (art. 17) and the right of the child to be protected (art. 19), the right to a name (art. 18), and the right to nationality (art. 20).

The consideration of *the right to nationality* as unrelated to public emergencies presents some problems.[47] It is clear that, in principle, the mere possession of nationality cannot threaten the security of the State, but on the other hand, as nationality is the prerequisite for the enjoyment of all the rights and guarantees recognized in the legislation of the State, deprivation of nationality has been used by governments as a means of depriving political opponents to the regime of all the rights and benefits to which nationality entitles them.

The prohibition of the suspension of the right to nationality in the American Convention does not affect the power of the State in emergencies to take two common measures relating to this right, namely, to expel some of its nationals or to refuse admittance to the country. These two measures, which would be against the rights recognized in article 22(5) of the American Convention (freedom of movement), could in principle be taken because this article is a derogable one. Moreover, what article 20(3) (the right to nationality) prohibits is the *arbitrary* deprivation of one's nationality; therefore, if the deprivation is not arbitrary (i.e. if it is carried out for genuine security reasons) one could be deprived of one's nationality in emergencies even if the article is a non-derogable one.

The use of the deprivation of nationality as a penalty against political opponent has been condemned by the IACHR in the *Third Report on Chile*.[48] Dut to the fact that Chile was not a party to the American

---

[45] Ibid., pp. 41–2, para. 155.

[46] Ibid., p. 20, para. 75(i).

[47] See also art. 15 of the UDHR. For an analysis of this right in HR treaties, see R. Lillich, 'Civil Rights', in Meron (ed.), *Human Rights in International Law*, pp. 153 ff. For a general approach to nationality in international law see Brownlie, *Principles*, pp. 381–420.

[48] IACHR, *Report on Chile*, 1977, Ch. 9, pp. 77–81. See also the use of this measure by the Greek Military Junta in 1967, see International Commission of Jurists, *States of Emergency*, pp. 143–5. See also the finding of the IACHR in its 1983 *Report on Cuba*,

Convention, the Commission referred to article 19 of the American Declaration, which also recognizes the right to nationality. The Military Junta, by means of Decree No. 175 (3 Dec. 1973), established a new reason for the loss of nationality for those 'seriously damaging from abroad the essential interests of the State during the state of emergency'. Nine persons were deprived of their nationality in the first four years of the Military Government, five of whom had held important posts as ministers of state, senators, and ambassadors in Allende's Government; three were trade union leaders and one a general in the Air Force. The IACHR has considered that these constitute an arbitrary deprivation of nationality. It is interesting to note that even though article 19 of the American Declaration, unlike article 20 of the Convention, does not expressly prohibit the arbitrary deprivation of nationality, the IACHR has found that the same standard was applicable under the American Declaration.[49]

The Commission found the measure to be arbitrary because it was a severe punishment imposed for reasons of political intolerance and used against political opponents who had no effective right to appeal.[50] The Commission has also noted the contradiction inherent in the application of a permanent penalty on the basis of a state of emergency, which, by definition, is of a temporary nature. The IACHR has insisted on the extraordinary importance of the right to nationality, 'one of the most important rights of man, after the right to life itself, because all the prerogatives, guarantees and benefits man derives from this membership in a political and social community, the State, stem from or are supported by the right'.[51] This measure has not been resorted to by the Government since the finding by the Supreme Court in 1977 that one of the nine deprivations of nationality was unlawful. However, the other eight persons have not, so far, been able to recover their nationality. The 1980 Constitution has suppressed the power of the President to impose that penalty in states of emergency.[52]

Even though most of the rights mentioned above have no direct relation with states of emergency, the applicant, in the *Macias case*, in the context of the contention of the Nicaraguan Government that it had a right to suspend human rights in emergencies, pointed out that the right to recognition before the law (art. 3), the right to a name (art. 18), and

which considers the deprivation of nationality of Cuban nationals by the mere fact of having emigrated to be unjust and incompatible with the observance of human rights (p. 180).

[49] IACHR, *Report on Chile*, 1985, p. 145, para. 1.

[50] Art. 8(4) of the 1961 UN Convention on the Reduction of Statelessness requires for a lawful deprivation of nationality the right to a fair hearing by a court or other independent body.

[51] IACHR, *Report on Chile*, 1985, p. 146, para. 6; see also 1977 *Report*, p. 80, para. 8.

[52] IACHR, *Report on Chile*, 1985, p. 81, para. 10.

the right to nationality (art. 20) were non-derogable according to article 27(2) of the Convention, and yet had been violated by the Government. Article 3 was construed by the applicant as including the right to 'access to law', which was denied to him by the acts of the Government. Article 18, establishing the right to a name, was construed as including the right to reputation attached to the person who bears the name. Finally, the right to nationality was allegedly violated because the applicant was forced by the Government to leave the country, and therefore his nationality had effectively been taken away from him, leaving him in the position of a stateless person.[53] The IACHR did not express any opinion on these points but recommended that the Government should submit the case to the Inter-American Court. If all these contentions were to be proved by the applicant, there is no doubt that the Court would find that these interferences in the applicant's rights are not justified by the emergency.

(b)   Rights which May have an Impact on Public Emergencies

Leaving aside the right to nationality, two other rights which may affect the security of the State are made non-derogable; the first is the right of freedom of conscience and religion, which is non-derogable under the Covenant and the American Convention, and the second (only non-derogable under the latter), the right to participate in government.

The *right to freedom of conscience and religion* is entrenched from derogation in both treaties, and the only problem which it may present in emergencies is that the definition of the right includes the freedom to profess or disseminate one's religion or belief either individually or together with others in public or in private.[54] This may sometimes affect the security of the State, e.g. during public disturbances and serious riots following religious ceremonies with political connotations. However, the same article in both treaties contains a 'limitation clause' authorizing restrictions of the right when necessary to protect public safety, order, health, morals, or the rights or freedoms of others.[55] This limitation clause could therefore have a similar effect to that of the derogation clause under certain conditions.[56]

---

[53] *Macias case*, IACHR, Case No. 9102, Res. No. 29/86 (10 Apr. 1986), in IACHR *Annual Report, 1985–6*, pp. 57–98.

[54] ICCPR, art. 18(1); ACHR, art. 12(1); similar provision is contained in art. 9 of the European Convention but it is a derogable right.

[55] ACHR, art. 12, para. 3; ICCPR, art. 18, para. 3.

[56] See the *travaux préparatoires* of the Covenant and the reasons in favour and against the entrenchment of the provision, A/5655, para. 52. See also the 'Siracusa Principles' on the principles applicable to limitations; some of them are similar to those of the derogation clause (i.e. principles of proportionality, non-discrimination, etc.).

The *right to participate in government* is only entrenched in the American Convention.[57] During the *travaux préparatoires* of the European Convention, the Consultative Assembly proposed at a very late stage of the drafting (August 1950) to include in the Convention an article guaranteeing those political rights which the ministers themselves agreed should be guaranteed even in time of war or civil disturbances.[58] The article was not accepted by the Committee of Ministers and was not included in the Convention.

The article in the American Convention contains important provisions concerning the right to take part in the conduct of public affairs, directly or through freely chosen representatives (para. 1 (*a*)); the right to vote and to be elected in genuine periodic elections (para. 1 (*b*)), and the right to have access to public services (para. 1 (*c*)). Even if in theory the prohibition of derogation from this provision seems to introduce an important guarantee (which does not exist in the other two treaties), in practice the two main measures relating to this right taken by States in situations of emergency seem also to be compatible with the American Convention, namely, the suspension of political rights by delaying general elections and the imposition of restrictions in the access to public services. The first measure could be compatible with the American Convention provided that the conduct of public affairs would continue through freely chosen representatives and that the election would be held as soon as the state of emergency, which by its very nature is temporary, ended. On this last point, the IACHR in its 1981 *Report on Nicaragua* seems to confirm this position. The Commission takes into account the fact that due to the state of emergency, the Government had postponed the elections until 1985, and did not consider that this fact was *per se* against article 27(2) in conjunction with article 20, but only that 'the date proposed for calling the elections was too far off'.[59]

4. OTHER RIGHTS AND PRINCIPLES WHICH ARE HELD NON-DEROGABLE BY IMPLICATION

According to the three human rights treaties under consideration, it seems that in addition to the rights expressly entrenched in paragraph 2 of the derogation clause, other provisions could be deemed non-derogable by implication.

[57] Art. 23. Similar provisions: European Covenant, art. 25 and First Protocol of the European Convention art. 3, but these are derogables.

[58] ECHR, *Collected Edition of the* Travaux préparatoires, vii. 36 (declaration by Sir D. Maxwell-Fyfe); art. 17 of the proposal is on p. 250.

[59] IACHR, *Report on Nicaragua*, 1981, p. 139. See another restriction on political rights in the *Landinelli case* (see below, pp. 157–8).

At an early stage of the drafting of the European Convention and of the Covenant, the original derogation clause was worded in such a way as to permit derogations only of those provisions which were contained in the substantive part of both treaties, in other words, in that part which contained the enumeration of the fundamental rights recognized in those treaties.[60] However, the final wording of the derogation clause in both treaties and also in the American Convention authorizes States parties to take measures derogating from their obligations under the treaty, without specifying any part of it.

This change could be inerpreted in the way that Uruguay actually did, when it claimed that 'while there was to be no derogation from certain provisions, derogation could be made from the rest of the Covenant, including the measures of implementation'.[61] As the comment of the UN Secretary-General to the draft Covenant pointed out, this 'might have far-reaching consequence'.[62] As will be seen below, such interpretation would run against the correct interpretation of human rights treaties. It will be shown that some of the provisions which are not expressly included in the list of non-derogable rights cannot in fact be derogated from. These provisions can be classified in three main categories:

provisions related to the exercise of non-derogable rights, i.e. the right to an effective remedy and the prohibition of discrimination;

provisions which contain general exceptions;

provisions related to the machinery of implementation, i.e. inter-state applications, the right to individual petition, and the report procedure under the UN Covenant.

(a)   Provisions Related to the Exercise of Non-Derogable Rights

In addition to articles on the definition of the rights protected, human rights treaties contain some provisions which specify basic principles relating to the enjoyment of those rights. Two of these provisions recognize first, the right to an effective remedy before domestic courts in the case of violation of those recognized fundamental rights and freedoms[63] and, second, the right to enjoy those rights without discrimination.[64] These two principles are applicable to all the rights and freedoms recognized in the human rights treaties, and States parties to the treaties

---

[60] For the UN Covenant see the text of para. 1 approved by the HR Commission at its 5th Session (1949), doc. E/1371, art. 4(1), and also the one approved at its 6th Session (1950), doc. E/1992, art. 2(4). For the European Convention see the text proposed by the UK, ECHR, *Collected Edition of the* Travaux préparatoires, iii: 186 and 190 (art. 4).

[61] A/2929, para. 45.

[62] Ibid.

[63] ECHR, art. 13; ICCPR, art. 2(3); ACHR, art. 25.

[64] ECHR, art. 14; ICCPR, art. 26; ACHR, art. 24.

should guarantee them. However, in situations of emergency, these two principles, when applied to non-derogable rights, are themselves entrenched from derogation. Thus, a State cannot either suppress all the effective remedies for the protection of the non-derogable rights,[65] or discriminate in the suspension of the non-derogable rights simply because it is not allowed to suppress those rights. However, as far as the derogable rights are concerned, the State may derogate from these two principles provided that the other conditions for derogations are met. Ironically enough, one of the main conditions for taking measures of derogation, as will be seen in Chapter 6, is, according to the express terms of the UN Covenant and the American Convention, that those measures do not involve discrimination.[66] The European organs have given to the general provision on non-discrimination similar effect to the specific non-discrimination provision in emergencies found in the other treaties.[67]

At the same time, the American Convention in its article 27(2) *in fine* expressly prohibited the derogation of 'the judicial guarantees essential for the protection of such (non-derogable) rights'. In two advisory opinions, the IA Court HR has reaffirmed this important principle.[68]

(b) Provisions which Contain General Exceptions

The second kind of provisions which raise the question of their compatibility with the derogation clause are those containing general exceptions. These general exception clauses are:

the prohibition preventing the State from engaging in any activity aimed at destroying or limiting to a greater extent than is provided the rights and freedoms recognized in the treaties;[69]

the prohibition on applying permitted restrictions for any purpose other that the one prescribed by the treaties;[70]

the prohibition on using the treaties as an excuse for limiting or derogating from the human rights recognized in municipal law or in other treaties.[71]

---

[65] See *inter alia* ILA Paris Report (1984), pp. 92–3; 'Kingston Seminar', para. 237(i); E/CN 4/Sub. 2/1989/30, Add. 2, Rev. 1; S. R. Chowdhury, *Rule of Law in a State of Emergency* (London, 1989), pp. 259–64.

[66] ICCPR, art. 4(1); ACHR, art. 27(1).

[67] See Ch. 6.

[68] IA Court HR, Ser. A: Judgments and Opinions, advisory opinion of 30 Jan. 1987, 'Habeas Corpus in Emergency Situations (arts. 27(2), 25(1), and 7(6) American Convention HR)'; advisory opinion of 6 Oct. 1987, 'Judicial Guarantees in States of Emergency (arts. 27(2), 25 and 8 American Convention on Human Rights)'. See below, pp. 112, 121, 160.

[69] ICCPR, art. 5(1); ECHR, art. 17; ACHR, art. 29(*a*).

[70] ECHR, art. 18; ACHR, art. 30 *in fine*. The European Convention does not contain an express provision on this principle.

[71] ICCPR, art. 5(2); ECHR, art. 60; ACHR, art. 29(*b*).

The conflict between these provisions and the derogation clause can be presented as follows: on the one hand, these provisions prohibit, for instance, any act aimed at destroying or limiting the rights, but, on the other hand, the derogation clause allows the State to derogate from the obligations of the treaty, one of which is precisely that contained in that general exception clause. The problem is then which of these clauses has primacy. The logical conclusion is to give priority to the general exception provisions over the derogation clause, and consequently to prohibit the derogating State in a public emergency from committing any act aimed at destroying or limiting the right or freedom recognized, more than is necessary according to the derogation clause. Therefore, these provisions containing general exceptions should be deemed non-derogable in states of emergency.

This has also been the position of the European Commission. In the *Greek case*, the applicant Governments contended that the measures of derogation by the Greek Government under article 15 were excluded by articles 17 and 18, articles which contain the first two general exception provisions mentioned above, read together with the Preamble of the Convention.[72] According to article 18, measures of derogation were only permitted in exceptional cases for the specific purpose of protecting these rights and freedoms. Article 17, on the other hand, excludes derogations which are aimed at the destruction of the rights and freedoms of the Convention; by the introduction of a totalitarian regime and by the destruction of the rights and freedoms of the Convention, the respondent Government was prevented from invoking article 15 as a legal justification for the derogations.[73] The respondent Government did not contest this construction, but rather stated that the measures of derogation taken were aimed at saving democracy and parliamentary life.

The Commission did not rule against the right of a so-called revolutionary Government to use the right of derogation of article 15. However, the Commission, in its approach to articles 17 and 18 in relation to article 15, seemed to follow the same reasoning as was advocated above. The Commission did not say that articles 17 and 18 are derogable because of the functioning of article 15; nevertheless, it found that, due to the fact that the main condition of justification for derogations, namely, the existence of a public emergency threatening the life of the nation, was not satisfied in this case, it was not called upon 'to express a view on the further question whether the respondent Government's derogations

---

[72] *Greek case*, opinion of the European Commission on HR in *YBECHR* 12 (1969), 111, para. 221.

[73] Ibid., p. 112, paras. 223–4.

under article 15 were also excluded by articles 17 and 18'.[74] One may assume that in the case of a lawful public emergency within the meaning of article 15, the application of derogations for any purpose other than that of overcoming the emergency and protecting human rights would be found by the Commission and the Court to be in violation of articles 17 and 18.

This was what happened in the *Lawless case*, in which the Court found that there was a public emergency threatening the life of the nation and, at the same time, that the preventive detention of Mr Lawless, as a derogation from article 5, was strictly required by the exigencies of the situation. However, the Court assessed, according to article 18, whether this measure was applied for a purpose other than the one for which the power of detention was granted. It found that it was not and therefore that there had been no violation of article 18.[75]

Finally, as far as the third clause is concerned, i.e. the prohibition of using the Convention as an excuse for limiting or derogating from the human rights granted in domestic law, the operation of this clause in conjunction with article 15 has been discussed by Velu in the particular case of the Belgian legislation and the specific right to respect for private life. In Belgian public law, emergency measures derogating from fundamental human rights cannot be taken except in time of war. Therefore, article 15 could not be interpreted as conferring on the organs of the Belgian State the right to take measures of derogation from the right to private life, in any emergency threatening the life of the nation other than in time of war.[76]

(c)   Provisions Related to the Machinery of Implementation of the Treaties

As was seen above, during the *travaux préparatoires* of the Covenant, the representative of Uruguay held the opinion that, leaving aside certain provisions which were made expressly non-derogable, the rest of the Covenant, including the measures of implementation, could be derogated from.[77] However, to hold this position would in fact defeat the main objective of the systems of guarantees established in human rights treaties.

---

[74] Ibid., para. 225.

[75] *Lawless case*, European Court of HR, Ser. A: Judgments, p. 59, para. 38. Precisely because the Greek Government did not show that the derogating measures were ultimately aimed at restoring democratic rights and freedoms, Mr Bussutil and Mr Ermacora denied the right of derogation to the Greek Government, *Greek Case*, opinion of Mr Bussutil, *YBECHR* 12 (1969), 113–19; dissenting opinion of Mr Ermacora, ibid., pp. 102–5.

[76] J. Velu, 'The European Convention on Human Rights and the Right to Respect for Private Life, the Home and Communications', in A. H. Robertson (ed.), *Privacy and Human Rights* (Manchester, 1973), p. 68.

[77] A/2929, para. 45.

The recognition by States parties of the right to derogate from the machinery of implementation of the treaties, especially from the right to individual petition and from the right of inter-State complaints, would preclude any check by the bodies established by the treaties of the lawful application of the derogation clause according to its conditions and principles.

In fact, there has been no indication so far that the States parties to the treaties have ever attempted to claim the right to derogate from the implementation provisions. At the same time, the practice of the monitoring bodies has constantly maintained, according to the functions of the bodies under each treaty, the right to review the correct application of the derogation clause by States parties, through either inter-State or individual applications.

5. STUDY OF SEVERAL PROPOSALS TO MAKE A PRIORI NON-DEROGABLE SOME MINIMUM GUARANTEES AGAINST ARBITRARY DETENTION AND OF DUE PROCESS OF LAW

(a) Introduction

There is no doubt that the right to be free from arbitrary arrest or detention and the right to due process of law are two of the most fundamental rights for the protection of human beings at all times. Therefore, it is not surprising to find in the *travaux préparatoires* of the three human rights treaties that there were several proposals for the addition of these two rights to the list of non-derogable rights. The evolving nature of human rights law and the experience of gross violations of the most fundamental human rights, have provoked several international organizations, in the light of the jurisprudence of international monitoring organs, to propose that at least some of the most fundamental guarantees against arbitrary detention and some minimum rights of due process should be made non-derogable. In fact, the violation of fundamental human rights including the non-derogable ones i.e. life and freedom from torture, has been made possible very often because of the lack of guarantees against arbitrary detention and the derogation of all provisions of due process.

Another reason for making these guarantees non-derogable is that the international machinery of implementation of human rights treaties and the control exercised by the international bodies (with the exception of the European organs, perhaps) over the correct application of the conditions of derogation by States parties—for instance, the principle of proportionality—are far from perfect. Therefore, rather than insisting only on scrupulous compliance with those conditions, international organizations and experts on international law have proposed to entrench some of the basic guarantees relating to administrative detention and fair trial, which would make the violation of non-derogable rights more difficult. These guarantees are so fundamental and so closely related to non-derogable rights that it would be hard to maintain that their derogation could be strictly required by the exigencies of the situation. The formal entrenchment of these guarantees by means of a Protocol to the existing human rights treaties has been suggested, but it seems politically unworkable for the moment.[78] However, the important function that the identification of these standards may fulfil will be to provide clear guide-lines on those guarantees which are so fundamental for the protection of human beings in emergencies that they should in principle be considered non-derogable. These guide-lines would be useful not only for States which may use the right to derogation, but also for the international bodies which monitor the strict compliance of the right with the principle of proportionality. The departure from these standards by the derogating States should be subjected to a higher standard of proof, as being made absolutely necessary by the exigencies of the situation. The international bodies should take into account the special importance of these fundamental guarantees as a priori non-derogable. In fact, it seems that this has been the practice of the European organs.

In the *Lawless case* and in the *Ireland* v. *UK case*, it was pointed out that these two articles (article 5, the right to be free from arbitrary arrest, and article 6, the right to a fair trial and due process of law) were, after the four non-derogable rights, the most important rights of the Convention. Consequently, their derogation should be subjected to the strictest control and to some fundamental safeguards against possible abuses.

These standards seem more pertinent to states of emergency arising from internal disturbances falling short of civil war, which perhaps have been the most common of all over the last years, and which have produced the gravest violations of human rights. Having said that, they could *mutatis mutandis* be applied to the gravest states of emergency, such as international or civil wars. It is interesting to note that most of the standards of due process proposed by these international organizations have been taken from the laws of war, especially from the 1949 Geneva Conventions and their two Protocols, thus justifying the assertion that, even in the gravest situations, these standards should be considered prima-facie non-derogable. The standards formulated in those Conven-

---

[78] See T. Meron, 'On a Hierarchy of International Human Rights', *AJIL* 80 (1986), 11 ff. See also Meron, *Human Rights in Internal Strife: Their International Protection* (Cambridge, 1987), pp. 135 ff.

tions, together with their Protocols, could serve as indicators of the feasible standards applicable in the gravest situations and therefore at all times.

(b)   Guarantees against Arbitrary Detention in Emergencies

All major studies on human rights in states of emergency produced by the UN Sub-Commission Special Rapporteur Mme Questiaux, the International Commission of Jurists, the International Law Association (ILA), the Group of Experts gathered in Siracuse, the UN 'Kingston Seminar', etc., have pointed to the importance of the non-derogability of some fundamental guarantees relating to administrative detention in states of emergency.[79] These bodies have produced several standards to protect human beings in emergencies in particular cases; many of these standards are common to all of these studies, and have been the product not only of the experience of dealing with the topic, but also of the application of these fundamental guarantees by the international monitoring bodies in emergency cases. These fundamental guarantees obviously call for examination.

(i) *General Guarantees.* The International Commission of Jurists has insisted on the importance of some fundamental guarantees applicable according to the general requirements of states of emergency. The introduction of the measure of administrative detention, because of its exceptional character, can only be resorted to in a state of emergency formally proclaimed according to domestic law, and on its authorization by a democratic parliament, which should periodically review the need for it (i.e. every six months). Once the emergency has terminated, the authority to make administrative detentions should cease and all administrative detainees be freed. Administrative detention should be used only when it is absolutely necessary to protect national security and public order, and not for other kinds of crimes, which should be submitted to the ordinary criminal procedures. Furthermore, the Constitution or the pertinent legislation should spell out the grounds for detaining a person administratively.[80]

(ii) *Minimum Rights of Detainees.* The rights of detainees in emergencies according to the various proposals have been formulated as follows:

---

[79] At the same time, the UN General Assembly has recently adopted by consensus a Body of Principles for the Protection of All Persons under Any Form of Detention or Imprisonment (UNGA Res. 43/173 (9 Dec. 1988)). These principles, which are in line with the minimum standards explained below, are non-derogable in states of emergency.

[80] Int. Commiss. of Jurists, *States of Emergency: Their Impact on Human Rights* (Geneva, 1983), p. 461, Nos. 16–20.

(a) The right to be informed of the reasons of the detention. The normal requirement in human rights treaties that every detainee should be informed at the time of arrest or promptly thereafter of the reasons for the arrest and of the charges against him[81] should be adapted to the situation of emergency, but should not disappear altogether. The ILA proposes that the detainee be informed within seven days of the grounds for his detention. However, disclosure of facts which could endanger the security or the public order of the State need not be made.[82] The International Commission of Jurists, on the other hand, recommends the issue of detention orders containing the grounds for detention, and statements concerning the facts and circumstances justifying it. The detention orders should be issued before arrest or at least within 24 hours of arrest, and one copy should be given to the detainee. Under the emergency regulations in Northern Ireland the practice of not informing the persons arrested of the reasons for their arrest was declared unlawful by the domestic courts.[83]

(b) Guarantees against detention incommunicado. One of the main factors which has facilitated torture has been the holding of prisoners incommunicado for a long time.[84] In the light of this, these bodies have proposed several guarantees, namely:

to authorize incommunicado detention only for a brief period of time; the 'Siracusa Principles' suggested no longer than a few days (three to seven days).[85] In the recent *Brogan case*, the European Court of HR found that the detention of suspects of terrorism for four days and six hours was in violation of article 5(3) of the ECHR. In this case, even though the UK had not sent a notice of derogation, the Court took into account the emergency situation in Northern Ireland.[86]

to guarantee access to a lawyer after arrest and at any time thereafter[87]; access to the family which would be allowed to make regular

---

[81] ICCPR, art. 9(2); ECHR, art. 5(2); ACHR, art. 7(4).

[82] ILA Paris Report (1984), p. 75; 'Kingston Seminar', p. 52 (from the moment of detention).

[83] *Ireland* v. *UK*, Ser. A: Judgment of the Court, p. 76, para. 198 (cases quoted by the Court: *Mc Elduff case; Kelly case,* and *Moore case*).

[84] In fact the UN Special Rapporteur on Torture considered that 'every effort should be made to declare incommunicado detention illegal', E/CN. 4/1988/17.

[85] 'Siracusa Principles', p. 54, No. 70(2); see also Questiaux, *Study of the Implications*, p. 45.

[86] *Brogan case*, European Court of HR, Ser. A: Judgment of 29 Nov. 1988, vol. 145-B. The Court reached the decision by 12 votes to 7 (see the dissenting opinions, especially that of Sir V. Evans). As a consequence of this finding, the UK has sent a formal notice of derogation for the relevant provisions under the ECHR and the ICCPR. For the UK position see *Netherlands Quarterly of HR*, 7 (1989), 255–7.

[87] Int. Commiss. of Jurists, *States of Emergency*, p. 462, No. 23; ILA Paris Report (1984), p. 75; 'Kingston Seminar', p. 52.

visits[88]; and access to a doctor soon after arrest, and thereafter to periodical examinations of which records should be kept.[89]

(c) The right to be treated with humanity. This right included the guarantee that administrative detainees cannot be held in worse conditions than convicted persons.[90] Even if this right is contained in the list of non-derogable rights under the American Convention and is applicable to all kinds of prisoners, it is important to insist upon its relevance in administrative detention.[91]

(d) The period of detention. No one should be detained for a period longer than 30 days, unless the reviewing authority, which must be an independent and impartial tribunal or committee with at least one judge, considers that there is sufficient cause for such detention.[92] A similar guarantee was established by the emergency laws in Northern Ireland.[93]

(e) Guarantees against indefinite detention. The practice of indefinite detention should be prohibited even in emergencies. The Special Rapporteur, Mme Questiaux, in her well-known report on emergencies, expressed the view that administrative detention of unlimited duration should be prohibited. The 'Siracusa Principles' expressed the same view.[94] The ILA Paris Minimum Standards proposed that detention should not last longer than one year; however if the circumstances warrant the continuation of detention, the detaining authority should issue a new detention order subject to all the guarantees prescribed since the beginning of the procedure.[95] It is interesting to note that in the US Restatement, 'prolonged arbitrary detention' is deemed to be against customary international law.[96] The IACHR has firmly pointed out that this kind of detention is against human rights standards even in situations of emergency. It said that

no domestic or international legal norm justifies, merely by invoking this special power, the holding of detainees in prison for long and unspecified periods, without any charges being brought against them for violation of the Law of

---

[88] Int. Commiss. of Jurists, *States of Emergency*, p. 462, No. 26; ILA Paris Report, p. 76(2)(*d*).

[89] Int. Commiss. of Jurists, *States of Emergency*, p. 462, No. 29.

[90] Ibid., p. 463, No. 35; ILA Paris Report, p. 76(2)(*g*).

[91] Meron considers this right which is contained in art. 10(1) of the Covenant as customary international law: *Human Rights and Humanitarian Norms*, p. 96.

[92] Int. Commiss. of Jurists, *States of Emergency*, p. 462, No. 24; ILA Paris Report, p. 75(2)(*d*).

[93] *Ireland* v. *UK*, European Court of HR, Ser. A: Judgments, pp. 34–9 (see the guarantees established for detention and internment).

[94] Questiaux, *Study of the Implications*, p. 41, para. 189; 'Siracusa Principles', p. 54, No. 70(2).

[95] ILA Paris Report, pp. 75–6(2)(*e*).

[96] US *Restatement*, p. 161, para. 702.

National Security or another criminal law, and without their being brought to trial so that they may exercise the right to a fair trial and to due process of law.[97]

The IACHR has urged all member States of the OAS 'to limit detentions carried out in states of emergency to a brief period of time and always subject to judicial review'.[98]

(f) Judicial or quasi-judicial guarantees for detainees. Two different subjects arise under this heading: the possibility of challenging the lawfulness of the administrative detention in emergencies before ordinary courts according to a habeas corpus (amparo) remedy, and the establishment of review procedures under *ad hoc* committees or tribunals of a quasi-judicial character.

In peacetime, the writ of habeas corpus or its equivalent is recognized in almost all legal systems as a fundamental remedy in order to challenge the lawfulness of a detention. The three human rights treaties under consideration also recognized this remedy.[99] However, the problem is whether this remedy should be maintained in states of emergency. The writ of habeas corpus can have several different functions, namely, to present the detainee before the judge in order to ascertain that he is alive, in good health, and not subjected to torture; to check whether all legal procedures have been complied with; and to test the legality of the individual's detention. Moreover, the judge, when he is not satisfied with the lawfulness of the detention, has the power to order the prisoner's release. It seems that the question is not so much whether or not the remedy of habeas corpus could be suspended in states of emergency, but which functions could be assigned to this remedy in emergencies. Thus it seems to be a matter of balancing the right of the individual against arbitrary detention and the need of the State to overcome the emergency.

Most of the studies in the field have pointed out that the complete suspension of the remedy has been one of the major causes of the complete lack of defence of detainees in states of emergency, and of the gross violations of the right to life and the right to be free from torture and ill-treatment. In many countries, including those of Latin America, the writ of habeas corpus has been completely suspended in emergencies without any procedure being established for challenging the possible arbitrariness of the detention. Consequently, as a major contribution in this delicate area, international organizations have proposed the entrenchment of the writ of habeas corpus, at least for limited purposes.[100]

---

[97] OAS, *The IACHR: Ten years of Activities (1971–1981)* (Washington, 1982), p. 337.
[98] Ibid., p. 318.
[99] ICCPR, art. 9(4), ECHR, art. 5(4); ACHR, art. 7(6).
[100] The ILA proposes that it should be available only in order to check the lawfulness of the order of detention (or in the case in which the order was made *mala fides*), the identity

However, one of the controversial questions is whether the courts should also be competent to look at the grounds for detention and to adjudicate on the merits—in other words, to look at the reasonableness of the decision taken by the executive branch when ordering the detention of suspects. In general, many studies in this area do not seem to contemplate this possibility.[101] However, the IA Court HR, in its recent advisory opinion on this very issue, seems to interpret the prohibition of derogation of the writ of habeas corpus in its widest possible sense.[102] The writ of habeas corpus does not only have the limited purpose of guaranteeing the judicial review against possible violations of the right to life, and to freedom from torture or ill-treatment, and against the possibility of formally unlawful detentions, but seems to extend the power of the courts to challenge the substance of the detention orders made by the executive.[103]

There is no doubt that this advisory opinion of the IA Court HR, which has been described as 'theoretically sound'[104] is of great importance. However, it remains to be seen whether or not the American States will make their practice in states of emergency conform to the finding of the Court. The advisory opinion could have a wider effect in the evolution of international human rights law outside the context of the American system. However, it seems that the opinion is based to some extent on the special circumstances of the American system. First of all, becaue the opinion is based on article 27(2) *in fine* of the American Convention, which expressly entrenched from derogation 'the judicial guarantees essential for the protection of non-derogable rights'. Secondly, because it provides a strong response to the dreadful violations of the right to life and physical integrity occurring in the area, linked to states of emergency and facilitated by the abrogation of the right to habeas corpus. Finally, because it seems that States parties to the ACHR have not explored the possibility of establishing other safeguards or guarantees against the possible abuses of arbitrary detention, outside the framework of this judicial procedure.

In Northern Ireland, for instance, the writ of habeas corpus before ordinary courts could be used for very limited purposes and did not amount to the necessary guarantees according to the terms of article 5(4).

---

of the person detained, and the implementation of the minimum rights of the detainee. ILA Paris Report, p. 76, No. 3.

[101] See *inter alia* Questiaux, *Study of the Implications*, p. 45; 'Kingston Seminar', p. 51, para. 237. See also the limited use of this remedy in Northern Ireland: *Ireland* v. *UK*, Ser. A: Judgments, p. 33, para, 82 and p. 77, para. 200 *in fine*.

[102] IA Court HR, advisory opinion of 30 Jan. 1987, 'Habeas Corpus in Emergency Situations'.

[103] Ibid., paras. 39–43.

[104] ILA Warsaw Report (1988), p. 11.

However, the European Court unanimously found that the system of guarantees established by the emergency legislation through the review procedure by quasi-judicial bodies was not in violation of article 5 taken together with article 15, because the measures adopted by the UK, even if they derogated from some aspects of article 5(4), did not exceed the extent strictly required by the exigencies of the situation.[105]

(g) Extra-judicial guarantees. The last point leads on to the question of the safeguards and guarantees against arbitrary arrest which provide an alternative to those of a judicial character. Thus, the ILA Paris Minimum Standards established that detention should be reviewed within 30 days by a judicial or quasi-judicial body constituted in accordance with the procedures designed to make such guarantees effective.[106] At the same time, a detention cannot continue beyond a period of one year; if it is necessary to continue the detention, the whole procedure should start again.[107] The 'Siracusa Principles' also insist on the need for the periodic review by an independent tribunal of cases of continued detention.[108] In the *Lawless case*, the Court gave weight to the safeguards that the Irish Government had set up (a Detention Commission, to which any person detained could refer his case, and the decision of which was binding on the Government) in order to assess whether the 'detention without trial' met the principle of proportionality.[109]

(h) Guarantees against interrogation abuses while in detention. The International Commission of Jurists has recommended several guarantees against possible abuses in the process of interrogation. First, all persons involved in interrogation should be accountable for the well-being of the detainee. Second, specific guide-lines regarding interrogation procedures should be issued and made public by governments.[110]

(i) The law of evidence should not be altered. The International Commission of Jurists has pointed out that one of the principal conditions which facilitates torture and ill-treatment while a person is in detention and facing an eventual criminal charge is precisely the changes that may occur in the application of the law of evidence. These changes encourage greater reliance on confession and limit the defendant's right to contest

---

[105] *Ireland* v. *UK*, European Court of HR, Ser. A: Judgment, p. 84, para. 220.

[106] ILA Paris Report, p. 75, No. 2(*c*).

[107] Ibid., pp. 75–6, No. 2(*e*).

[108] 'Siracusa Principles', para. 70(*d*). Several of these guarantees were applicable in Northern Ireland during the state of emergency from 1971 to 1975 and were reviewed by the European organs; see *Ireland* v. *UK*, European Court of HR, Ser. A: Judgment, paras. 81–90. See also the proceedings suggested by the 'Kingston Seminar', p. 52, para. 237; pp. 47–8, paras. 218–19. See also Meron, 'Draft Model Declaration on Internal Strife', *International Review of the Red Cross*, 262 (1988), 59–76, esp. art. 11.

[109] *Lawless case*, European Court of HR, Ser. A: Judgment, para. 37.

[110] See these guide-lines in Int. Commiss. of Jurists, *States of Emergency*, pp. 462–3; some of these measures were also applicable in Northern Ireland (ibid., p. 431).

evidence collected during the investigatory stage of proceedings.[111] The HR Committee, in its general comment on article 7 (torture and inhuman treatment), has included as a safeguard a provision making confession, or any other evidence, extracted in violation of article 7, inadmissible in court. Even though the HR Committee does not say that this applies to states of emergency, it should be deemed so due to the fact that it is a safeguard against a non-derogable right.[112]

(j) Central Register of Detainees. Finally, the IACHR, the HR Committee, and other non-governmental organizations, due to the frequency of disappearances or unacknowledged detentions, have stressed the importance of having a central register with all the names of the detainees, the date of detention, and the date of release. The names of the detainees and their release should also be published in an 'official gazette'.[113] Furthermore, the authorities should inform the families of the detainees as soon as possible. In the 1973 *Report on Chile*, the IACHR considered that the failure to notify the family of the detainee within 24 hours was not justified, due to the modern means of communication and due to the fact that the arrest occurred within the town or city where the family of the arrested person resided.[114]

(c) Guarantees of Due Process of Law in Emergencies

After a thorough analysis of articles 14 and 15 of the Covenant, the International Commission of Jurists has identified twenty different rights related to due process of law. In the light of the standards applicable to armed conflicts of an international and non-international character contained in the Laws of War, the International Commission of Jurists has proposed that some of these rights should be considered non-derogable. As Professor Jimenez de Arechaga has pointed out, guarantees that are considered non-derogable in time of war might *a fortiori* be considered non-derogable in times of lesser threats to the nation.[115] There is no doubt that the norms of due process play an important role in the Laws of

---

[111] Ibid., p. 462, No. 28; see also p. 430.

[112] HR Committee, General Comment 13(21)d/art. 14, para. 14 in CCPR/C/21/Add. 3, p. 6.

[113] OAS, GA Resolution 510, No. 13 (in IACHR, *Annual Report, 1980–1*); OAS, *The IACHR: Ten Years of Activities*, p. 317; Int. Commiss. of Jurists, *States of Emergency*, p. 463, Nos. 34 and 36; ILA Paris, p. 76(*h*); Questiaux, *Study of the Implications*, p. 45; 'Siracusa Principles', p. 70(*a*); HR Committee, General Comment 7(16), para. 1.

[114] IACHR, *Report on Chile* (1977), p. 43. The Int. Commiss. of Jurists also recommends that States accept regular visits by the International Committee of the Red Cross. See *States of Emergency*, p. 463, No.37.

[115] Final Recapitulation of the General Rapporteur of the Inter-American Seminar on State Security, Human Rights, and Humanitarian Law, San José, Costa Rica (quoted in Int. Commiss. of Jurists, *States of Emergency*, p. 427).

War.[116] It is interesting to note that in the discussion of the possible incorporation of the right to fair trial in the list of non-derogable rights under the Covenant, Mrs Roosevelt opposed the inclusion as unnecessary because a list of due process rights were already contained in the 1949 Geneva Conventions.[117] This would mean that at least those rights should also be guaranteed in emergencies according to the Coventant.

(i) *The Minimum Rights of Due Process.*

The International Commission of Jurists, the ILA Paris Minimum Standards, the 'Siracusa Principles', and the 'Declaration on Internal Strife' prepared by Professor Meron have established the following 'minimum rights' based on articles 75 (Protocol I) and on article 6 (Protocol II):

1. The right to be informed promptly and in detail of the charges.[118]
2. The right to 'all the rights and means of defence necessary'.[119]
3. The right to be present at one's trial.[120]
4. The presumption of innocence.[121]
5. The right not to be forced to give incriminatory evidence or to confess.[122]
6. The right to a tribunal 'which offers the essential guarantees of independence and impartiality'.[123]
7. The right to appeal.[124]
8. The principle of non-retroactivity of penal laws.[125]

---

[116] See Y. Dinstein, 'Human Rights in Armed Conflicts: International. Humanitarian Law', in Meron, *Human Rights in International Law*, p. 349. See the Geneva Convention III of 1949, arts. 82–108. For certain due process guarantees for protected persons in the Convention IV, see arts. 43, 65, 67, 71–6, 78, 117, 126. See also the more comprehensive catalogue of due process rights in art. 75 of Protocol I and art. 6 of Protocol II. The list in art. 75 is considered by US Government experts as enumerating rules of customary international law (quoted in Meron, *Human Rights and Humanitarian Norms*, p. 68).

[117] E/CN. 4/SR. 196, p. 6.

[118] Protocol II, art. 6. 2(a); Protocol I, art. 75. 4(c); Int. Commiss. of Jurists, *States of Emergency*, p. 427, No. 1; ILA Paris Report, p. 82, No 1; 'Siracusa Principles', g. 1.

[119] Protocol II, art. 6. 2(a); Protocol I, art. 75. 4(a); Int. CJ, p. 427, No. 2; ILA Paris Report, p. 83, No. 2; 'Siracusa', g. 2; Meron, 'Declaration', a.

[120] Protocol II, art. 6, 2(e); Protocol I, art. 75. 4(e); Int. CJ, p. 427, No. 3; ILA Paris Report, p. 83, No. 3; 'Siracusa', g, 4; Meron, d.

[121] Protocol II, art. 6. 2(d); Protocol I, art. 75. 4(d); Int. CJ, p. 428, No. 4; ILA Paris Report, p. 83, No. 5; 'Siracusa', g; Meron, c.

[122] Protocol II, art. 6. 2(f); Protocol I, art. 75. 4(f); Int. CJ, p. 428, No. 5; ILA Paris Report, p. 83, No. 9; 'Siracusa', g. 5; Meron, c.

[123] Protocol II, art. 6. 2; Protocol I, art. 75; Int. CJ, p. 428, No. 6; ILA Paris Report, p. 83, No. 11; 'Siracusa', (f)(e); Meron, a.

[124] Protocol II, art. 6. 3; Int. CJ. p. 428, No. 7; ILA Paris Report, p. 83, No. 12; 'Siracusa', g, 8.

[125] Protocol II, art. 6. 2(c); Protocol I, art. 75(4); Int. CJ, p. 428, No. 8; ILA Paris Report, p. 83, art. 15; Meron, g.

9. The right to obtain the attendance and examination of defence witnesses.[126]

10. The right not to be retried after a final judgment.[127]

11. The right to a lawyer of one's choice.[128]

12. The right to free legal assistance if necessary.[129]

There is another group of rights which is recognized in article 14 of the Covenant which do not have a very clear status, and which have little influence in states of emergency, due to their very general character or because their implications are post-emergency. Among those are: the general principle of equality before the courts; the right to a free translator; the right to have the final judgment made in public; and the right to compensation for unlawful conviction. Perhaps the most relevant of these, departure from which would be justified in certain cases, is the right to a free translator in the case of an indigent defendant who does not understand the language of the court. The right to compensation would normally be raised once the state of emergency is over, and the convictions are reviewed. In these circumstances, the right to compensation does not threaten the security of the State. The right to a public judgment is a recognized non-derogable right in article 75 of the First Protocol.

(ii) *The Rights whose Derogation would be Justified in Principle*. On the other hand, there are some other rights relating to fair trial and due process of law whose derogation in states of emergency seems to be justified once the principle of proportionality is satisfied. These are: the right to a public trial, the right to a trial without undue delay, and the right to examine prosecution witnesses.

The right to a public trial in states of emergency seems to be accepted as a derogable right. Moreover, even in normal circumstances, this right may be limited for reasons of public order or national security, according to the wording of the human rights treaties.[130] However, the ILA has pointed out that such restrictions shall not apply at least to the members of the family of the defendant.[131] Mme Questiaux added that, as an important safeguard, legal observers appointed by non-governmental organizations should be allowed to attend the trial.[132]

The right to a trial without undue delay: in states of emergency, longer

---

[126] Protocol I, art. 75. 4(*g*); Int. CJ p. 428; ILA Paris Report, p. 83, No. 13.

[127] Int. CJ, pp. 427–8; ILA Paris Report, p. 83, No. 8; 'Siracusa', *i*; Meron, *f*.

[128] Int. CJ, p. 428; ILA Paris Report, p. 83, No. 2; 'Siracusa', *g*. 3.

[129] Int. CJ, p. 428; ILA Paris Report, p. 83, No. 2; 'Siracusa', *g*. 3.

[130] ICCPR, art. 14(1); ECHR, art. 6(1); ACHR, art. 8(5).

[131] ILA Paris Report, p. 83, No. 4.

[132] Questiaux, *Study of the Implications*, p. 45.

delays than normal in the proceedings will be permitted. However, the IACHR has pointed out in the exposition of its doctrine that 'the indefinite prolonging of trials for crimes against public order and state security, which give rise in some instances to deprivation of the freedom of the accused for a longer period than the longest sentence he could receive, is a violation of the right to a fair trial . . .'.[133]

The right to examine prosecution witnesses is a delicate matter. In the three human rights treaties and in Protocol I, the rights to examine prosecution and defence witnesses are incorporated into the same provision.[134] However, the right to obtain the appearance and the examination of prosecution witnesses must sometimes be derogated from because of the justified fear that terrorist organizations may take revenge against those witnesses. Otherwise, this fact would discourage other potential witnesses from appearing in court and giving testimony. Consequently, as is the case in Northern Ireland, this right could be suspended in exceptional circumstances even if article 75 of Protocol I made it non-derogable. The International Commission of Jurists and the ILA have, nevertheless, pointed out that, as a safeguard, it should be accepted that the defence has the right to test the veracity of the evidence of the prosecution witnesses.[135] On the other hand, it seems that there is no reason to justify the derogation of the defendant's right to present and examine his own witnesses.

Some of the rights proposed for entrenchment have been the object of controversy, such as the *right to a lawyer of one's choice*. Sometimes it has been argued that the State has a right to derogate from this guarantee because some lawyers could play the role of intermediary with terrorist organizations and would therefore be a danger to the security of the State. In Germany, for example, the law providing for curtailment of contacts between an accused detained in custody and his legal counsel was strictly an emergency measure taken in response to a series of terrorist acts, and could only be imposed in order to avert imminent danger to life, limb, or freedom of persons and when the suspicion that such danger emanated from a terrorist organization was based upon hard evidence.[136] However, as the International Commission of Jurists has pointed out, the alternative is not necessarily the complete derogation of the right, but could be the elaboration of a list of lawyers who have been

---

[133] OAS, *The IACHR: Ten Years of Activities*, p. 320.
[134] ICCPR, art. 14(3)(*e*); ECHR, art. 6(3)(*d*); ACHR, art. 8(2)(*f*); Protocol I, art. 75(4)(*g*).
[135] Int. CJ, *States of Emergency*, p. 429; ILA Paris Report, p. 83, No. 13.
[136] HR Committee, A/33/40 (1978), p. 59, para. 360.

security vetted, rather than the imposition by the State of a military lawyer.[137]

(iii) *The Right to be Tried by an Independent and Impartial Tribunal.* Another controversial point is the meaning of the words 'to be tried by an independent and impartial tribunal'. This has been interpreted as requiring that courts be structurally independent from the other branches of government. The importance of preserving the independence of the judiciary in emergencies has been stressed by the IACHR. This raises the question of the legitimacy of two practices very often resorted to by the executive in public emergencies, namely, the creation of *ad hoc* tribunals to judge crimes against the security of the state; and the transference of the trial of civilians from the jurisdiction of ordinary courts to the military courts. The lack of independence from the executive, and the absence of the minimum guarantees of fair trial and due process of law from these tribunals have been pointed to in many cases by the relevant international bodies.[138] The best view seems to be that if the ordinary courts are operating normally, the creation of *ad hoc* tribunals, or the transference of the trials of civilians to the military jurisdiction, is hardly justified. However, in the *Lawless case*, which occurred at a time when the courts were operating normally, some of the dissenting opinions in the Commission thought that it would have been better to have instituted a Special Criminal Court for dealing with terrorist offences, and for such a Court to have judged Mr Lawless. This would have been a lesser restriction of freedom than the detention without trial that actually took place.[139] The case of Northern Ireland shows that even in a very grave public emergency, trial of civilians by ordinary courts is possible, with some modifications of the rules of procedure. However, as a last resort, and provided that the measures are strictly required by the exigencies of the situation, it could be argued that the present human rights standards in emergencies do not in theory prohibit the extreme measure of transferring the trial of civilians to military courts, provided that the minimum rights of due process as described above are respected.[140] This position seems to have been confirmed in two cases.

In the *Fals Borda case*, the applicant contended that the use of military courts in Colombia to judge civilians in offences against the security of

---

[137] Int. CJ, *States of Emergency*, p. 428.

[138] See OAS, *The IACHR: Ten Years of Activities*, p. 331; see also *Report on Nicaragua* (1981; on Anti-Somocistas Tribunals), *Report on Chile* (1985; military tribunals). The same position has been taken by the HR Committee, General Comment 13(21)d/art. 14, para. 4.

[139] *Lawless case*, Ser. B: Report of the Commission, opinions of Mr Eustathiades, p. 137; Mr Susterhenn, p. 153, No. 10; Mr Dominedo, p. 154; Mme Janssen-Petvschin, p. 154.

[140] See the position of the HR Committee (n. 138 above).

the State nullified the right to a competent, independent, and impartial tribunal as established in the Covenant and that the only competent tribunal to try civilians should be the ordinary courts. The Committee, however, did not consider that the mere fact of being tried by a military court, without any submission of facts on violations of judicial guarantees, was contrary to article 14 of the Covenant.[141]

In the *Greek case*, the European Commission considered that even if it could be said that there existed a public emergency threatening the life of the nation in Greece, the maintenance of extraordinary courts-martial and the denial of the right of re-hearing before the Court of Appeal for offenders against national security or public order were not strictly required by the exigencies of the situation, and that, therefore, there had been a violation of article 6 of the Convention. The executive in fact had a great deal of influence over the decisions of the military courts. Moreover, some of the heavy sentences imposed were later suspended by a general amnesty, which suggested to the Commission that many of them were of an arbitrary nature. At the same time, there was no right to appeal, and defence lawyers found it very difficult to carry out their duties; some of them were even arrested.[142]

(iv) *The Practice of the HR Committee.* The HR Committee has not so far had the opportunity to express a clear opinion on the general question of non-derogability of fundamental guarantees of fair trial and due process of law in states of emergency. However, the practice of the Committee seems to confirm the position advanced earlier that some of these guarantees are non-derogable in emergencies. Three cases will illustrate this point.

The Committee, when adopting its views under the Optional Protocol, has found in a number of cases from *Uruguay* that there had been clear violations of almost all the rights which form part of article 14 of the Covenant.[143] The situation in Uruguay was characterized by the application in all these cases of 'prompt security measures' (emergency laws) aimed at dealing with the alleged terrorist threat. According to Uruguayan laws, political opponents were charged in many cases with crimes such as 'subversive association' and 'conspiracy against the Constitution', and tried by military tribunals without the guarantees of article 14 of the Covenant. From the legal point of view, Uruguay was in fact in a state of

---

[141] *Fals Borda* v. *Colombia*, Comm. No. 46/1979, Adoption of Views, 27 July 1982, in Human Rights Committee, *Selected Decisions under the Optional Protocol*, i. *Second to Sixteenth Sessions, 1977–1982*, UN (New York, 1985), pp. 142 and 144, paras. 9.2 and 13.3.

[142] *Greek case*, p. 149, para. 328.

[143] See a list of all the cases related to art. 14 in HR Committee, 'Selected Decisions', i (up to the 16th Session in 1982), 159; see also ii (up to the 32nd Session in 1988), 230.

'public emergency', and on 30 July 1979 the Government had notified the emergency to the UN Secretary-General according to article 4(3). However, in the notice of derogation the Government did not state which due process rights had been derogated from. Moreover, in the proceedings befores the Committee, it did not make any submission of fact or law to justify such derogations, although it argued that there was a public emergency. The Committee has stated in these cases:

the HR Committee has considered whether acts and treatment, which are *prima facie* not in conformity with the Covenant, could for any reasons be justified under the Covenant in the circumstances. The Government has referred to provisions of Uruguayan law, including the 'prompt security measures'. The Covenant (art. 4) allows national measures derogating from some of its provisions only in strictly defined circumstances, and the Government has not made any submission of fact or law to justify such derogation. Moreover, some of the facts referred to above raise issues under provisions from which the Covenant does not allow any derogation under any circumstances.[144]

These views of the HR Committee need to be put in perspective. It is true that the Committee did not explicitly say that the measures of derogation from due process standards taken by the Uruguayan Government were not strictly required by the situation, and because of this in violation of the Covenant. However, the fact that the Committee did not accept a general justification (the existence of an emergency) in order to derogate from certain due process guarantees, together with the requirement that the Government should comply with a high standard of proof in order to show that those derogating measures were required by the situation, underlines the importance of these guarantees in emergencies. Furthermore, the fact that the Committee found violations of article 14, even in the emergency, is a sound precedent for the Committee in the eventual monitoring of other cases in which governments may argue about the necessity of similar measures of derogation.

In another interesting communication concerning *Colombia*, the Committee found that, even though the Government pleaded that the derogation of article 14(5) was justified because of the state of emergency in the country, it did not accept this defence and found a violation of article 14(5) of the Covenant, 'because Mrs Consuelo Salgar de Montejo was denied the right to review of her conviction by a higher tribunal'.[145] In reaching this decision, the Committee took into account that the notice of derogation by the Colombian Government did not mention article 14,

---

[144] *Inter alia Weinberger* (v. *Uruguay*), Comm. No. 28/1978, Adp. Views, 29 Oct. 1980, in HR Committee, *Selected Decisions*, i. 60, para. 14.

[145] *Salgar de Montejo* (v. *Colombia*), Comm. No. 64/1979, Adp. Views, 24 Mar. 1982, in HR Committee, *Selected Decisions*, i. 127–30.

but only articles 19(2) and 21 as derogated from, and that the mere invocation of the emergency by the State concerned without giving details of the relevant facts could not be used to evade its obligations under the Covenant.

Finally, when reviewing the *Report of Syria*, Sir Vincent Evans underlined the great importance in emergencies of the due process guarantees in article 14 of the Covenant. He mentioned in particular the rights to a defence counsel, to adequate time and facilities for the preparation of the defence, and the right to have the conviction reviewed by a higher tribunal, and added that 'it was difficult to see what could justify derogations from the safeguards mentioned'.[146]

(v) *The Practice under the ACHR*. In a recent advisory opinion, the IA Court HR has emphasized the extraordinary importance of the due process rights contained in article 8 of the American Convention for the exercise of remedies, such as habeas corpus and amparo, which are themselves not suspendable in situations of emergency when they have the aim of protecting non-derogable rights.[147] The Court stated that the essential guarantees which are not derogable according to article 27(2) 'should be exercised within the framework and the principles of due process of law, expressed in article 8 of the Convention'.[148] Moreover, the principles of due process cannot be suspended in states of emergency, insofar as they are the conditions necessary for the exercise of the remedies considered by the Convention to be non-derogable.

The practice of the IACHR also confirms the extraordinary importance of these fundamental guarantees of fair trial and due process, and considers them to be non-derogable. The assessment of the findings of the IACHR is more difficult because, as has been pointed out already, the procedure in elaborating country reports is not adversarial and therefore States do not present elaborate legal arguments in defence of their position. The IACHR acts mainly as a fact-finding body. As far as the rights to fair trial and to due process of law are concerned, the Commission has not answered the question of which of the due process rights are non-derogable. The Commission, when studying the situation of human rights in the respective States, many of them in a state of emergency, has found deficiencies and violations of some of the fundamental

---

[146] *Year Book, HR Committee* (1979–80), i. 167. See also the interesting comments made by members of the Committee in reviewing the *Report on El Salvador*, which seem to assume that even in emergencies some minimum guarantees of due process are non-derogable (A/39/40 (1984), p. 15).

[147] IA Court HR, advisory opinion, 'Judicial Guarantees', para. 30. In para. 29 the Court considered that these guarantees should also be applied to all remedies in force in emergencies.

[148] Ibid., para. 41(3).

rights related to due process. However, the reports of the Commission do not explain why some measures of derogation from due process standards were not legitimate in those states of emergency. It is true that in several country reports, the Governments themselves have not contended that the derogation of these rights was justified according to article 27, but in some other cases, even though the Government, as in the case of Nicaragua in 1981, has insisted on its right of derogation from these guarantees of article 8 due to the emergency, the Commission has found that they were in violation of due process standards, without analysing the compliance or non-compliance with the principle of proportionality.[149]

On at least two occasions, the IACHR has pointed out that the right to due process of law is a non-derogable right. In the 1980 *Report on Argentina*, it says that 'certain fundamental rights can never be suspended as in the case, among others, of the right to life, the right to personal safety, or the right to due process . . . in other words, under no circumstances, may governments employ . . . the denial of certain minimum conditions of justice as the means to restore public order'.[150] This text was included in the statement of the doctrine of the Commission. It is interesting to note that Argentina was not party to the American Convention and therefore the statement was based either on the standards applicable under the American Declaration (perhaps under the influence of the 1968 IACHR Resolution on States of Emergency in which due process is a non-derogable right)[151] or on a rule of customary international law.[152] At the same time, in the 1983 *Report on Guatemala*, after enumerating the procedural deficiencies of the Courts of Special Jurisdiction, the Commission recommended the modification of the law creating these courts, to ensure the conformation of 'the text (of the law) to the judicial guarantees indispensable to due process included in the ACHR, which should not be suspended even in emergency situations'.[153] In this case, Guatemala was party to the Convention.

It seems therefore that the IACHR considers the standards of due process, or at least most of them, to be the standards applicable to those countries in states of emergency within the Inter-American System both to States parties to the American Convention (art. 8), and to those which are not parties and to which the American Declaration applies (mainly articles 18 and 26).[154]

---

[149] IACHR, *Report on Nicaragua* (1981). See below, Ch. 5.
[150] IACHR, *Report on Argentina* (1980), p. 26; see also OAS, *The IACHR: Ten Years of Activities*, p. 341.
[151] IACHR, *Resolution on the Protection of Human Rights in Connection with the Suspension of Guarantees or 'State of Siege'* (12 Sept. 1968), para. *e*.
[152] For the discussion of this issue see below, Ch. 10.
[153] IACHR, *Report on Guatemala* (1983), p. 18.
[154] See IACHR, *Report on Chile* (1985), p. 149 (the right to fair trial is embodied in art. 18 and to due process in art. 26 of the American Declaration); *Report on Cuba* (1983), p.

The Commission has found the following to be violations of funda-mental rights of due process:

1. The right to be informed promptly and in detail of the charges.[155]
2. The right to 'all the rights and means of defence necessary'.[156]
3. The right to be present at one's trial.
4. The presumption of innocence.[157]
5. The right not to be forced to give incriminatory evidence or to confess.[158]
6. The right to a tribunal 'which offers the essential guarantees of independence and impartiality'.[159]
7. The right to appeal.[160]
8. The principle of non-retroactivity of penal laws.[161]
9. The right to obtain the attendance and examination of defence witnesses.[162]
10. The right not to be retried after a final judgment.[163]
11. The right to a lawyer of one's choice.[164]
12. The right to free legal assistance if necessary.[165]

The IACHR has stressed the importance of an independent judiciary for the protection of human rights. The Commission has complained about the lack of independence of the judiciary in many of the Latin American States.[166] Especially strong has been its criticism in cases in which, due to the state of emergency, States have introduced military and special courts. The Commission has indicated not only the lack of

---

51; *Panama* (1978) and *Paraguay* (1978). Other articles of the Declaration adduced by the Commission: art. 25 (non-retroactivity) and art. 28 (right to be tried within a reasonable time). See *Report on Argentina* (1980), p. 224; *Cuba* (1962), p. 4.

[155] IACHR, *Report on Cuba* (1983), p. 57.

[156] IACHR, *Report on Panama* (1978), p. 116; *Colombia* (1981), p. 181; *Nicaragua* (1981), p. 168; *Guatemala* (1983), p. 95; *Chile* (1985), pp. 171, 193; *Suriname* (1983), p. 68; *El Salvador*, in *Annual Report, 1985–6*, p. 154.

[157] IACHR, *Report on Argentina* (1980), p. 224; *Nicaragua* (1981), pp. 88–9, 93, 168; *Suriname* (1983), p. 68.

[158] IACHR, *Report on Guatemala* (1983), pp. 95 and 280.

[159] IACHR, *Annual Report, 1977*, pp. 27–9; *Report on Cuba* (1962), p. 4; *Paraguay*, in *Annual Report 1979–80*, p. 117; *Argentina* (1980), p. 224; *Haiti* (1979), p. 79; *Nicaragua* (1981), pp. 88–9, and *Annual Report, 1982–3*, p. 18, *1983–4*, p. 118; *Cuba* (1983), pp. 51, 58, 177; *Chile* (1985), p. 192.

[160] IACHR, *Report on Panama* (1978), p. 116; *Chile* (1985), p. 191; *Nicaragua* (1981), pp. 168 ff.

[161] IACHR, *Report on Cuba* (1962), p. 4; *Haiti* (1963), p. 325; *Argentina* (1980), p. 224.

[162] IACHR, *Report on Guatemala* (1983), p. 95.

[163] IACHR, *Report on Cuba* (1983), p. 74; *Cuba* (1979), pp. 32–3.

[164] IACHR, *Report on Paraguay* (1978); *Argentina* (1980), pp. 222–3, 265; *Guatemala* (1983), p. 95; *Cuba* (1983), p. 56; *Suriname* (1983), p. 68.

[165] IACHR, *Report on Paraguay* (1978) *Panama* (1978), p. 116; *Argentina* (1980), pp. 222–3; *Guatemala* (1983), p. 95; *Suriname* (1983), p. 68.

[166] IACHR, *Annual Report, 1977*, pp. 27 and 29(2). See also *Annual Report, 1984–5*.

independence and impartiality of many of these courts, but also the lack of fundamental guarantees of due process.[167]

6. CONCLUSION

1. The principle of non-derogability of certain rights has been considered to be one of the most important principles of the derogation clause. According to this principle, even in situations of emergency threatening the life of the nation there are some rights which can never be suspended; consequently, this principle establishes a clear limitation on the right of the State to take measures derogating from human rights standards when it faces an emergency.

2. In the drafting of the three treaties under consideration, there was complete unanimity on the necessity of including this principle in the derogation clause. However, the elaboration of a concrete list of non-derogable rights presented more difficulties. From the *travaux préparatoires* of the treaties, three conclusions can be drawn. First, that there was a lack of a profound and careful examination of the category of 'non-derogable rights' and of all the implications that the exclusion of some rights might have. Second, it is not clear what the main criteria actually were for including some rights in the list and excluding others. In fact, it would seem that two different criteria were used, but not consistently: one was to include those rights which are absolutely fundamental and indispensable for the protection of human beings; the other was to include those rights from which derogation by the State would never be justified because they have no relation with an emergency. Third, in the elaboration of the list, especially under the Covenant, the case of war, which undoubtedly is the gravest of emergencies, was very much in the mind of the drafters when they refused to include in the list some important rights. This fact could be of some importance when assessing, in the light of the *travaux préparatoires* of the treaties, the need for derogations in situations of less gravity (i.e. internal disturbances).

3. The result of these problems in the drafting process is the existence

---

[167] IACHR, *Report on Cuba* (1962), p. 4; *Panama* (1978), 'night courts', pp. 67–8; *Argentina* (1980), pp. 222–3; *Colombia* (1981), p. 181; *Nicaragua* (1981), Special Courts, pp. 93, 168; *Annual Report, 1982–3*, 'Anti-Somocists Tribunals', p. 82; *Guatemala* (1983), Special Courts, pp. 43–61 and 132; *Chile* (1985), Military Tribunals, p. 193. See also IACHR, *Report on Colombia, 1981*, pp. 219–22; and also p. 181. In this case the Commission failed to examine whether the introduction of military courts in trying civilians was absolutely necessary in view of the fact that the ordinary judicial system was operating normally and with all the guarantees of due process.

of three different lists of non-derogable rights. These lists can be criticized for two reasons. First, because they do not include some fundamental guarantees which need to to be entrenched in emergencies (i.e. minimum guarantees for detainees and due process). Secondly, because the decision of enlarging the list with the inclusion of rights which are not particularly at risk in emergencies could have an adverse psychological effect; this is especially clear in respect to the ICCPR and the ACHR. Even if in principle it is true to say that there is no reason for not including these rights in the list, there is a case for maintaining that a short list of rights, containing only those rights which must be protected in emergencies because they are so fundamental and because there is a high risk of them being violated, should be preferred. A short list would have two effects: first, it would give more strength to the category of non-derogable rights, and, secondly, there would be a greater prospect of it being accepted and implemented by States.

4. A comparative analysis of the three lists shows that there are four common non-derogable rights, namely, the right to life, the right to be free from torture and other inhuman or degrading treatment or punishment, the right to be free from slavery or servitude, and finally, the right to be free from a retroactive application of penal laws. These four rights are so fundamental that they are not only customary international law, but also norms of *ius cogens*, and no derogation whatsoever can be permitted in emergencies. However, international monitoring organs have documented gross violations of these fundamental rights in emergencies.

5. A detailed analysis of the other non-derogable rights in the ICCPR and the ACHR confirms the impression that the inclusion of these rights in the list was due to the fact that most of them are not directly related to emergencies and therefore there is no need to derogate from them. This is especially true in respect to the right to be recognized as a person before the law (ICCPR, art. 16; ACHR, art. 3), the right to be free from imprisonment merely on the ground of inability to fulfil a contractual obligation (ICCPR, art. 11; ACHR, art. 7(7)), the right to marry and to build a family (ACHR, art. 17), the right of the child to be protected (ACHR, art. 19), and finally the right to a name (ACHR, art. 18).

6. As far as those non-derogable rights which may have an impact on emergencies are concerned, such as the right to nationality, the right to freedom of conscience and religion and the right to participate in government, the existence of important qualifications within these rights means that their entrenchment against derogation is not so important.

7. In addition to the rights expressly entrenched in the derogation clause, other provisions in the three treaties could be deemed non-derogable by implication; these provisions are those related to the exercise of the non-derogable rights (remedies and non-discrimination), those

containing general exceptions, and those related to the machinery of implementation of the treaties.

8. There is no doubt that the right to be free from arbitrary arrest and the right to due process of law are two of the most fundamental rights for the protection of human beings at all times. These rights, however, have not been made non-derogable in the treaties, even though there was a long discussion about the need to include at least some of their guarantees. Two facts have caused several international organizations, in the light of the jurisprudence of human rights organs, to propose that at least some of the most fundamental guarantees contained in these rights should be made in principle non-derogable. The first is the fact that in the last decades gross violations of the most fundamental rights (the rights to life and to freedom from torture) have been made possible in part because of the lack of minimum guarantees. The second is the relative weakness of the international machinery of implementation of human rights (especially under the UN Covenant and the American Convention) when monitoring the correct application of the conditions of derogation by States parties to the treaties (i.e. the principle of proportionality). Therefore, rather than insisting only on scrupulous compliance with these conditions, legal experts have proposed the entrenchment of some of the basic guarantees relating to administrative detention and fair trial in order to make the violation of the non-derogable rights more difficult.

9. Among the main guarantees against arbitrary detention, the minimum rights of detainees should be mentioned first; these include: the right to be informed of the reasons for detention, safeguards against incommunicado detention, the right to be treated with humanity, a limit on the period of detention unless it is reviewed by a competent tribunal, judicial or quasi-judicial guarantees for detainees, such as the possibility of challenging the lawfulness of the detention before ordinary courts according to habeas corpus (amparo) remedy, and the establishment of review procedures under *ad hoc* committees or tribunals of a quasi-judicial character. The IA Court HR, in two important advisory opinions, has given great importance to the non-derogable character of these remedies in emergencies. Other guarantees which should be mentioned are those against interrogation abuses while in detention, those against a change in the application of the law of evidence, and finally, the necessity of keeping a central register of detainees.

10. As far as the guarantees of due process are concerned, the minimum rights the derogation of which, in principle, would not be justified are the right to be informed of the charges, the right to all the rights and means of defence, the right to be present at the trial, the presumption of innocence, the right not to be forced to give incriminatory evidence or to confess, the right to an impartial and independent tribunal, the right to

appeal, the right to examine defence witnesses, the right not to be retried after a final judgment, the right to a lawyer of one's choice, and finally, the right to free legal assistance if necessary. The importance of respecting these minimum guarantees, which have been based on the minimum standards of due process contained in the 1949 Geneva Conventions on the Laws of War and its two Protocols, has been stressed by the UN HR Committee and the IACHR.

Appendix

Reservations to Non-Derogable Rights in the Three Human Rights Treaties

An interesting problem is whether or not a reservation to a non-derogable right can be held to be valid. As has been already seen in this chapter, human rights treaties prohibit States from derogating from certain human rights which are considered fundamental even in states of emergency. One way in which States might seek to restrict their international obligations under the treaties is through reservations. However, due to the extraordinary importance of the principle of non-derogability of some rights, it would be prima facie against the spirit of the treaties to make reservations to such fundamental rights. As Imbert has pointed out, it seems reasonable in principle to assume that if derogations are not permitted, reservations should not be permitted either.[168] The aim of these reflections, therefore, is to study whether or not it is possible to make reservations to those rights, and to ascertain which are the main principles regulating this matter according to the law of treaties.

The IA Court HR addressed this question in its advisory opinion on the validity of Guatemala's intention to extend its reservation to article 4(4) to paragraph 2 of the same article 4 of the American Convention, which deals with the right to life.[169] The facts were as follows: when Guatemala ratified the Convention, on 25 May 1978, it made a reservation to article 4(4), because article 54 of the Guatemala Constitution only excludes political crimes, not common crimes related to political ones, from the application of the death penalty.[170] However, on 1 July 1982, after a *coup d'état*, the new regime of general Rios Montt created Courts of

---

[168] P. H. Imbert, 'Reservations and Human Rights Conventions', in Maier (ed.), *Protection of Human Rights in Europe*, p. 93.

[169] IA Court HR, advisory opinion of 8 Sept. 1987, 'Restrictions to the Death Penalty (arts. 4(2) and 4(4) American Convention)'.

[170] Ibid., para. 10.

Special Jurisdiction to deal with subversive activities. Decrees Law 46–82, which created the Special Courts, prescribed the death penalty for eighteen crimes that were not previously so punishable; previously, only six offences had been punishable with the death penalty.[171] When several executions were decreed and carried out, the IACHR intervened, due to the legal monstrosity of this practice, taking into account the clear and absolute prohibition of article 4(2) *in fine* of the American Convention: 'the application of such punishment (death penalty) shall not be extended to crimes to which it does not presently apply.'

The Government of Guatemala justified the practice on two grounds: firstly, on the basis of the sovereign power of the State to modify its internal criminal legislation in exceptional circumstances as a protective measure for society, and secondly, on the basis of an extensive interpretation of its reservation to article 4(4).[172] The IACHR requested an advisory opinion of the Court and worded the question in general terms.

The Court in the dispositive part of the judgment held that article 4(2) imposed 'an absolute obligation on the extension of the death penalty and that, consequently, the Government of a State Party cannot apply the death penalty to crimes for which such a penalty was not previously provided for under its domestic law', and that a reservation restricted by its wording to article 4(4) cannot justify the introduction of legislation contrary to article 4(2).[173]

The Court, dealing with the question of the validity of reservations to non-derogable rights, established an important principle. The regime of reservations under the American Convention should be interpreted according to the Vienna Convention on the Law of Treaties (art. 75, American Convention). Therefore, reservations are valid provided that they are not incompatible with the purpose and object of the treaty (art. 20(1) of the Vienna Convention).[174] The Court said:

it would follow therefrom that a reservation which was designed to enable a State to suspend any of the non-derogable fundamental rights must be deemed to be incompatible with the object and purpose of the Convention and, consequently, not permitted by it. The situation would be different if the reservation sought merely to restrict certain aspects of a non-derogable right without depriving the right of its basic purpose.[175]

---

[171] C. Moyer and D. Padilla, 'Executions in Guatemala as Decreed by the Courts of Special Jurisdiction in 1982–3', in OAS, *Human Rights in the Americas: Homage to the Memory of Carlos A. Dunshee de Abranches* (Washington, 1984), p. 281.

[172] IACHR, *Report on Guatemala* (1983), pp. 54–5.

[173] IA Court HR, advisory opinion of 8 Sept. 1987, 'Restrictions', para. 76.

[174] IA Court HR, advisory opinion of 24 Sept. 1982, 'The Effect of Reservations on the Entry into Force of the American Convention (arts. 74 and 75)', para. 35.

[175] IA Court HR, advisory opinion of 8 Sept. 1987, 'Restrictions', para. 61.

Therefore, a reservation which aims substantially to suspend a non-derogable right would be incompatible with the Convention, whereas the restriction of certain aspects of it could be deemed compatible. The Court considered that the Guatemala reservation fell within the latter category, since it did not deny the right as such, and was therefore compatible with the object and purpose of the treaty.[176] As Professor Buergenthal has pointed out, 'the opinion constituted the first unambiguous judicial articulation of a principle basic to the application of human rights treaties that links non-derogability and incompatibility'.[177]

It seems, therefore, that the decisive criterion for the validity of reservations is compatibility with the object and purpose of the treaty; but because the non-derogable rights are considered to be so fundamental, a general reservation to one of the rights will be deemed incompatible *ipso facto* with the object and purpose of the treaty, whereas a reservation to a particular aspect of it will in principle be held to be valid provided that it passes the test of compatibility.

This construction of the Inter-American Court seems to be sound and could be applied to the other two treaties. The UN Covenant contains no clause whatsoever on reservations. It seems that the intention of the parties was to abide by the rules adopted at that time by the International Law Commission in the drafting of the Law of Treaties Convention. As several representatives pointed out during the drafting of the Covenant, 'the right of a contracting party to make reservations to a multilateral treaty was now an accepted principle, subject to the provision that such reservations were not incompatible with the object and purpose of the treaty'.[178] This principle, formulated for the first time in the Advisory Opinion of the International Court of Justice in the *Genocide case*, was incorporated into article 19(c) of the Vienna Convention on the Law of Treaties, and can be said to constitute a norm of customary international law.[179]

The European Convention contains a specific clause on reservations (art. 64) which establishes a more restrictive regime in the sense that, while it prohibits reservations of a general character, it allows reservations to any particular provision of the Convention to the extent that any law in force in the territory of the reserving State party is not in con-

---

[176] Ibid.

[177] T. Buergenthal, 'Judicial Interpretation of the American Human Rights Convention', in OAS, *Human Rights in the Americas*, p. 258.

[178] Report of the Third Committee to the General Assembly (1966), A/6546, paras. 142–3.

[179] See the application of this test by the HR Committee in the case of Trinidad and Tobago with express reference to art. 19(c) of the Vienna Convention, UNGAOR. A/40/40 (1985), p. 22, para. 112. See also *'Reservations to the Genocide Convention'*, *ICJ Reports* (1951).

formity with the provision at the time of signing or ratifying the Convention. This provision aimed to encourage a great number of States to become parties to the Convention. Even if some of their domestic legislation was not in conformity with some provisions of the Convention, it was implicitly understood that they would bring that legislation into line with the Convention as soon as possible. However, even under the European Convention, as Professor Jacobs points out, 'any far-reaching reservation would in any event be illegal, as being incompatible with the object and purpose of the Treaty'.[180] This position seems to have been confirmed recently by the European Court of Human Rights in the *Belilos case*, where it states that article 64 'expressly prohibited reservations of a general character and prohibited by implication those which were incompatible with the Convention'.[181]

It is interesting to note that the ILA approved a resolution in its Fifty-Fourth Conference at The Hague which took the same line: the only reservations permitted in multilateral treaties on human rights are those which are not incompatible with the object and purpose of the treaty.[182] In the following pages, some reflections will be made on the reservations made by States on the principle of non-derogability of fundamental rights.

1. Reservations of a General Character

Before dealing with particular reservations to some aspects of the non-derogable rights, due consideration should be given to reservations of a general character to the whole principle of non-derogable rights established in paragraph 2 of the derogation clause.

The first to be considered is the *Mexican reservation* made at the time of Mexico's signature to the American Convention. As has been seen in the *travaux préparatoires* of the Convention, Mexico took a very strong stand against the principle of non-derogability of certain rights embodied in paragraph 2 of the derogation clause, because it conflicted with its fundamental law. The aim of Mexico was to give governments a virtually unlimited authority to suspend the Convention's guarantees in time of emergency. Mexico's proposal to delete the principle did not succeed, and as a result, Mexico entered into a reservation to article 27(2) of the Convention included in the Final Act of the Conference, which states that:

---

[180] F. G. Jacobs, *The European Convention on Human Rights* (Oxford, 1975), p. 212.
[181] *Belilos case*, European Court of HR, Ser. A: Judgment of 29 Apr. 1988, para. 52. See also G. Cohen-Jonathan, 'Les Réserves à la Convention européenne de droits de l'homme: A propos de l'arrêt [Belilos du 29 avril 1988],' *Revue generale de droit international public*] (1989), pp. 273–315.
[182] ILA, Report of the 54th Conference at the Hague, 1970 (London, 1971), p. 625.

the Constitution of the United Mexican States authorizes in a general way the suspension of any rights that may be obstacles in order to cope, rapidly and easily, with situations of extreme emergency. The Mexican delegation consequently expresses a reservation to point 2 of article 27, which restricts this authorization of suspension with respect to certain given rights.[183]

It is important to note that Mexico did not include this reservation at the time of its accession to the Convention. Therefore, according to article 23(2) of the Vienna Convention, which requires that 'a reservation must be formally confirmed by the reserving State when expressing its consent to be bound by the treaty', this reservation would not be valid because it was not confirmed at the moment of accession to the Convention.[184] Interestingly enough, the only other express reservation included in the Final Act of the Conference was confirmed by Uruguay at the time of ratifying the Convention.[185] The fact that Mexico did not confirm its reservation may well indicate that it was withdrawn. In any case, in the light of the above-mentioned advisory opinion of the Inter-American Court, the Mexican reservation, in view of its general aim to limit all non-derogable rights, would not be compatible with the object and purpose of the treaty.

Another general reservation is the one made by *Trinidad and Tobago* in 1978 on ratifying the UN Covenant. It reads as follows:

the Government of the Republic of Trinidad and Tobago reserves the right not to apply in full the provision of paragraph 2 of article 4 of the Covenant since section 7(3) of its Constitution enables Parliament to enact legislation even though it is inconsistent with sections (4) and (5) of the said Constitution.[186]

This reservation would authorize Trinidad and Tobago, in the case of a public emergency threatening the life of the nation, to take measures derogating from any of the rights which the Covenant declared non-derogable. There is no doubt that this reservation should be deemed incompatible with the Covenant, due to its general and sweeping character. It would allow the State in question to derogate from fundamental rights like the right to life, or to freedom from torture and ill-treatment. The Federal Republic of Germany and the Netherlands raised a formal objection to the reservation. In view of the text and the history of the

---

[183] Statements included in the Final Act of the San José Conference (1969). Reservation of Mexico to art. 27(2), in Buergenthal and Norris (eds.), *The Inter-American System*, vol. i, booklet 3, p. 27. See above, pp. 92–3.

[184] OAS, *Handbook of Existing Rules Pertaining to Human Rights in the Inter-American System* (Washington, 1985), p. 94.

[185] Ibid., pp. 64 and 68.

[186] UN, *Multilateral Treaties Deposited with the Secretary-General (Status as at 31 Dec. 1986)* (New York, 1987), p. 135.

Covenant, they considered the reservation to be incompatible with the Covenant's object and purpose.[187]

The HR Committee, in the review of the Trinidad and Tobago periodic reports, has also insisted on the incompatibility of this reservation, and has asked the Government to withdraw it.[188] Professor Higgins pointed out that

> the reason why certain rights were regarded as being non-derogable was because there was a feeling that it was never necessary to derogate from them, even in the most difficult circumstances. For example there was no excuse for a Government to allow arbitrary deprivation of life or the adoption of criminal laws having retroactive effect. It would seem inconsistent that Trinidad and Tobago might wish to act in such a way, even if a state of emergency was proclaimed.[189]

The validity of the *French reservation* to the derogation clause both to the European Convention and the UN Covenant is very controversial.[190] In any case, if one takes into account the express terms of the reservation and the statement of the French representative in the presentation of the periodic report before the HR Committee, such a reservation does not affect the second paragraph of the derogation clause, which deals with the non-derogable rights. Therefore, the President of the French Republic may take any measure considered necessary to overcome the emergency, apparently even in derogation of the principle of proportionality provided that it does not affect any of the non-derogable rights.[191] The French reservation states that the principle of proportionality established in paragraph 1 of the derogation clause 'cannot limit the power of the President to take the measures required by the circumstances'. At the same time, as far as the principle of exceptional threat is concerned, the reservation considers that the circumstances enumerated in article 16 of the French Constitution are to be understood as meeting the purpose of the derogation clause.[192]

---

[187] Ibid., pp. 137–8.

[188] CCPR/C/SR. 555 (1982), p. 2, Mr Tomuschat; CCPR/C/SR. 765, p. 6, Mr Cooray and Mr Lallach. See also Imbert, 'Reservations', p. 95 n. 24.

[189] CCPR/C/SR. 765, p. 5, para. 16.

[190] The validity of the French reservation has been seriously put in doubt by several writers, including: Jacobs, *The European Convention on Human Rights*, pp. 212 and 209; P. Van Dijk and F. Van Hoof, *Theory and Practice of the ECHR* (Deventer, 1984), p. 453; Higgins, 'Derogations under Human Rights Treaties', p. 317.

[191] For the French reservation to the ECHR, see *YBECHR* 17 (1974), P. 4; for the reservation to the Covenant, see UN, *Multilateral Treaties Deposited with the Secretary-General*, Status as at 31 Dec. 1986 (New York, 1987), p. 131; for the declaration of Mr Guillaume before the HR Committee, see A/38/40 (1983), p. 75, para. 320. Also CCPR/C/SR. 445, p. 8, para. 32.

[192] There is no room here for a long discussion of the French reservation, See, however, J. F. Villevieille, 'La Ratification par la France de la Convention européenne des droits de l'homme', *AFDI* 29 (1973), 922–7; V. Coussirat-Coustère, 'La Réserve française à l'article

Another interesting prima-facie reservation of a general character, but this time only to one of the non-derogable rights, was the *Congo's reservation* to article 11 of the UN Covenant, which states that 'no one shall be imprisoned merely on the grounds of inability to fulfil a contractual obligation'. As was mentioned before, this article is non-derogable only under the UN Covenant. The first part of the Congo's reservation says in plain terms that 'it does not consider itself bound by the provisions of article 11'. In fact, according to the Congolese legislation, 'in matters of private law, decisions or orders emanating from conciliation proceedings may be enforced through imprisonment for debt when other means of enforcement have failed, when the amount due exceeds 20,000 CFA francs and when the debtor between 18 and 60 years of age, makes himself insolvent in bad faith'.[193]

Belgium, supported by the Netherlands, raised a formal objection to this reservation on the grounds of the non-derogable character of the article. This article has a particularly restricted sphere of application: it prohibits imprisonment only when there is no reason for resorting to it other than the fact that the debtor is unable to fulfil a contractual obligation. Therefore, any reservation to that article would destroy its effects and would contradict the letter and the spirit of the Covenant. The toleration as a matter of principle of this reservation would create a precedent which would have a considerable effect. Thus the Belgium Government hoped that it would be withdrawn.

Even if this was the position of the objecting Governments as a matter of principle, they found, after a careful examination of the content of the reservation to article 11 in the light of the *travaux préparatoires*, that there was no incompatibility between the article as it stands and the Congolese legislation. They believed that imprisonment can be imposed under article 11 when the debtor, by acting in bad faith or through fraudulent manoeuvres, has placed himself in the position of being unable

---

15 de la Convention européenne des droits de l'homme', *Journal du droit international*, 2 (1975), 269–75; K. Vasak (ed.), 'L'Histoire des problèmes de la ratification de la Convention par la France', *HRJ* 3 (1970), 558–66. A. Pellet, 'La Ratification de la Convention européenne des droits de l'homme', *Revue de droit public* (1975), pp. 1319–79; Questiaux, 'La Convention européenne des droits de l'homme et l'article 16 de la constitution du 4 octobre 1958', *HRJ* 3 (1970), 668; Cohen-Jonathan, 'La Reconnaissance par la France du droit de recours individuel devant la Commission européenne des droits de l'homme', *AFDI* 27 (1981), 280 ff. For a good exposition of Turkey's reservation on the application of the derogation clause of the ECHR when making a declaration under art. 25 [accepting the right to individual petition, see I. Cameron, 'Turkey and Article 25] of the ECHR', *ICLQ* 37 (1988), 887–925. See also C. Zanghi, 'La Declaration de la Turquie relative à l'article 25 de la Convention européenne des droits de l'homme', *Revue general de droit international public*, 93 (1989), 69–95.

[193] *Multilateral Treaties*, p. 130.

to fulfil his obligation. This is precisely the condition in the Congolese legislation, and therefore the reservation would be unnecessary.[194]

2. Reservations to Particular Aspects of the Non-Derogable Rights

(a)   Reservations to the Right to Life

The Malta reservation.   When Malta ratified the European Convention in 1967, it made a reservation to the right to life, in the sense that the principle of lawful defence admitted in article 2(2)(*a*) as an exception for the deprivation of life should apply in Malta also to the defence of property in the light of its Criminal Code (arts. 237–8). According to these articles, 'actual necessity of lawful defence' included three cases:

repelling an invasion of one's property at night;

acts in defence against any person committing or attempting to commit theft or plunder;

the defence of one's chastity or of the chastity of another person.[195]

The standard of article 2(2)(*a*) is 'the use of force that is no more than absolutely necessary . . . in defence of any person from unlawful violence'. Unfortunately, neither the Court nor the Commission has interpreted this provision. In fact, in their legislations, many States accept the deprivation of life when it is necessary to protect property, honour, or national security. Moreover, the legislation of several newly independent States has adopted the same wording of article 2 of the European Convention, with the addition of protection of property as a justification for a lawful deprivation of life. However, it seems that the evolving standard in international law justifies the deliberate killing of another person only as a response to an imminent threat of injury or death to the victim. It is rightly argued that the value of life cannot be put at the same level as other values, such as property.[196]

Malta's reservation has not raised either objections from other states or opinions from the Court or the Commission. With the recent acceptance by Malta of the right of individual petition under article 25 of the Convention, some cases may arise in the future, and the Strasbourg bodies would have to pronounce on the validity of this reservation to one of the most important, if not the most important, right of the Convention.

The Barbados reservation.   As was seen before, the Guatemalan reservation to one aspect of the right to life (article 4(4)) under the

---

[194] Ibid., pp. 137–8.

[195] Council of Europe, *Collected Texts of the ECHR*, p. 605.

[196] K. Boyle, 'The Concept of Arbitrary Deprivation of Life', in Ramcharan (ed.), *The Right to Life in International Law*, p. 241.

American Convention has been considered by the IA Court HR to be a valid reservation. By this reservation, Guatemala excluded political crimes from the application of the death penalty, but not common crimes related to political ones.[197] By parity of reasoning in relation to the Court's opinion in this case, the reservation made by Barbados to the same provision of the American Convention, by which it reserves the right to apply the death penalty in the case of crimes of treason, which are sometimes regarded as political crimes, should also be deemed valid due to its very concrete character. However, the second Barbados reservation to article 4 could present some problems. This reservation was made to paragraph 5, which prohibits the imposition of the death penalty on persons who at the time the crime was committed were under 18 or over 70 years of age, because Barbados legislation authorizes in principle the execution of persons over 16 or over 70 years of age.[198] It seems that there have been no objections to this reservation up until the present.

As far as the reservation to the prohibition of the execution of juveniles under 18 is concerned, two comments should be made. On the one hand, it might be argued that because the reservation is a very concrete one, and not one of a general character, and it does not aim to restrict the right to life as a whole, but only a partial aspect of it, the reservation could be considered to be valid. On the other hand, there are some grounds for saying that there is a rule of customary international law which prohibits the execution of persons under 18 years of age. The IACHR in the *Roach and Pinkerton case* has said that, even if this norm does not exist yet in customary international law, it is emerging.[199] However, Professor Shelton has criticized this opinion and has shown, using much evidence based on State practice, that the norm has already emerged.[200] If this is the case, the compatibility of the reservation with the object and purpose of the treaty will be more difficult to maintain. In any case, the imposition by Barbados of the death penalty on juveniles under 18 will be a violation of article 6(5) of the UN Covenant, ratified by Barbados without any reservation.[201] In fact, in reviewing the Barbados report, some of the members of the HR Committee asked the representative to bring the law concerning the imposition of the death penalty on persons under 18 in line with article 4(5) of the Covenant.[202]

---

[197] IA Court HR, advisory opinion, 'Restrictions', para. 61; for the Barbados reservation, see *Handbook of Existing Rules*, p. 86.

[198] Ibid., pp, 86–7.

[199] IACHR, *Roach and Pinkerton (v. USA)*, Case No. 9647. Resolution No. 3/87 (27 Mar. 1987). para. 60.

[200] D. Shelton, 'Application of Death Penalty on Juveniles in the US: Violation of Human Rights Obligations within the Inter-American System', *HRLJ* 8 (1987), 345–61.

[201] *Multilateral Treaties*, p. 129.

[202] HR Committee, A/43/40 (1988), p. 131.

There is no corresponding provision in the Covenant with respect to the reservation to the provision prohibiting execution of persons over 70. In the improbable case of Barbados imposing the death penalty on persons over 70, and wanting to carry it out despite eventual appeals on humanitarian grounds, it is uncertain that the American organs would find a violation of the Convention because the reservation would be held to be incompatible with the treaty, due to its very limited extent.

Finally, *Mexico*, when depositing the instrument of adhesion to the American Convention, made an interpretative declaration to article 4(1) in relation to the issue of abortion. It considered that the expression 'in general' 'does not constitute an obligation to adopt or maintain in force legislation that protects life "from the moment of conception" in as much as this matter belongs to the domain reserved to the States'.[203] In the light of the Commission's opinion in the *Baby Boy case*, even though the Commission has taken a very flexible approach which allows the expression 'in general' to be construed as allowing States parties a wide margin of appreciation in establishing permissive legislation on abortion, an absolute liberalization of abortion by one of the States parties could hardly be considered compatible with the object and purpose of the treaty.[204]

(b) Reservations to the Principle of Non-Retroactivity of Penal Laws

Three reservations have been made to paragraph 1 *in fine* of the principle of non-retroactivity of penal laws, recognized in article 15 of the Covenant, which states: 'If subsequent to the committing of the offence, provision is made by law for the imposition of a lighter penalty, the offender shall benefit thereby.'

In its reservation, *the Federal Republic of Germany* declares that 'when provision is made by law for the imposition of a lighter penalty, the hitherto applicable law may for certain exceptional categories of cases remain applicable to criminal offences committed before the law was amended'.[205] Taking into account the limited scope of the reservation, the fact that only 'exceptional categories of cases' are exempted from the application of article 15(1), the reservation seems to be compatible with the treaty. *Italy* and *Trinidad and Tobago* made the same reservation, in as much as the provision of article 15(1) should be applied exclusively to cases in progress and not to cases in which a person has been convicted by a final decision.[206] It seems that this construction states the correct inter-

---

[203] IACHR, *Handbook of Existing Rules*, p. 94.
[204] IACHR, *Baby Boy (v. USA)*, Res. No. 3/87 (27 Mar. 1987).
[205] *Multilateral Treaties*, p. 132.
[206] Ibid., pp. 133 and 135; see also p. 149 n. 9.

pretation of the provision; in fact, Trinidad and Tobago call it an interpretative declaration rather than a reservation.[207]

In ratifying the ECHR on 8 November 1978, *Portugal* made a reservation to article 7, This provision would be applied subject to article 309 of
the Constitution of the Portuguese Republic, which provides for the
indictment and trial of officers and personnel of the State Police Force.
This article of the Constitution states that Law No. 8/75 of 25 July 1975
with its amendments shall remain in force; its provisions state that some
offences 'may be further defined by law', and that the exceptional exten-
uating circumstances may be specifically regulated by law. The above-
mentioned 1975 Law established the penalties applicable to the referred
persons and also that the military courts had jurisdiction in those cases.[208]

The exact extent and use of this law authorizing the further definition by law of offences which were to apply retroactively has not been
explained by Portugal; therefore it is difficult to judge the compatibility of
the reservation with the object and purpose of the treaty. It is interesting
to note that Portugal did not enter into the same reservation when
ratifying the UN Covenant almost on the same date.[209]

The Federal Republic of Germany and *Argentina* made a similar reservation to paragraph 2 of the article recognizing the principle of non-
retroactivity; Germany made a reservation to article 7(2) of the ECHR,
and Argentina to article 15(2) of the Covenant.[210] Paragraph 2 of this
provision has a similar wording in the two treaties, and it says that 'the
article shall not prejudice the trial and punishment of any person for any
act or omission which, at the time it was committed, was criminal according to the general principles of law recognized by the community of
nations'.[211]

The German reservation in fact does not restrict Germany's obligations
under the Convention.[212] The Argentine reservation subjects the application of the provision to the principles laid down in article 18 of the
Argentine Constitution, by which 'on no inhabitant of the Nation can
be inflicted punishment without previous trial, based on an earlier law
than the deed of the process, nor tried by special committees, nor taken

---

[207] See Council of Europe, *Problems Arising from the Coexistence of the UN Covenants
on Human Rights and the ECHR*, Report of the Committee of Experts (Strasbourg, 1970),
Doc. H (70) 7, p. 41, para. 152.
[208] *YBECHR* 21 (1978), 14.
[209] *Multilateral Treaties*, pp. 129–37.
[210] For the German reservation see Council of Europe, *Collected Texts of the ECHR* (8th
edn. 1974), p. 604. For the Argentine reservation see *Multilateral Treaties*, p. 129.
[211] Art. 15(2) of the Covenant. See also ECHR art. 7(2).
[212] See the opinion of the European Commission on HR, App. No. 1063/61, Decision on
Admissibility, 16 July 1962. See also Jacobs, *The European Convention on Human Rights*,
p. 125.

away from the judges designated by law before the deed of the trial'.[213]
Perhaps the explicit reference to the principle of non-retroactivity of
national law can be understood in the light of the difficulties faced by the
Argentine Government in the trial of the Military Junta officers for
violation of human rights, after the return to democracy. The law applied
in those trials has been exclusively national law, and therefore the sentences have been exclusively based in Argentine law without any reference to the general principles of law recognized by the community of
nations, as authorized by paragraph 2 of article 15 of the Covenant.[214]

(c) *Reservations to the Right to Freedom of Conscience and Religion*

In depositing the instrument of accession to the UN Covenant and the
American Convention, the Mexican Government made the same interpretative declaration in respect of this right recognized in article 18 of the
Covenant and article 12 of the American Convention. According to the
Mexican Constitution, the freedom of every person to profess religious
beliefs and to practice them has a limitation, in the sense that public
religious acts can only be performed in places of worship.[215] The Government considered that this limitation is included in paragraph 3 of the
above-mentioned articles. At the same time, the interpretative declaration to the Covenant includes a second provision by which 'studies carried
out in establishments designed for the professional education of ministers
of religion are not officially recognized'.[216] Due to the historical background of Mexico, the authorities might reasonably show that the first
limitation is necessary for the protection of public order, and to prevent
religious services from becoming political demonstrations. However, this
interpretation, which is not strictly speaking a reservation, is not necessarily binding upon the organs established by the treaties.

(d) *Reservations to Political Rights*

Mexico has also made an express reservation to article 23(2) of the
American Convention and to article 25(b) of the Covenant, in the sense
that according to its Constitution, 'ministers of religion shall have neither

---

[213] Constitution of the Argentine Republic, 1853 (Ministerio de relaciones exteriores.
Buenos Aires. 1926, English trans.
[214] See Argentina, National Appeals Courts (Criminal Division), 'Judgement on Human
Rights Violations by former Military Leaders (30, 12, 1986)', in *ILM* 26 (1987), 317–72. See
also Amnesty International, *Argentina: The Military Junta and Human Rights: Report of the
Trial of the Former Junta Members* (London, 1987).
[215] *Multilateral Treaties*, p. 133; OAS, *Handbook of Existing Rules*, p. 94.
[216] Ibid.

an active nor a passive vote, nor the right to form associations for political ends'.[217]

Even though the wording of the reservation says that it affects only paragraph 2, which lays out the limitations of the right, in fact it affects the whole set of rights of a group of citizens. The prohibition of the exercise of all political rights for life, and of also the right of political association, recognized in article 22 of the Covenant and 16 of the American Convention, for a group of citizens just because of their profession (religious ministry) seems to go too far, and if challenged before the monitoring organs could be deemed incompatible with the object and purpose of both treaties. This is especially true under the American Convention, where the right to participate in government as expressed in article 23 is a non-derogable right.

Finally, *Uruguay* has made another reservation to article 23(2) of the American Convention in so far as its legislation authorizes the suspension of a person's citizenship, and therefore his political rights, not only when he is under a prison sentence, as foreseen in article 23(2) but also when the person is 'under indictment on a criminal charge'.[218]

THE THREE SUBSTANTIVE CONDITIONS FOR THE DEROGATION OF RIGHTS IN STATES OF EMERGENCY

The derogation clause in the three main treaties establishes in its first paragraph three substantive conditions for taking measures of derogation from human rights obligations in situations of emergency.[219] Thus, the right of the State to take such measures is conditioned by the *principle of proportionality*, which states that the measures must be strictly required by the exigencies of the situation, by the *principle of non-discrimination*,[220] which states that the measures must not involve any discrimination, and finally, by the *principle of consistency*, which states that the measures should not be inconsistent with the States' other obligations under international law. These three principles will be dealt with in the next three chapters.

---

[217] Ibid.

[218] IACHR, *Handbook of Existing Rules*, p. 64.

[219] See e.g. the formulation of these conditions in respect of the Covenant in A/2929, para. 42.

[220] However, the derogation clause of the European Convention does not expressly mention this requirement. The consequences of this omission will be seen below in Ch. 6.

5

The Principle of Proportionality

I. INTRODUCTION

The principle of proportionality is not a principle that applies solely to the derogation clause or to the realm of human rights. The principle of proportionality can be deemed to constitute a general principle of international law;[1] and its application within different branches of international law seems to support this assertion. The principle of proportionality was first applied in the customary international law of reprisals and self-defence.[2] Nowadays, it has found application in several areas of international law, *inter alia* in humanitarian law of war,[3] in the law of counter-measures,[4] in cases of delimitation of the continental shelf between adjacent or opposite States,[5] and in the area of human rights.

One of the areas in which the principle of proportionality has found a major application has been in human rights. Moreover, from the historical point of view, proportionality and human rights have always been linked. In fact, in the Age of Enlightenment the problem of balancing the rights and freedoms of individuals with the public interest became one of the main problems in political and legal philosophy. Thus proportionality became one of the main legal principles available by which to determine the legality of States' interference in individual rights and freedoms. The same rationale lies behind its applicability in modern human rights law; the rights and freedoms recognized to individuals are not absolute or without limits; however, such limits must be propor-

---

[1] For a brief history of the notion of proportionality and its evolution in domestic and international law, see J. Delbrück, 'Proportionality', in R. Bernhardt (ed.), *Encyclopedia of Public International Law* (1984), vii. 396–400. Also H. Mosler considered it as a general principle of law, see *Recueil* 140. 4 (1974), 147–8.

[2] I. Brownlie, *International Law and the Use of Force by States* (Oxford, 1963), 261–4. Professor Brownlie has pointed out that the principle of proportionality has received little attention by jurists.

[3] See e.g., art. 57(2) (*b*) of Protocol I, Geneva Conventions.

[4] O. Y. Elagab, *The Legality of Non-Forcible Counter-Measures in International Law* (Oxford, 1988), pp. 83 ff.

[5] In this area the principle was applied for the first time by the ICJ in the *North Sea Continental Shelf cases* (1969). Afterwards, it was referred to by the Court of Arbitration in the *Anglo-French Continental Shelf case*, and by the ICJ again in the *Tunisia–Libya* and *Malta–Libya cases*. For a short summary of the application of the principle in these cases see Elagab, *Legality*, pp. 81–3. See also D. M. McRae, 'Proportionality and the Gulf of Maine Boundary Dispute,' in *Can. YBIL* 19 (1981), 287–301.

tionate to the legitimate aim pursued by the limitation. This has become a well-established principle in domestic as well as in international law. In international law, the principle of proportionality has been applied in three main areas: in the area of non-discrimination, in the area of the limitation clauses, and in the legal regime of derogations.

As far as the principle of *non-discrimination* is concerned, the jurisprudence of the international tribunals holds that not all distinction or differentiation of treatment constitutes discrimination. There are two criteria used to distinguish whether a treatment is discriminatory or not. The principle of non-discrimination is violated either if the differentiation of treatment lacks an objective and reasonable justification, or if there is no proportion between the means used and the legitimate end pursued.[6]

Concerning *limitation clauses*, the criteria applied by the European organs—the best system created so far for monitoring States' compliance with the terms of a human rights treaty—are very similar to those criteria just mentioned in relation to the principle of non-discrimination. Most of the limitation clauses in human rights treaties authorize States parties to limit the exercise of some rights when necessary in a democratic society for reasons of public order, national security, public morals, and the rights and freedoms of others. In order to assess the legitimacy of a limitation, the European organs have found that, first of all, the limitation should pursue one of the aims admissible in the clause; and secondly, that the 'necessity' of any limitation in a democratic society should make reference to the proportionality between the limitation and the aim pursued. In other words, the concept of proportionality is inherent in that of necessity.[7]

However, it seems that it is in the legal regime of the *derogation clause* that the principle of proportionality acquires paramount importance, being the main substantive criterion employed to assess the legality of the derogating measures taken by States in situations of emergency. Thus the derogation clause of the three main treaties stipulates that derogations from the obligations of the treaties are allowed 'to the extent (and for the period of time) strictly required by the exigencies of the situation'.

Some writers have suggested that two theoretical bases form the foundation of the principle of proportionality in the derogation clause. The first one is derived from the principles laid down in articles 29(2) of

---

[6] For a good study of the jurisprudence of the European organs on the principle of proportionality see M. Salvia, 'La Notion de proportionalité dans la jurisprudence de la Commission et de la Cour européenne de droits de l'homme,' *Diritto communitario e degli scambi internazionali*, 17 (Milan, 1978), 463–93.

[7] Salvia, 'La Notion', p. 493. The notion of proportionality as implicit in that of 'necessity' was also affirmed in the 'Siracusa Principles'; see para. 10(*d*) of the General Interpretative Principles relating to the justification of limitations.

the Universal Declaration, 5(1) of the ICCPR, and 17 of the ECHR, which embody a fundamental theory of limitation and imply that the extent of every limitation or derogation should be strictly proportionate to the need of defending the higher interest of society. The second can be found in the principle of self-defence in international law.[8]

Four separate cases. It is important from the outset to distinguish four separate cases relating to this topic which are not always clearly identified.

1. The first case refers to a declaration of emergency made by a Government in bad faith; in other words, when there is no public emergency whatsoever which would justify derogations. In this case, the issue of proportionality does not arise at all.

2. In the second case, the main question is whether the emergency, which actually exists, is so grave as to amount to what the treaties have defined as 'public emergency threatening the life of the nation'. The main question here concerns, therefore, the threshold of the emergency. In this case, one can talk about certain proportionality in the sense that the emergency must reach the degree (or proportion) required by the concept 'public emergency'. However, this is not what can be called 'proportionality' *stricto sensu*.

3. The third case refers to the principle of proportionality *stricto sensu*. This occurs only once the emergency existing in the country has been considered to be a 'public emergency' within the meaning of the concept in the treaties. In this case, what the monitoring organs have to do is to assess whether the derogating measures taken by a Government are proportionate to the threat.

4. Finally, the fourth case, which is in fact a particular application of the previous one, refers to the change of circumstances within the emergency. Emergencies can have a dynamic nature, in the sense that the gravity of the circumstances can vary over a certain period of time. In these cases, the measures to deal with the emergency must also vary in accordance with the different degree of gravity of the circumstances. There must therefore be a certain proportionality in each of the phases of the emergency.

This chapter will deal mainly with the last two cases which concern the principle of proportionality *stricto sensu*.

---

[8] ILA Paris Report (1984), p. 66, para. 5; see also Questiaux, *Study of the Implications*, para. 60.

(a) The Legislative History of the Treaties

The legislative history of the three treaties is of little assistance in the interpretation of this principle. Commenting on this principle of the derogation clause in the *travaux préparatoires* of the UN Covenant, the Secretary-General of the UN laconically commented that there was a general agreement on the inclusion of this requirement in the Covenant.[9] On the other hand, for the Special Rapporteur on Derogations of the IACHR, the inclusion of the principle in the Draft American Convention constituted a 'great conquest' for the hemisphere.[10]

A comparative analysis of the wording of this principle in the three treaties shows only one significant difference. Whereas the UN Covenant and the European Convention demand that the measures of derogation should be strictly required by the exigencies of the situation, the American Convention adds that they should also be limited to the period of time of the emergency. The addition of this element to the American Convention came from the 1968 IACHR Resolution on States of Emergency, which was incorporated in the IACHR Draft Convention.[11] The remote origin of this addition can be found in the modification made by the Seminar of Uruguay (1959) to the Draft Convention of the Inter-American Council of Jurists. This modification was later incorporated into the important study by Martins, which greatly influenced the latest draft of the American Convention.[12]

In any case, the fact that this additional element is not explicitly included in the other two treaties does not mean that a derogating measure under these two treaties can last once the emergency has ended. In fact, as has been pointed out in Chapter 1, states of emergency are essentially temporary; in other words, they are only justified as long as the emergency lasts.[13] Accordingly, all measures of derogation are justified only for the life-span of the emergency; thus a derogating measure should not be strictly required by the situation if it continues in force once the emergency has ended. The doctrine of the European organs and the Human Rights Committee in the application of the principle of proportionality confirms this position.[14] The construction of the principle of proportionality by the international monitoring organs of the three treaties will be studied in the next Section.

---

[9] A/2929, para. 43.
[10] Martins, *Protection of Human Rights*, p. 145.
[11] IACHR, Resolution on States of Emergency (1968).
[12] Martins, *Protection of Human Rights*, pp. 146–7.
[13] See above, Ch. 1, p. 30.
[14] For the doctrine of the European Commission in the *De Becker case*, see below, p. 151. For the doctrine of the HR Committee, see below, pp. 153–4.

2. THE APPLICATION OF THE PRINCIPLE OF PROPORTIONALITY BY THE SUPERVISORY BODIES UNDER THE THREE TREATIES

(a)   The Principle as Applied by the European Organs in Cases of Emergency

The application of the principle of proportionality—in other words, the assessment of whether the measures of derogation were strictly required by the exigencies of the situation—was one of the main issues in the *Lawless case* and the *Ireland* v. *UK case*. The European organs have always declared themselves to be competent to check the fulfilment of this substantive condition of derogation since the *First Cyprus case*.[15] However, in the *Lawless case*, the Irish Government contended that 'it was for a government, and for that government alone, to determine not only whether a state of emergency existed but also what measures were required by the exigencies of the situation'.[16] The European Court of Human Rights in the *Ireland* v. *UK case* expressed in plain terms its position on this point: it falls in the first place to each contracting State which has responsibility for the life of the nation to determine how far it is necessary to go in attempting to overcome the emergency. In principle, the government is in a better position than the international judge to decide on the nature and scope of derogating measures. In this connection, article 15(1) leaves the government with a wide margin of appreciation. However, States do not enjoy an unfettered discretion; the European organs, which are responsible for ensuring the observance of the States' obligations, have the power to rule on whether the requirement of proportionality has been met. Thus the domestic margin of appreciation is accompanied by a European supervision.[17]

The construction of the principle of proportionality provoked some disagreement within the Commission in the *Lawless case*. Some members of the Commission favoured a strict interpretation of the principle, giving weight to the expression of the derogation clause that the measures should be 'strictly required' and defending an objective interpretation of it. According to this construction, the European organs should analyse objectively whether each measure of derogation is strictly required by the

---

[15] *First Cyprus case*, p. 176. This competence was also confirmed in the *Lawless case*, Judgment of the Court, para. 22; *Ireland* v. *UK*, Judgment of the Court, para. 207; *Greek case*, para. 217; *Turkey case*, p. 32, para. 47.

[16] *Lawless case*, Ser. B: Report of the Commission, p. 104, para. 98(a), Counter-Memorial of the Respondent Government.

[17] *Ireland* v. *UK*, Ser. A: Judgment of the Court, para. 207. See the reflections made above in Ch. 2.

situation; consequently, if the State could have taken an alternative measure that would have been less prejudicial to individual rights, the European organs should find a violation of the principle of proportionality.[18] In other words, the government has to show clearly that it had no other measure available with which to deal with the emergency. On the other hand, some other members of the Commission defended a less strict interpretation of the provision. They accorded the government a wide margin of appreciation, because the government is in a better position to know the best course of action that should be taken to deal with the emergency. Therefore, the government has a certain choice of means. The function of the European organs is, consequently, not to replace the government's assessment of the situation by its own, but to review the lawfulness of the measures, and to be sure that the government has not exceeded its margin of appreciation. In this second interpretation, notions like good faith and reasonableness play an important role in assessing the proportionality of the measures.[19]

There is no doubt that, as Mr Sorensen pointed out, the application of the principle of proportionality in particular cases constitutes one of the most difficult and delicate tasks entrusted to the European organs.[20] In fact, in the *Lawless case*, there was a sharp division between the majority of the Commission (comprising eight members) who considered that the Irish Government had not exceeded its margin of appreciation and therefore the detention without trial of Mr Lawless was strictly required by the situation, and the minority (six members) who considered that other alternatives less harmful for the applicant should have been applied by the Government.[21] Surprisingly, the Court found *unanimously* that the requirement of proportionality had been met. This position of the Court, which was confirmed as a matter of principle in the *Ireland* v. *UK case*, supports the less strict interpretation of the principle as construed by the majority of the Commission in the *Lawless case*.[22]

From the jurisprudence of the Court and the Commission in the main cases on derogation, especially the *Lawless case* and *Ireland* v. *UK*, some

---

[18] *Lawless case*, Ser. B: Report of the Commission, opinions of Mr Eustathiades (pp. 134–42); Mr Susterhenn (pp. 142–54); Mr Ermacora (pp. 155–6). The same strict construction of the principle by the applicant; see p. 108 (oral hearings on 17 to 19 Apr.).

[19] Ibid., opinions of Mr Waldock (pp. 114–30) and Mr Sorensen (pp. 130–3). The same rationale can be found in the judgment of the Court in *Ireland* v. *UK*, para. 214.

[20] Ibid., opinion of Mr Sorensen (p. 130). See also the opinion of Mr Eustathiades in the *Greek case*, p. 106.

[21] For the majority: Messrs Waldock, Berc, Faber, Beaufort, Petren, Sorensen, Crosbie, Skarphedisson. For the minority: Messrs Eustathiades, Dominedo, Susterhenn, Mme Janssen-Pevtschin, Messrs Erim, Ermacora.

[22] See *Ireland* v. *UK*, Ser. A: Judgment, pp. 78–82. See also the comments on this question by Fawcett, *The Application of the European Convention*, pp. 309–12. See also the reflections made above in Ch. 1.

general principles on the application of the principle of proportionality can be formulated.

1. A succinct enunciation of the principle was made by Susterhenn in the *Lawless case*: 'the severity of the counter-measures must be proportionate to the gravity of the threat'. Thus the appreciation of the necessity of any measure to combat the threat depends essentially on the seriousness of the threat. Consequently, very drastic measures will be necessary when the threat to the life of the nation is very grave, but not when the emergency is less so.[23] Or, as Ermacora put it, 'measures which may be validly resorted to in time of war should not be taken when the situation bears no resemblance to a state of war'.[24]

2. The legitimacy of taking derogating measures arises only when the ordinary provisions of the law and the legitimate limitations foreseen for peacetime are not enough to deal with the emergency.[25]

3. The mere existence of a public emergency threatening the life of the nation, within the meaning of the derogation clause, does not justify all kinds of derogations from human rights standards. Each measure of derogation taken in a lawfully declared emergency should be necessary and proportionate to the threat. This principle was also applied by the UN Human Rights Committee in reviewing the *Report of Nicaragua* in 1981. Nicaragua was under a state of emergency, duly notified according to article 4(3). The Government in the relevant decree of suspension indicated that, for the Somocistas, all human rights were suspended except those mentioned by the ICCPR as non-derogable (art. 4(2)). The HR Committee seemed to accept the legitimacy of the state of emergency. However, as Sir Vincent Evans pointed out, even if the state of emergency was lawfully declared, notified, and recognized to be a genuine emergency, it does not necessarily follow from this fact that all the derogations from the Covenant will be 'strictly required by the exigencies of the situation'.[26] The Nicaraguan Government recognized that, due to its lack of legal experience in presenting derogation submissions, it had included all the derogable rights in the relevant decree, when in fact many of them had not been derogated at all.[27]

4. Each measure of derogation has to bear some relation to the threat; in other words, there must be a link between the facts of the emergency and the measures taken.[28] This can be called a 'qualitative proportionality'.

---

[23] *Lawless case*, Ser. B: Report of the Commission, opinion of Mr Susterhenn, p. 143.

[24] Ibid., opinion of Mr Ermacora, p. 156.

[25] *Lawless case*, Ser. A: Judgment of the Court, para. 36. Also *Ireland* v. *UK*, Ser. A: Judgment of the Court, para. 212.

[26] CCPR/C/SR. 420.

[27] A/38/40 (1983), p. 55, para. 244.

[28] *Ireland* v. *UK*, Ser. B: Report of the Commission, opinion of the Commission, p. 119.

This principle was referred to by a member of the HR Committee as being logically inherent in the derogation clause. The Government of Peru seemed to have suspended political rights in a state of emergency due to a natural disaster; this suspension would have no relation whatsoever with the emergency.[29]

5. The measures of derogation taken by the government should be potentially able to overcome the emergency. This, nevertheless, does not mean that the judgment on the strict necessity of the measures will depend on the fact that the measures have actually overcome the emergency. This was precisely one of the problems in the *Ireland* v. *UK case*. The Irish Government contended that the measures taken by the UK Government—in particular, the several different types of extra-judicial deprivation of liberty—were not strictly required by the situation because in fact they did not overcome the emergency, but, on the contrary, worsened it.[30] The Court did not accept this argument: 'the Court must arrive at its decision in the light, not of a purely retrospective examination of the efficacy of those measures, but of the conditions and circumstances reigning when they were originally taken and subsequently applied'.[31]

6. At the same time, the fact that a government did not take preventive measures before the emergency arose or at an earlier stage does not affect its right to take derogating measures once the emergency has arisen. The necessity and the proportionality of the measures according to article 15 should be judged in the light of the current state of emergency.[32]

7. As far as the rights recognized in the human rights treaties are concerned, not all rights have the same importance. Those rights which are more important need a closer and stricter scrutiny when the necessity for derogation and the proportionality to the threat is assessed. Mr Susterhenn classified the rights in the light of the derogation clause in three categories. These rights are safeguarded in varying degrees as far as the possibility of state interference with them is concerned:

(a) There are some rights which are immune from any State interference; these are the non-derogable rights, the rights most vigorously guaranteed.

(b) There are some other rights with a general limitation clause (i.e. articles 8, 9, 10, 11 of the ECHR), which are covered by less rigid guarantees against interference by States; there is no need to

---

[29] A/38/40 (1983), p. 60, para. 283. See remarks of Mr Prado Vallejo in CCPR/C/SR. 430.
[30] *Ireland* v. *UK*, Ser. A: Judgment of the Court, para. 214.
[31] Ibid.
[32] *Lawless case*, Ser. B: Report of the Commission, opinion of Mr Waldock, p. 125.

declare a state of emergency or to notify them in order to operate the limitations.

(c) Finally, the rest of the articles can be derogated from in the very special circumstances foreseen in the derogation clause, i.e if there is a genuine state of emergency, and subject to the principle of proportionality; these are articles 4(2), 5, 6, 12, 13, 14. Among these rights, articles 5 and 6 (the right of freedom from arbitrary arrest and the right to a fair trial) are especially important. Therefore, the need for their derogation should be subjected to the most rigorous scrutiny and their derogation should be allowed only in the gravest emergencies.[33]

8. In assessing whether a derogating State has complied with the principle of proportionality, the monitoring organs have to take into account not only the need for bringing into operation the derogating measures, but also the manner in which the derogating measures have been applied in practice. Thus in the *Ireland* v. *UK case*, the Commission examined the practice of arrest and detention on extraordinary grounds in order to check for compliance with the principle.[34]

9. In analysing the principle of proportionality, the monitoring bodies should take into account not only the need for and scale of a given measure, e.g. administrative detention, but also the need for and proportion of the suspension of some of the guarantees linked with the derogated right, e.g. the guarantee of habeas corpus.[35]

10. In order to assess the proportionality of the derogating measures, the monitoring bodies should analyse the other less grave alternatives open to the government in dealing with the emergency. In the *Lawless case*, this was a major issue; the Commission spent much time in the study of the possibilities open to the Irish Government in dealing with the terrorist threat in 1959. In this respect, the Commission analysed the alternatives to detention without trial: namely, trial by ordinary courts, Special Criminal Courts, Military Courts or Tribunals, and also the closing of the border. The different evaluation of the suitability of these possibilitites led to a split in the Commission over this issue.[36] Analysis of

---

[33] Ibid., opinion of Mr Susterhenn, pp. 143–5.

[34] *Ireland* v. *UK*, Ser. B: opinion of the Commission, pp. 120 ff. See also opinion of Mr Sorensen, p. 130.

[35] *Ireland* v. *UK*, Ser. A: Judgment of the Court, paras. 211, 215–21; Ser. B: Report of the Commission, pp. 122–6.

[36] *Lawless case*, Ser. B: Report of the Commission, pp. 102–55. See also the opinion of Mr Eustathiades on this point in the *Greek case*, pp. 104–10.

the alternatives was also the method followed by the Commission in the *First Cyprus case*.[37]

11. In assessing compliance with the principle of proportionality, special importance should be attached to the necessary safeguards taken by governments in order to avoid abuses. As was pointed out in the *Ireland v. UK case*, 'the derogation from normal guarantees could become excessive if no other safeguards were put in their place'.[38] In fact, this was also one of the most important principles established by the European organs in the *Lawless case*. In this case, the derogating measure from article 5 of the ECHR, detention without trial, was accompanied by a number of safeguards, namely, constant supervision by Parliament, and a Detention Commission to review detentions and with the power of ordering the release of detainees if the detention was not longer justified; moreover, the Irish Government offered to release the detainees if they gave an undertaking to respect the Constitution and the Law and not to engage in any illegal activity against the Republic. Thus, the European Court concluded:

the detention without trial provided for by the 1940 Act, subject to the above mentioned safeguards, appears to be a measure strictly required by the exigencies of the situation within the meaning of article 15.[39]

12. The safeguards mentioned by the Court in the *Lawless case* are not normative in all cases; in other words, not in every emergency do the same safeguards have to be established by the government when derogating from the right to liberty and in order to comply with the principle of proportionality. Each case is different, and in confronting an emergency the State has to choose its own means of action to overcome the threat and to establish its own safeguards to avoid abuses.[40]

13. In assessing the proportionality of the measures, account must be taken not only of the nature of the threat, but also of its intensity at a given moment. During a state of emergency, the intensity of the threat could vary over a long period of time, therefore the proportionality of the measures should be assessed according to the variations in intensity. If the threat decreases, the measures should also be softened or even ended. In the *Greek case*, Mr Eustathiades maintained that measures which at the beginning of the emergency could have been justified had no justification after a certain period.[41]

---

[37] *First Cyprus case*. The Commission's report remains confidential, but Mr Eustathiades in the *Greek case* (p. 108) disclosed this point.

[38] *Ireland v. UK*, Ser. B: Report of the Commission, p. 124.

[39] *Lawless case*, Ser. A: Judgment of the Court, para. 37.

[40] *Ireland v. UK*, Ser. B: Report of the Commission, p. 124.

[41] *Greek case*, Report of the Commission (1969), opinion of Mr Eustathiades, pp. 104–10. See also the opinion of Mr Busuttil, pp. 117–18, in which he justifies the necessity

The emergency in Northern Ireland has also shown that the threat against the life of the nation can be maintained for a long time, although the intensity and gravity of the threat can vary in each phase. In its fight against terrorism, the British Government introduced different measures and safeguards during the whole period considered by the European organs in the *Ireland* v. *UK case* (1970–5).[42] Although, as a matter of principle, the different phases with their different measures and safeguards dependent upon the intensity of the threat have to be examined separately, this approach should not be conducted too rigorously. One of the main contentions of the Irish Government in the case was that in the first stages of the emergency, the safeguards established by the British Government to prevent abuses in the operation of the extra-judicial deprivation of liberty were not satisfactory. The Court established an important principle in answering this position of the applicant: the examination of the situation should be an overall one, leaving room for the government to learn from experience and to establish the necessary safeguards according to the evolution of the situation. Consequently, instead of defending what one may call a 'static' approach, what the Court advocates is a 'dynamic' approach which would leave room for progressive adaptations of the derogating measures and safeguards. The Court said:

When a State is struggling against a public emergency threatening the life of the nation, it would be rendered defenceless if it were required to accomplish everything at once, to furnish from the outset each of its chosen means of action with each of the safeguards reconcilable with the priority requirements for the proper functioning of the authorities and for restoring peace within the community.[43]

It is interesting to note that the only dissenting opinion on this point (that of Judge O'Donohue) related precisely to this fact. He did not consider that the safeguards established by the UK Government in the first phase of the extra-judicial deprivation of liberty were sufficiently satisfactory. In the light of the *Lawless case* judgment he thought that the safeguards mentioned there were also required in this case.[44]

Although the Court and the Commission recognized that in the first phase some of the guarantees were not completely satisfactory, they found that the legislation and the practice of the British Government during the whole period 'evolved in the direction of increasing respect for

and proportionality of some derogating measures taken by the Greek Government just for a short period of time after the military coup of 21 Apr. 1967.

[42] For an account of the different legislative and administrative measures and safeguards introduced by the UK Government, see *Ireland* v. *UK*, Ser. A: Judgment of the Court, paras. 80–240.

[43] Ibid., para. 220.

[44] Ibid., para. 246 (16 votes to 1). See separate opinion of Mr O'Donohue, p. 108.

individual liberty', even establishing quasi-judicial guarantees.[45] The Court, therefore, having analysed each of the different phases and the whole period, and having taken that 'dynamic' approach, found that the British Government did not exceed the limit strictly required by the exigencies of the situation.[46]

14. As has been pointed out in Chapter 1, states of emergency are essentially temporary and are only justified as long as the emergency lasts.[47] Consequently, all measures of derogation taken are also justified only as long as the emergency lasts. Therefore, a derogating measure should not be strictly required by the situation if it continues to be in force once the emergency has ended. This principle, which follows from the very nature of states of emergency, was explicitly incorporated into the derogation clause of the American Convention. (art. 27(1)). Even if this requirement is not explicitly included in the other two treaties, it derives from the very essence of the derogation clause. Thus, the opinion of the European Commission in the *De Becker case* supports this assertion. The Commission pointed out that

the measures of derogation are . . . only justified in the circumstances defined in paragraph 1 (article 15) with the result that if they remain in force after those circumstances have disappeared, they represent a breach of the Convention.[48]

In this case, the Belgian Government tried to justify an almost complete lifetime suspension of the right to freedom of expression in areas not related to the political offence for which Mr De Becker was convicted as a Nazi collaborator in the Second World War; and this suspension was still in force 15 years after the end of the war. The Belgian Government submitted that 'wartime measures, originally justified under article 15, could not automatically cease to be effective when the war was ended'.[49] Although one can agree as a matter of principle with the formulation of the Belgian Government that the maintenance of some wartime measures could be justified for a short time after the end of the war in order to restore peace and order, the Commission rightly did not accept this argument in the case, due to the disproportionate length and the far-reaching extension of that suspension.[50]

---

[45] *Ireland* v. *UK*, Ser. A: Judgment of the Court, para. 220; Ser. B: Report of the Commission, pp. 122–5.

[46] Ibid., Ser. A: Judgment, para. 220 *in fine*. See also the conclusion of the Commission: Ser. B, p. 126.

[47] See above, Ch. 1, p. 30.

[48] *De Becker case*, Ser. B: Report of the Commission (8 Jan. 1960), p. 133.

[49] See some remarks made by the Commission which seem to accept as a matter of principle this reasoning (ibid., p. 127). See also the opinion of Mr Bussutil in the *Greek case*, pp. 117–18.

[50] *De Becker case*, Ser. B: Report of the Commission, p. 133.

15. Finally, States enjoy a margin of appreciation in deciding which measures are necessary to deal with an emergency. They have a certain choice of means aimed at overcoming the emergency and restoring peace and order. Concepts like good faith and reasonableness on the part of the government will play an important role in the assessment by the European organs of the legality and proportionality of the measures.

(b)   The Principle as Applied by the UN Human Rights Committee

The Human Rights Committee, in assessing whether the derogating measures taken by States parties to the Covenant were 'strictly required by the exigencies of the situation' (art. 4(1)), has followed an approach similar to that of the European organs. This can be seen in the doctrine of the Committee in reviewing States' reports and in the proceedings under the Optional Protocol.

(i) *The Review of States' Reports.* In order to assess whether a derogation is necessary and proportionate to the emergency, the monitoring organs obviously need detailed information both about the threat and about the extent of the derogating measures; without detailed knowledge of the facts it is impossible to put forward any view on this issue. The lack of the detailed information which should be furnished by States has been one of the main problems faced by the Committee when having to express an opinion on the proportionality of the measures. Many States have not given the Committee any concrete information either about the facts of the emergency and the gravity of the threat, or about the exact extent of the measures. This information has not been furnished either through the notice of derogation in accordance with article 4(3), or through the periodic reports sent to the Committee in accordance with article 40. In the chapter on the principle of notification (Chapter 3), it was pointed out that this principle requires States fully to inform the other States parties of all measures taken in derogation of the provisions of the treaty and the reasons thereof. The widespread lack of compliance with this requirement was also pointed out in that chapter.[51] At the same time, the reports sent by many States which found themselves subject to public emergencies do not give sufficient information. The Committee has therefore asked for more information on restrictions in order to assess the necessity and proportionality of the measures.[52] Due to the lack of detailed information

---

[51] See Ch. 3, pp. 77 ff.
[52] See *inter alia* A/34/40 (1979), Report of Chile, p. 19, para. 78 and p. 23, para. 108; A/39/40 (1984), Report of Chile, II, p. 90, paras. 449–50. See also the review of the reports of Colombia, Uruguay, El Salvador, Iran, Nicaragua, and Peru.

the Committee has been unable in many cases to give an opinion on the merits of those cases; in other cases, the Committee has relied on other information at the disposal of the international community. Nevertheless, many of the principles of interpretation concerning proportionality formulated by the European organs can also be found in the practice of the Committee.

1. The Committee has affirmed its competence to pronounce on States' compliance with the principle of proportionality. This position has not been challenged by any State.[53]

2. The legitimacy of resorting to derogating measures is justified only in the case where ordinary legislation and the normal means of maintaining public order are not sufficient to deal with the emergency. In reviewing the Uruguayan report in 1982, it appeared from the evidence before the Committee, and from that government's own admission, that the threat posed by the Tupamaros movement had dropped substantially. Consequently, members of the Committee asked the Uruguayan Government to repeal a number of provisions derogating from the Covenant, because 'a clear and present danger to the public good could be controlled by appropriate means while still observing the rights provided for in the Covenant'.[54]

3. The measures of derogation are justified only as long as the emergency lasts. This is in consonance with the temporary nature of states of emergency. Thus the HR Committee in its 'general comment' on the derogation clause stated: 'measures taken under article 4 are of an exceptional and temporary nature and may only last as long as the life of the nation concerned is threatened.'[55] Therefore, once the threat has decreased or even disappeared, the measures must end. This was the case in Uruguay (as already mentioned), when the threat by the Tupamaros had diminished substantially and the Committee asked the Government to repeal the derogating measures.[56]

Similarly, when on 22 August 1984 the UK notified the termination of its derogation from several provisions of the Covenant, members of the Committee in reviewing the UK report asked whether the rights derogated from had been fully restored.[57] The UK representative explained in great detail before the Committee that there was no need to

---

[53] See on the competence of the Committee the statement made by Mr Dieye in CCPR/C/SR. 356, p. 2, para. 2 (Report of Uruguay). On the different opinions within the Committee regarding its competence in reviewing states of emergency see above, pp. 48–51.

[54] Report of Uruguay, opinion of Mr Tomuschat, CCPR/C/SR. 357, P. 5.

[55] HR Committee, General Comment 5/13, in CCPR/C/SR. 21, p. 5, para. 3.

[56] See n. 54 above. Also CCPR/C/SR. 356, p. 5 (opinion of Mr Bouziri) and p. 6 (Mr Ermacora); and CCPR/C/SR. 357, p. 3 (Mr Tomuschat).

[57] A/40/40 (1985), p. 100, para. 528.

continue the derogations of those articles, even though the emergency in
Northern Ireland had not disappeared.[58]

One of the main violations of the principle of temporariness of the
derogating measures is the 'institutionalization' of states of emergency,
or, in other words, the indefinite duration of the derogating measures
once the original threat has disappeared. As Mr Dieye noted, 'the letter
and spirit of article 4 provided that when a country took measures to
suspend human rights, those measures must be temporary, they could not
be institutionalised'.[59] The Committee has found violations of this principle in several cases. In the case of Chile, several members of the
Committee pointed out that the restrictions of rights which, according to
article 4, were 'intended to be limited in space and time, had been
transformed into institutional restrictions in force through the country for
an indefinite period'.[60] In the case of Colombia, the state of emergency
had been in force in different forms for 30 years.[61]

Totalitarian governments are often caught in a dilemma. On the one
hand, for political reasons and with the aim of strengthening the legitimacy of the government in the eyes of public opinion, they declare
themselves to be in full control of the country and minimize the importance of the threats of insurgents or opposition groups. On the other hand,
they have to show the gravity of the threat in order to justify the adoption
of derogating measures otherwise unacceptable in times of normalcy.

4. Although it has not been an easy thing to do, due to the insufficiency
of detailed information, the Committee has examined in several cases the
extent of each measure of derogation in order to assess the necessity and
the proportionality. Thus, in reviewing the UK report, members of the
Committee expressed 'concern at the continued derogation from articles
9, 10, 12, 14, 17, 21 and 22 of the Convention and requested clarifica-
tions as to the reasons for and extent of such derogations, bearing in mind
the conditions laid down in article 4'.[62]

In the case of Uruguay, the impression given by the notice of derogation was that all rights had been suspended. Therefore, members of the
Committee requested detailed information on the specific rights suspended
and the extent to which those derogations were required by the situation.[63] In the case of Nicaragua, although there was a genuine state of

---

[58] Ibid., para. 533.

[59] CCPR/C/SR. 556, p. 2.

[60] A/34/40 (1979), p. 19, para. 78; A/37/40 (1982), para. 449.

[61] A/35/40 (1980), p. 54, para. 247. This fact was recognized by the Colombian
representative (para. 241). In Jordan, the state of emergency seemed to have been in force
for 25 years. A/37/40 (1982), para. 172. However, the representative before the Committee
denied this fact; see para. 187.

[62] A/34/40 (1979), p. 55, para. 235.

[63] A/37/40 (1982), pp. 59–60, para. 270. Also CCPR/C/SR. 356, p. 6, para. 26 (Mr

emergency, the Government included in the corresponding decree of suspension all the rights except those declared non-derogable in the Covenant. A member of the Committee rightly pointed out that, as a matter of principle, that construction of the derogation clause was inconsistent. Even if there is a public emergency according to article 4, the requirements of this article are very strict, and only those derogating measures which are strictly required by the situation can lawfully be taken.[64]

5. The Human Rights Committee has tried to examine not only the extent of the measures but also the way in which they are applied in practice. One of the main difficulties in this task has been the fact that some States have presented in their reports an abstract and idealistic picture of the situation of the country with exclusive reference to the legislative and administrative framework, but with very little reference to the way in which the restrictions on human rights actually operate in the country. Members of the Committee strongly contested some of these reports and asked for more objective information. In reviewing the report of Afghanistan, one of the members very unusually qualified it as a monumental distortion of reality' and a 'deluge of lies, augmented by pure propaganda'.[65]

6. The Committee has also insisted that the derogating measures should be related to the emergency. Thus, in reviewing the Peruvian report, Mr Prado Vallejo pointed out that there is 'an inherent logic to article 4, namely, that the rights suspended must bear some relation to the kind of emergency declared'.[66] In fact, Peru seemed to have suspended political rights in an emergency which was in reality due to a natural disaster.[67]

7. Finally, the Committee has also insisted on the importance of safeguards to prevent abuses and violations of human rights in emergencies.[68] The UK representative gave an account of the procedural safeguards introduced by the UK in the 'Diplock courts', which established a special procedure for the trial of suspected terrorists.[69] The Spanish representative also explained before the Committee the special safeguards intro-

Ermacora). The Committee took the same attitude when reviewing the report of Iran: A/37/40 (1982), p. 68, para. 306.

[64] CCPR/C/SR. 420, remarks by Sir Vincent Evans.

[65] CCPR/C/SR. 604 (1985) p. 13, para. 18 (Mr Bouziri). See also the remarks of members of the Committee on reviewing the Chilean report: A/39/40, p. 88, para. 473.

[66] CCPR/C/SR. 430 (1983).

[67] A/38/40 (1983), p. 60, para. 263.

[68] A/37/40 (1982), Report of Uruguay, pp. 59–60, para. 270; A/39/40 (1984), El Salvador, p. 14, para. 75; ibid., Chile, p. 88, paras. 450 ff.

[69] A/40/40 (1985), UK, pp. 101–2, para. 532.

duced by the Anti-Terrorist Law of 1984, which dealt with the restriction on the individual rights of those suspected of terrorism.[70]

The opinion of the Committee on some derogating measures. The HR Committee has also on several occasions expressed its opinion on the proportionality of some measures of derogation. Analysing some of these measures, members of the Committee have pointed out that they were not strictly required by the exigencies of the situation. Thus, on reviewing the Colombian report, members of the Committee requested an explanation concerning 'the disquieting extension of military jurisdiction affecting a number of Covenant rights'.[71] The procedure followed by military courts in the emergency in El Salvador was also of concern to the Committee; a number of fundamental rights of due process were not followed by these tribunals; for the Committee these derogations seemed not to be required by the situation.[72] On reviewing the Uruguayan report, the Committee pointed out that the measures against the Tupamaros were especially repressive and out of proportion to the threat: e.g. the imprisonment and deprivation of human rights for 15 years.[73] In respect of the suspension in Chile of all political rights embodied in article 25 of the Covenant for nine years, members of the Committee found it hard to see how the measure was necessary and found it hard to reconcile the proportionality of this measure with the alleged internal disturbances. It was pointed out that article 25, which is a derogable right under the Covenant, could be subjected to reasonable restrictions. However, in Chile it was subjected to a general suspension for nine years, which appears to be unreasonable according to any standards.[74] In a similar case, when the military *coup d'état* took place in Greece in 1967, and parliamentary life and elections were suspended indefinitely, the European Commission found that 'even if it be said that there has been a continuing public emergency threatening the life of the nation . . . there is no indication that the situation has been and still is such as to require the suspension of parliamentary life or that elections could not be held'.[75]

The principle of proportionality has also been applied by the Committee in assessing the legitimacy of the use of force by law enforcement officials in quelling riots, effecting arrests, etc.[76]

---

[70] Ibid., Spain, p. 86, para. 476 (Organic Law No. 8/1984).

[71] A/35/40 (1980), Colombia, p. 54, para. 247.

[72] A/39/40 (1984), p. 14, para. 78.

[73] CCPR/C/SR. 356, p. 3, para. 5 (Mr Dieye).

[74] CCPR/C/SR. 528, p. 4 (Sir V. Evans), p. 5 (Mr Tomuschat).

[75] *The Greek case*, Report of the European Commission (1969), p. 179, para. 418. See also the opinion of Mr Delahaye, ibid, para. 419.

[76] See e.g. the *Suarez de Guerrero case* (v. *Colombia*), Comm. No. 45/1979, Adoption of Views (31 Mar. 1982), in HR Committee, *Selected Decisions*, i. 112–18 (excessive use of force by police resulting in the death of 7 people).

(ii) *Under the Optional Protocol.* In the same way as under the reporting procedure of article 40, the Committee has found that some States did not furnish enough information to enable the Committee to assess the necessity and the proportionality of the derogating measures taken in emergencies. However, the proceedings under the Optional Protocol are, to a certain extent, of an adversarial nature. Therefore, the Committee has had at its disposal a good deal of information provided by the alleged victim of human rights violations. The Committee has taken the salutary and logical position that in the absence of any submission of fact or law justifying derogating measures by the government concerned, that government cannot apply article 4, and therefore must be accountable for the application of normal standards of the Covenant.

This is what has happened in a number of cases concerning Uruguay.[77] On 30 July 1979, Uruguay notified the Secretary-General of the United Nations, according to the procedure of article 4(3), of the existence of a state of emergency in the country. However, it did not indicate either the derogating measures taken, or the reasons for them.[78] Although the notice of derogation of the Uruguayan Government was defective by any standards, the Committee took a flexible approach and left open the possibility of showing the need for those derogating measures in the proceedings under the Optional Protocol. In most of these cases, Uruguay has not made any submission proving the necessity for the derogations; in others, it has tried to justify the measures with general and abstract allegations.

The *Landinelli case* is a good example of this approach of the Committee.[79] The facts were undisputed. By Institutional Act No. 4 of 1 September 1976 (art. 1(a)), the authors of the application together with a considerable number of other political activists were deprived of their right to engage in any activity of a political nature, including the right to vote for a period of 15 years.[80] The main contention of the applicants was that this measure was clearly incompatible with article 25 of the Covenant. The defence of the Uruguayan Government was based on the derogation clause, and especially on the fact that article 25 was not mentioned among the non-derogable rights in the Covenant. The Government, therefore, assumed that it had the right temporarily to derogate from some provisions relating to political parties.[81] The Com-

---

[77] See above, p. 51.

[78] For the Uruguayan notice of derogation see UN, *Multilateral Treaties Deposited with the Secretary-General*, Status as at 31 Dec. 1986 (New York, 1987), pp. 84–5.

[79] *Landinelli et al. (v. Uruguay)*, Comm. No. 39/1978, Adoption of Views (8 Apr. 1981, in HR Committee, *Selected Decisions*, i. 65–6.

[80] Ibid., para. 2.

[81] Ibid., para. 6.

mittee, however, was unable to accept merely on this information that the requirements of article 4(1) had been met. The Committee recalled the importance of the principle of proportionality as a fundamental condition for derogation.

However, the Uruguayan Government did not furnish any detailed information of the relevant facts in order to justify the measures of derogation either through the notice of derogation, or through the proceedings of the case. The Committee, when assessing the compliance with the principle of proportionality, needs full and comprehensive information, and consequently the State party is duty-bound to give that information. Otherwise, the Committee 'cannot conclude that valid reasons exist to legitimize a derogation from the normal legal regime prescribed by the Covenant'.[82] Even assuming that there was a public emergency, which had not in fact been shown by the Government, the Committee could not find any reason whatsoever which could justify the deprivation of all the applicants' political rights for such a long time. Therefore, the Government failed to show that the measure was strictly required by the situation; in the eyes of the Committee, the measure unreasonably restricted their rights under article 25 of the Covenant.[83] In another case with a similar finding, the Committee explicitly stated that 'the principle of proportionality would require that a measure as harsh as the deprivation of all political rights for a period of 15 years be specifically justified.'[84]

The Human Rights Committee has adopted the same attittude when assessing the need for further restrictions in those articles which contain a limitation clause. In several cases dealing with freedom of expression (art. 19 of the Covenant), the Committee found that the arrest, detention, and trial of the applicants were not justified on any of the grounds included in paragraph 3 of the same article. The authors of the communications were in fact convicted for their prior political and trade union activities. The Government did not supply any detailed information on the nature of their offences; it only referred in abstract terms to 'subversive association and conspiracy to violate the Constitution, followed by preparatory acts thereto'.[85]

In the *Salgar de Montejo case*, the Government of Colombia did not

---

[82] Ibid., para. 8(1–3).
[83] Ibid., paras. 8(9), 9. See also *Weinberger case (v. Uruguay)*, Comm. No. 28/1978, Adoption of Views (29 Oct. 1980), in HR Committee i. *Selected Decisions*, 60, paras. 15–16.
[84] *Pietraroia case (v. Uruguay)*, Comm. No. 44/1979, Adoption of Views (27 Mar. 1981), in HR Committee, *Selected Decisions*, i. 79, paras. 16–17.
[85] *Pietraroia case*, para. 15. See also *Garcia Lanza* et al. *(v. Uruguay)*, Comm. No. 8/1977, Adoption of Views (3 Apr. 1980), in HR Committee, *Selected Decisions*, i. 49, para. 6 *in fine*; *Saldias de Lopez (v. Uruguay)*, Comm. No. 52/1979, Adoption of Views (29 July 1981), in HR Committee, *Selected Decisions*, i. 91, paras. 11(5), 13.

show that the derogation of article 14(5), the right to review by a high tribunal, was made in accordance with the principle established in the derogation clause. In particular, it did not show that there was a public emergency and that the derogation of article 14(5) was strictly required; consequently, the facts of the case disclosed a violation of that provision.[86]

(c) The Application of the Principle of Proportionality under the ACHR

The IACHR, in its reports on those countries under states of emergency, has insisted, in the same way as the European organs and the Human Rights Committee, on the importance of the principle that derogating measures are legitimate only if they are taken 'to the extent and for the period of time strictly required by the exigencies of the situation' (art. 27(1) of the American Convention). Commenting on this requirement, the Commission has pointed out that 'the measures adopted should be proportionate to the danger, both with respect to degree and duration; thus, once the danger that threatens the security of the States has been overcome, the special provisions should also be terminated.'[87]

It is interesting to note that this requirement contained in article 27(1) has been applied by the Inter-American Commission not only to those States parties to the American Convention, but also to those that are non-parties but members of the OAS.[88] For the Commission this principle, as well as some others included in the derogation clause, such as the principle of exceptional threat and the principle of non-derogability, constitutes 'the most accepted doctrine' on this topic of human rights in states of emergency, and is therefore applicable as a substantive criterion to assess whether the measures of suspension of rights can be considered justified and legitimate.[89] The legal status of these principles as forming part of a possible emergency rule of customary international law will be discussed in Chapter 10.

(i) *The General Principles Formulated by the Inter-American Court.* Before dealing with several cases in which the IACHR has applied the principle of proportionality in a concrete way, it is worthwhile recalling

---

[86] *Salgar de Montejo (v. Colombia)*, Comm. No. 64/1979, Adoption of Views (24 Mar. 1982), in HR Committee, *Selected Decisions*, i. 129, para. 10(2, 3).

[87] IACHR, *Nicaragua–Miskitos case*, p. 117.

[88] At the same time, this doctrine had been applied even before the American Convention entered into force.

[89] See IACHR, *Report on Chile*, 1985; *Report on Paraguay*, 1987. See also the formulation of this principle in what the IACHR called 'Doctrine': OAS, *The IACHR: Ten Years of Activities (1971–1981)* (Washington, 1982), pp. 341–2. See also the application of this principle in the *Report on Argentina*, 1980, pp. 24–7 and *Report on Colombia*, 1981, pp. 15–18.

the important remarks made as a matter of principle on the substantive requirements of the derogation clause by the IA Court HR in its advisory opinion on 'Habeas Corpus in Emergency Situations'. The Court starts by establishing an important principle in its interpretation of the derogation clause within the American Convention. Reacting against the implicit assumption that in states of emergency all rights can be suspended, the Court says:

rather than adopting a philosophy that favours the suspension of rights, the Convention, establishes the contrary principle, namely, that all rights are to be guaranteed and enforced unless very special circumstances justify the suspension of some, and that some rights may never be suspended, however serious the emergency.[90]

The expression 'very special circumstances' refers to the conditions laid down in paragraph 1 of the derogation clause *inter alia* to the principle of exceptional threat. With respect to the proportionality of the measures, the Court deduced from the derogation clause a principle already expressed by the European Commission: 'since article 27(1) envisages different situations . . . what might be permissible in one type of emergency would not be lawful in another.'[91] Referring to the criterion of legitimacy of derogating measures, the Court said:

the lawfulness of the measures taken to deal with each of the special situations referred to in article 27(1) will depend, moreover, upon the character, intensity, pervasiveness, and particular context of the emergency and upon the corresponding proportionality and reasonableness of the measures.[92]

Thus any action taken by the public authorities that goes beyond the limits of what is strictly required by the emergency (and these limits must be specified with precision in the decree of proclamation), would be unlawful even in a genuine state of emergency. *In concreto*, there would be a violation of these general principles on derogations if the derogating measures 'infringed the legal regime of the state of emergency, if they lasted longer than the time limit specified, if they were manifestly irrational, unnecessary or disproportionate, or if, in adopting them, there was a misuse or abuse of power'.[93] The Court finally insisted on the importance of a system both governed by the rule of law and with an independent judiciary which exercises control over the lawfulness of the measures of derogation.[94]

[90] IA Court HR, advisory opinion on 'Habeas Corpus', para. 21.
[91] Ibid., para. 22.
[92] Ibid., para. 22.
[93] Ibid., para. 39.
[94] Ibid., para. 40.

(ii) *The Application of the Principle by the IACHR.* The IACHR has not produced a clear legal analysis of the proportionality of the measures of derogation taken by States in emergencies. This is due to the nature of its country reports, in which, when it exercises its functions as a fact-finding body, the Commission describes the situation of human rights in the country, rather than strictly analysing the compliance with all the legal requirements of the Convention. What the Commission usually does in many of its reports is to describe the restrictions and limitations imposed by States on the enjoyment of rights, and it usually classifies those limitations as violations of human rights standards, but without a thorough analysis of the proportionality of the measure to the emergency. A good example of this position is the *Report on Argentina* of 1980.[95]

However, especially in the last years, some interesting applications of the principle of proportionality can be found in the reports of the Commission. In the *Nicaragua–Miskitos case*, the Commission assessed whether the compulsory relocation of 8,500 Miskitos Indians in five camps in derogation of the right to residence and movement (art. 22), was 'to the extent and for the period of time strictly required by the exigencies of the situation'.[96] The Commission found first of all that there existed a real and imminent threat to the security of the State in Nicaragua at the time, and therefore that there was a genuine state of emergency according to article 27(1). As the Commission pointed out in other cases, the derogation clause demands that the derogating measures should be proportionate to the danger, both with respect to the nature of the measure and to its duration. Analysing the reasons adduced by the Government for taking the measure (for military reasons and with the aim of protecting the life of the population caught in an area of heavy armed conflict), the Commission found that the measure could be considered to have met the requirement of proportionality in the absence of any other alternative to deal with the emergency and taking into account all the facts of the case. However, the Commission, on applying the principle of temporariness and giving the Government a margin of discretion, subjected the proportionality of the measure to one condition: once the emergency ended, the Miskitos should be free to return to their land if they wished to do so. The measure, due to its very temporary nature, cannot outlast the emergency, it must be limited to the duration of it. The Commission did not find the compulsory relocation of Miskitos for

---

[95] IACHR, *Report on Argentina*, 1980; see especially the chapters on the derogable rights: the right to freedom of expression (pp. 235–8), the right of labour (pp. 239–43), and political rights (pp. 245–50).
[96] IACHR, *Nicaragua–Miskitos case*, 1983.

economic reasons justified once the emergency was over, as had been planned by the Government before the emergency arose.[97]

From the legal point of view another interesting analysis of derogating measures according to the principle of proportionality was made by the Commission in relation to the measures taken by the *Nicaraguan Government* in 1985–6.[98] The Government declared a new state of emergency on 15 November 1985 for the period of a year, and through Decree No. 128 suspended a number of provisions of the Statute on the Rights and Guarantees of Nicaraguans.[99] In this case, the Commission recognized that there was a public emergency threatening the life of the nation and therefore in principle the Nicaraguan Government was entitled to take derogating measures. Afterwards, the Commission examined whether the measures taken were proportionate to the threat in terms of duration and in terms of the nature and extent of what was required to deal with the emergency. The Commission interpreted the aim of this requirement as seeking to limit the extent of the Government's discretionary power, in order to prevent excesses from being committed under the guise of an emergency.[100] The Commission, in fact, only made a judgment on the proportionality of the first two measures; in the other cases, it only described the measures taken.

Thus, as far as the first measure was concerned, the Commission considered that some of the measures taken against the Catholic Church were out of proportion to the threat. Consequently, the requirement of proportionality in article 27(1) was not satisfied. These measures were: the prolonged take-over of the COPROSA premises, the seizure of the newspaper *Iglesia*, the indefinite closing down of *Radio Catolica*, the harassment of priests, the conscription of seminary students, the confiscation of a printing press, and the expulsion from the country of two senior Church officials.[101]

The second measure scrutinized was the indefinite closure of the independent newspaper *La Prensa* on 26 June 1986, a paper which had criticized many of the measures taken by the Government. The Commission found this restriction on the freedom of expression difficult to justify, especially as the newspaper was already subjected to prior censorship. The Commission, therefore, considered that the principle of proportionality was not met in this case either.[102] Surprisingly, however, in

---

[97] Ibid., pp. 117–19.

[98] IACHR, *Annual Report, 1985–6*, Nicaragua, pp. 165–75.

[99] Ibid., p. 167.

[100] Ibid., pp. 168–9.

[101] Ibid., p. 172. COPROSA stands for (in translation) Archidiocesan Social Promotion Commission.

[102] Ibid., pp. 172–3. In the IACHR *Annual Report, 1982–3*, p. 23, the Commission

both cases the Commission does not explicitly mention any article of the American Convention as being the object of violation by the Nicaraguan Government.

In the rest of the report, the Commission describes some of the measures without explicit judgment on their necessity and proportionality, although they are presented implicitly as excessive. It would have been more suitable, however, to follow the same method used with the first and second measures, and to pass a judgment on their proportionality.[103]

Finally, the Commission stressed the importance of restoring the right of habeas corpus to those accused of having committed crimes against State security.[104] The Commission regarded the suspension of the right of habeas corpus as a violation of article 27(2) *in fine* of the American Convention; this right, as was later confirmed by the Inter-American Court, cannot be suspended even in situations of emergency.[105]

In the *Report on Bolivia*, the Commission made an overall assessment of the proportionality of the measures taken by the Government after the military coup of 17 July 1980. The rights suspended were the right to assembly, to freedom of association, to freedom of thought and expression, to personal freedom, and to movement and residence. In a rather unclear passage, the Commission considered that the principle of proportionality had not been met insofar as the measures exceeded the lawful limits:

the Commission finds no direct causal relationship between the acts of opposition that the military declaration occasioned and the actions and decisions the Bolivian authorities took, at times legally and at other times ipso facto. In the Commission's opinion the Bolivian authorities exceeded the limits of state action by disregarding the restrictions on the use of such measures stipulated in the American Convention both with respect to the gravity of the situation and the period of time such measures should remain in effect.[106]

It is interesting, however, that there is no assessment of the proportionality of each measure of restriction in the respective chapters.

The state of emergency in *Nicaragua* in 1979, in the aftermath of the Sandinistas' take-over of the country, provoked an illuminating debate between the Nicaraguan Government and the Inter-American Com-

considered the restrictions on the freedom of expression as arbitrary and going beyond those limits a government can reach, even in a state of emergency.

[103] The other measures analysed by the IACHR referred to the freedom of expression, action against opposition parties and independent trade-unions, and the mass detention of peasants in rural areas.

[104] IACHR, *Annual Report, 1985–6*, pp. 173–5.

[105] Ibid., p. 167. IA Court HR, advisory opinion on 'Habeas Corpus', para. 44.

[106] IACHR, *Report on Bolivia*, p. 23. See also p. 116, para. 3. Even if Bolivia did suspend these fundamental rights, it did not send a notice of derogation to the Secretary-General of the OAS in accordance with art. 27(1).

mission on the compatibility with the Convention of certain measures taken by the Government.[107] Perhaps the most important and serious measure was the derogation of the fundamental guarantees of fair trial and due process of law of those accused of crimes committed during the previous regime. Articles 7 and 8, the rights to personal freedom and to fair trial, are not among those rights which cannot be suspended in states of emergency; consequently, *stricto sensu*, they can be derogated from. However, for the derogations to be valid, they must meet the other conditions of article 27(1), one of which is the principle of proportionality.

The Commission, after a thorough analysis of the legal framework within Nicaragua, in particular of the Statute of the Rights and Guarantees of Nicaraguans, and the emergency laws which suspended the rights of the Somocistas, found numerous irregularities, incompatible with the American Convention, in the trial of the Somocistas by Special Tribunals. These irregularities were: the lack of opportunity given to the accused to exercise their rights; the length of time the detainess were kept in detention before being brought to trial; the vagueness and imprecision of many of the charges; the very short period of time the accused were given to prepare their defence and to present evidence; the lack of basis of the judgments; and finally, the lack of jurisdiction of the Appeals Court to review the facts established by the Special Tribunals.[108]

These measures were found by the Commission to be incompatible with the following rights contained in the Convention: the right to a hearing, the right to a fair trial within a reasonable time, the right to a competent, independent, and impartial tribunal, previously established by law, the right to presumption of innocence (all contained in art. 8(1)); the right to adequate time and means to prepare one's defence (art. 8(2)); the principle of non-retroactivity of penal laws (art. 9); and the right to appeal to a higher court (art. 8(2) and (4)).[109]

It may be noted that the Commission did not make any reference to the principle of proportionality and did not check whether each and all of these measures were strictly required by the situation.[110] It seemed to accept that in Nicaragua in 1979 there was a genuine state of emergency within the meaning of article 27, and it did not doubt the good faith of the Government; however, it failed to analyse the proportionality of the derogating measures.[111]

---

[107] IACHR, *Report on Nicaragua*, 1981. The Government of Nicaragua proclaimed and notified the state of emergency according to art. 27.

[108] Ibid., p. 93. For the exposition of the IACHR doctrine on the non-derogability of certain minimum guarantees of fair trial and due process of law, see above, Ch. 4.

[109] IACHR, *Report on Nicaragua*, 1981, pp. 69–93 and esp. 88–93.

[110] See the remarks made in Ch. 4, pp. 121 ff.

[111] Neither is there any mention of the theory of the margin of appreciation conceded to

Two explanations are possible for the Commission's attitude: either that it was not aware of the operation of the principle of proportionality, which is implausible, or that it considered at least some of those guarantees of fair trial and due process of law to be so fundamental that they could not be derogated from even in states of emergency. The second explanation seems to be more plausible. In fact, the Commission pointed out at the end of its analysis that those guarantees, which are called 'minimum guarantees', ought to be applicable without discrimination to all inhabitants of the country, including the Somocistas.[112]

In its observations on the Commission's report, the Nicaraguan Government tried to justify the measures derogating from articles 7 and 8 of the Convention.[113] The Government insisted on the grave situation of emergency in which the country found itself as a consequence of the bloody war of liberation, and the necessity of taking some extraordinary measures to deal with the emergency. The deprivation of liberty of the Somocistas and the creation of Special Tribunals to try them were measures aimed at quietening the anger of the people who were calling for vengeance against those guilty of the most egregious crimes. The Government concluded:

the measure taken by the Government namely the establishment and operation of special courts, somewhat dampened the rage of the masses and reduced the danger of a confrontation between the thousands of family members of the heroes and martyrs that died in the struggle, who wanted to take the law into their own hands, and the recently constituted authorities. It is virtually impossible, in a situation like that which existed, in which there was deep popular resentment and continuing calls by the masses for rapid justice, that the situation of the Somoza criminals could be dealt with by ordinary legislation.[114]

In fact, the Commission, in its report, recognized the gravity of the situation and the great difficulties of the Government which had to deal with a country lacking all the elements of public administration, police, and the administration of justice.[115] However, the course taken by the

States in assessing the need for derogating measures. See, however, the reference to this doctrine by the IACHR in the *Nicaragua–Miskitos case*, pp. 117–19.

[112] IACHR, *Report on Nicaragua*, 1981, p. 93. See the position of the IACHR on this point in Ch. 4, above.

[113] 'Observations and Comments of the Government of Nicaragua on the Report of the Inter-American Commission of Human Rights on the Status of Human Rights in that Country', OAS Doc. OEA/Ser. P, AG doc. 1369/81 (27 Oct. 1981. See Buergenthal, Norris, and Shelton (eds.), *Protecting Human Rights: the Americas*, p. 226. Even though these 'Observations' were sent to the Preparatory Committee of the General Assembly on 14 Oct. 1981, it seems that they are basically the same as those sent to the Commission on 15 June 1981, which the Commission took into consideration before publishing the report.

[114] 'Observation and Comments of the Government of Nicaragua', p. 203.

[115] IACHR, *Report on Nicaragua*, 1981, pp. 69–70, 92–3.

Government was considered by the Commission to be incompatible with the Convention, even in emergencies. The Government, against the opinion of the Supreme Court which favoured more trials by ordinary courts, had established Special Tribunals in order to speed up the trials and to calm the anger of the population, with the result that some fundamental due process guarantees were suspended. There is no doubt that the irregularities in those trials can hardly be justified, even in emergencies, as being strictly required by the situation.[116] The only element in which derogation seems to be justified is the delay in the trials, especially in the light of the observations of the Government on the speed of the trials and the number of releases through three successive pardons.[117] In the analysis of the Commission, it does not appear that the delay in the trials due to the emergency was justified.[118]

These Special Tribunals operated from December 1979 to February 1981 and convicted 4,331 of those accused. The suspension of such fundamental due process guarantees can hardly be defensible, some weeks after the end of the war, when the Government was in full control of the country. The defence made by the Commission of the fundamental guarantees of fair trial and due process of law underlines the importance of these guarantees, which cannot be derogated from even in emergencies. The preference of the Commission—in line with the advice of the Supreme Court—for a less grave alternative to the derogating measure, that is, for the trial of the accused by ordinary courts and with full procedural guarantees, seems to be reasonable in the circumstances.

As far as political rights are concerned, the Commission considered that the date proposed for the elections was too distant, even taking into account the special situation in Nicaragua in which the holding of elections immediately after the victory of the Sandinistas would have been impossible. According to the decision of the Council of State in August 1980, the elections would be held in 1985. This opinion of the Commission is reasonable, especially if one considers that all political activity by the parties was forbidden until 1984. Here again, the Commission considered the restrictions on political rights to be disproportionate because of their duration.[119]

---

[116] For a complete explanation of the procedure of these Special Tribunals, see ibid., pp. 74–93.

[117] 'Observations and Comments of the Government of Nicaragua' (original in Spanish), pp. 52–3. However, see the remarks of the Commission on the uncertain legal status of those detainees. IACHR, *Report on Nicaragua*, 1981, p. 89.

[118] Ibid.

[119] Ibid., pp. 139 and 171(a). For the case of Colombia and an assessment of the derogation of the right to fair trial and due process of law, see IACHR, *Report on Colombia*, 1981, pp. 220–1.

The application of the principle to States non-parties to the ACHR. The principle of proportionality of derogating measures has also been applied by the IACHR to States non-parties to the Convention as a part of what has been called 'the most accepted doctrine' on this subject.[120] Three cases will illustrate this application. The first two cases refer to Chile and Paraguay, and the third to Suriname.

Although in the two comprehensive *Reports on Chile* and *Paraguay*, the IACHR considered that there was no 'public emergency threatening the life of the nation' and therefore the main condition for the operation of the right of derogation was not met,[121] it examined the restrictions on rights made by the two Governments in the light of the principle of proportionality. In the two reports, the Commission concluded that the restrictions were out of proportion both in respect of the extent and the duration of the derogating measures. There is no need to conduct a detailed analysis of each measure taken by these Governments. This was done by the Commission in its two lengthy and comprehensive reports. However, a few remarks on its general approach would be useful.

As far as *Paraguay* is concerned, the Commission concluded that 'the practice of the Government of Paraguay in this respect has been characterised by a permanent lack of compliance with reasonable criteria in applying the provisions of the state of siege'.[122]

As far as *Chile* is concerned, the Commission focused on the measures restricting the right to personal liberty, the right to residence and movement, the right to nationality, the right to freedom of expression and opinion, labour union rights, and, finally, political rights. It also analysed in great length the serious violation of the right to justice, the violation which made possible those other gross violations of fundamental rights. In particular, the Commission refers, first, to the weakening of the remedies instituted for the protection of fundamental rights, because of the limitations imposed upon the judiciary both by the Government and by the judiciary itself; secondly, to the expansion of the jurisdiction of military courts, that lack independence and impartiality when judging breaches of the law; and finally, to the serious restrictions placed on the due process rights. As a result of all of these grave irregularities, the Commission concluded that the rule of law did not exist in Chile at that time.[123]

---

[120] See, however, the complete lack of judgment on the compliance with the principle of proportionality of the restrictions imposed by Argentina. The Commission only describes the violations of human rights that occurred in the country. IACHR, *Report on Argentina* (1980).

[121] See above, Ch. 1, p. 26.

[122] IACHR, *Report on Paraguay*, 1987, pp. 21–2.

[123] IACHR, *Report on Chile*, 1985, pp. 43–5, 285–90.

Finally, in the 1983 *Report on Suriname*, the Commission found that the provisions of Decree A–11 which governed the state of emergency were in flagrant violation of article 4 of the Covenant, because that decree did not require that the derogating measures must comply with the principle of proportionality.[124]

The few resolutions on individual cases in which the IACHR has dealt with rights considered derogable under the Convention contain no elements relevant to the issue of states of emergency, and therefore to the application of the principles of consistency, proportionality, or non-discrimination.[125]

3. CONCLUSION

The principle of proportionality, which has found application in several areas, can be deemed to constitute a general principle of international law. As far as our area of study is concerned, the principle embodies one of the most important conditions for derogation of human rights in states of emergency. The principle is contained in the three main human rights treaties. According to this principle, the derogating measures, in order to be valid, must be strictly required by the exigencies of the situation; in other words, they must be necessary and proportionate to the gravity of the threat. This principle refers not only to the extent but also to the period of time during which a derogating measure can be held to be justified; once the emergency has ended, or its gravity has diminished, the derogating measures have no justification.

In the application of the principle of proportionality, the monitoring organs of the three treaties have formulated certain general principles which constitute important guide-lines for its application. Although the European Commission and Court, the UN Human Rights Committee, and the Inter-American Court and Commission are bodies with different functions according to the terms of the respective treaties, there is a wide coincidence in the general interpretation of the principle of proportionality. The assessment of the proportionality of the measures of derogation is considered one of the most difficult tasks of the monitoring organs in emergencies.

[124] IACHR, *Report on Suriname*, 1983, p. 37, para. 82.

[125] See the list of these cases relating to the right to privacy, freedom of thought and expression, right of assembly, freedom of association, and the freedom of movement and residence, in Buergenthal and Norris (eds.), *The Inter-American System*, vol. iv, booklet 26, Appendices, pp. 13–14. It is interesting to note the sharp contrast between the great number of cases decided that involve violation of the non-derogable rights—plus personal liberty and fair trial—and the very few cases relating to the other rights of the American Convention.

The main principles governing the application of the principle of proportionality formulated by the monitoring organs can be summarized as follows:

1. Measures derogating from human rights standards can only be take when the ordinary provisions of the law and the limitations foreseen for peacetime are not enough to deal with the emergency.

2. The mere existence of a public emergency threatening the life of the nation, within the meaning of the derogation clause, does not justify *ipso facto* every derogation from human rights standards. Each measure of derogation taken in a lawfully declared emergency should be necessary and proportionate to the threat.

3. Each measure of derogation has to bear a relation to the threat; in other words, there must be a link between the facts of the emergency and the measures taken.

4. The measures of derogation taken by the government should potentially be able to overcome the emergency. This, however, does not mean that the judgment on the strict necessity of the measures will depend on the fact that the measures actually overcome the emergency.

5. At the same time, the fact that a government did not take preventive measures before the emergency arose or at an earlier stage does not affect its right to take derogating measures once the emergency has arisen. The necessity and the proportionality of the measures according to article 15 should be judged in the light of the current state of emergency.

6. As far as the rights recognized in the human rights treaties are concerned, not all rights have the same relevance. Therefore, those rights which are more important need closer and stricter scrutiny when the necessity for derogation and the proportionality to the threat are judged.

7. In assessing whether a derogating State has complied with the principle of proportionality, the monitoring organs have to take into account not only the need for bringing the derogating measures into operation, but also the manner in which the derogating measures have been applied in practice.

8. In analysing the principle of proportionality, the monitoring bodies should take into account not only the necessity and proportionality of a given measure e.g. administrative detention, but also the necessity and proportionality of the suspension of some of the guarantees linked with the derogated right, e.g. the writ of habeas corpus.

9. In order to assess the proportionality of the derogating measures, the monitoring bodies should analyse the other less grave alternatives open to the government in dealing with the emergency.

10. In assessing the compliance with the principle of proportionality, special importance should be attached to the necessary safeguards taken by governments in order to avoid abuses.

11. In order to analyse the proportionality of the measures and the sufficiency of the safeguards against abuses, attention should be paid to every different phase of the emergency.

12. As has been pointed out in Chapter 1, states of emergency are essentially temporary; in other words, they are only justified as long as the emergency lasts. Consequently, all measures of derogation are only justified as long as the emergency lasts. Therefore, a derogating measure is not strictly required by the situation if it continues to be in force once the emergency has ended.

6

The Principle of Non-Discrimination

I. INTRODUCTION

The second substantive condition for the derogation of rights in emergencies is that the derogating measures must not involve discrimination on the grounds of race, colour, sex, language, religion, or social origin. This condition is contained in the derogation clause of the UN Covenant and of the American Convention, but not in the European Convention.[1] The reason for the absence of this requirement in the latter Convention can be explained by referring to the legislative history of these treaties. The principle of non-discrimination did not appear either in the first UK proposal of a derogation clause to the UN Covenant, or in the first texts of the derogation clause as approved by the UN Commission of Human Rights at its Fifth (1949) and Sixth (1950) Sessions.[2] The principle was introduced by a UK proposal in 1952, and incorporated into article 4 of the Covenant in the Eighth Session of the Commission in 1952.[3] However, in 1950 the drafters of the European Convention borrowed from the draft UN Covenant the text of the derogation clause as it was then worded, that is, without any mention of the principle of non-discrimination.[4] As is well known, the European Convention was signed in Rome on 4 November 1950, and therefore it could not possibly take into account the further developments of the derogation clause in the *travaux préparatoires* of the Covenant, which finally incorporated the principle of non-discrimination. The consequences of the absence of this requirement for derogation in the European Convention will be studied below.

---

[1] See ICCPR, art. 4(1); ACHR, art. 27(1) and ECHR, art. 15(1). For a general view of this principle in international law, see W. McKean, *Equality and Discrimination under International Law* (Oxford, 1983).

[2] See the UK original proposal, art. 4. See also the text of the derogation clause approved at the 5th Session (1949), E/1371, art. 4(1); and 6th Session (1950), E/1681, art. 2(1).

[3] See the UK proposal in E/CN. 4/L. 139/Rev. 1, and also E/CN. 4/SR. 331, p. 6. For the text of the derogation clause approved at the 8th Session see E/2556, art. 3(1).

[4] See ECHR, *Collected Edition of the* Travaux préparatoires, iii. 186, 190, 280.

(a)   The Characteristics of the Provision

At first sight, the provision of non-discrimination in the derogation clause has two characteristics. First, the prohibition of non-discrimination in emergencies refers to a list of grounds that is less numerous than the one contained in the general non-discrimination provisions of the Covenant (articles 2(1) and (26).[5] In these articles, the discriminatory grounds mentioned are: race, colour, sex, language, religion, political or other opinion, national or social origin, property, and birth or other status. The derogation clause, which contains a 'closed' list of grounds, omits four of these, namely, political or other opinion, national origin, property, and birth or other status. The main reason for these omissions was that the drafters of the Covenant thought that in some cases, legitimate restrictions on these grounds could be imposed in states of emergency.[6] Secondly, the derogation clause contains the word 'solely', which could have important consequences for the interpretation of the provision; it seems to imply that it may be possible to take derogating measures which discriminate even on those grounds mentioned in the clause, provided that they are not *solely* discriminatory, but rather relate to other legitimate reasons, e.g. military necessity. If this is the right interpretation of the provision it would obviously weaken considerably the strength of the prohibition of non-discrimination in emergencies. Due to the importance of this issue, it is necessary to undertake a detailed analysis of the *travaux préparatoires* in order to find out whether or not this interpretation is correct. The introduction of the principle of non-discrimination as a condition for derogation in paragraph 1 of the derogation clause was probably due to the failure to include the non-discrimination provision (final article 26) among the non-derogable rights in paragraph 2 of the derogation clause. A closer examination of this point could help in interpreting the exact extent of the principle of non-discrimination.

(b)   The Debate on the Principle of Non-Discrimination as Non-Derogable

There is no doubt that the principle of equality before the law and non-discrimination is one of the most fundamental human rights;[7] hence it is not surprising that during the discussion about the rights which should be entrenched against derogation, it was proposed that this principle should be included in the list of non-derogable rights.[8] Moreover, this proposal

---

[5] These grounds are the same as those mentioned in art. 2(2) of the UDHR. See also ECHR, art. 14, and ACHR, art. 1(1).

[6] A/2929, para. 44.

[7] See Jimenez de Arechaga, *Recueil* (1978), pp. 174–7.

[8] The proposal came from Mr Malik (Lebanon), E/CN. 4/SR. 195 (29 May 1950), p. 23.

to make article 26 (then art. 20) a non-derogable right was approved by the HR Commission in May 1950 when it was listing these rights.[9] However, immediately after this decision, several members of the Commission realized the possible consequences of including article 20 among the non-derogable rights. René Cassin was afraid that the Commission had made 'a hasty decision', and, therefore it was decided to re-examine the matter. Mr Kirou (Greece) felt that the inclusion of article 20 among the non-derogable rights 'would constitute a serious blow to the draft Covenant and a virtual invitation to the Governments not to ratify it'.[10] After a lively debate, it was decided not to include article 20 in the list because the majority thought that it might well become necessary for a State in emergencies to draw certain distinctions between different categories of persons. Special reference was made to the necessity of taking derogating measures against enemy aliens, such as restrictions on the freedom of assembly and due process of law, or their segregation in detention camps.[11] Some members pointed out that it was impossible to treat enemy aliens in wartime in exactly the same way as citizens, and, therefore, the provisions of article 20 were not applicable in emergencies. Even those members who favoured the inclusion of article 20 interpreted the provision as not preventing States from making special arrangements for certain groups in emergencies.[12]

(c)   The Introduction of the Principle as a Condition for Derogation

As was pointed out above, it is very possible that the failure to include the general provision on non-discrimination among the non-derogable rights in paragraph 2 led to the introduction by the UK of a proposal to make non-discrimination into one of the substantive conditions of derogation in paragraph 1.[13] The UK proposal was introduced in the belief that

the right of derogation should not be so absolute as to permit discrimination solely on the grounds of race, colour, sex, language or religion. In time of war or public emergency, discrimination on the ground of national status might be essential, but even in time of crisis discrimination for the reasons set forth in the UK text should not be permitted.[14]

---

[9] The article was adopted by 8 votes to 4 with 1 abstention, E/CN. 4/SR. 195, para. 143.
[10] Ibid., p. 3.
[11] The Commission decided by 8 votes to 2 with 3 abstentions that art. 20 should not be included, E/CN. 4/196, p. 6, para. 19. See comments by Mr Kyrou (Greece), Mrs Roosevelt (USA), Mr Leroy Beaulieu (France), Mr Nisot (Belgium), Mr Metha (India), Mr Mendez (Philippines), at pp. 3–5.
[12] See the remarks by Mr Oribe (Uruguay), ibid., p. 4.
[13] The proposal was made by the UK in 1952, E/CN. 4/L. 139. See also E/CN. 4/SR. 330, p. 3.
[14] Ibid., p. 4.

In the debate following the UK proposal, there was general agreement that the grounds which do not justify derogation in emergencies should include 'social origin' but not 'birth'.[15]

There was also a debate on whether to include the word 'solely'. Mr Hoare, the UK representative, believed the word to be of great importance, because

during an emergency a State would impose restrictions on a certain national group which at the same time happened to be a racial group; that word would make it impossible for the group to claim that it had been persecuted solely on racial grounds.[16]

The inclusion of the word required a separate vote, and it was finally adopted.[17] It was pointed out that the evil that should be avoided with the introduction of the word was the adoption of discriminatory measures based exclusively on some of the grounds mentioned in the clause.[18]

(d) The Extent of the Provision

In the light of the *travaux préparatoires*, one may wonder what the practical implications and the extent of the non-discrimination provision of the derogation clause in the Covenant and in the American Convention actually are. This is so, first of all, because the non-discrimination provision allows derogating measures which discriminate on a series of grounds (those not explicitly mentioned in the clause); and secondly, because, even on the grounds mentioned, what the clause prohibits is a discriminatory measure *solely* based on those motives. Therefore, States could take derogating measures in emergencies even against a racial or ethnic group, provided that the main reason for this is not the fact of race but, for example, military necessity. The practice of States supports this position; two examples could illustrate this. The first example refers to the Miskitos in Nicaragua. Measures derogating from the rights to residence and free movement were taken by the Nicaraguan Government against the Miskito Indians living by the Coco River because of military necessity and the need to protect their lives and not for racial motives, even though the Miskitos constituted a racial and ethnic group different from the rest of the Nicaraguan citizens. This measure was considered by the IACHR to be non-discriminatory.[19]

The second example is found in the measures of relocation and ex-

---

[15] Ibid., pp. 4–11.
[16] Ibid., p. 10, Mr Hoare (UK).
[17] It was adopted by 9 votes to 7 with 2 abstentions, E/CN. 4/SR. 331, p. 6.
[18] A/2929, para. 44.
[19] *Nicaragua–Miskitos case.* See a detailed analysis on this point, pp. 187–8 below.

clusion taken by the USA Government against Americans of Japanese origin during the Second War World. There is no doubt that in principle these measures were directed against those people because of their racial or ethnic origin and the fact that they had links with a nation at war with the USA. It should be pointed out that even if some were aliens resident in the USA, many of the 120,000 involved were American citizens born in the USA, although of Japanese origin: Therefore, the different treatment could not be justified on the grounds of citizenship or birth, because they were not strictly speaking 'enemy aliens'.

Although such an *a posteriori* analysis of the derogating measures indicates that there was no justification for their adoption, the Special Commission set up by the US Congress did not rule out the possibility that military necessity could compel States in some cases to take measures derogating from human rights standards, establishing different treatment in respect of certain groups.[20] Furthermore, in the case of the evacuation and relocation of the indigenous population of the Aleutian Islands following a Japanese attack, the same Commission considered the measures to be justified for reasons of military necessity, although it criticized the manner in which those measures were carried out.[21]

Therefore, derogating measures which make a distinction between different groups of people, even on racial grounds, could be considered legitimate when they have not been taken exclusively on these grounds, but because the measures were necessary and reasonable in the circumstances, and proportionate to the emergency. Thus, the inclusion of a non-discrimination provision in the derogation clause of the Covenant and the American Convention would not preclude the possibility of taking these kinds of measures lawfully. In the light of this interpretation, the presence of this requirement in the two above-mentioned treaties does not seem to restrict the powers of States to take derogating measures any more than does the European Convention, for example, whose clause of derogation does not explicitly contain this requirement of non-discrimination, but has a general non-discrimination provision (art. 14).

In its comparative study of the Covenant and the European Convention, the Committee of Experts of the Council of Europe confirms this position. The Committee considered that the inclusion of the non-discrimination provision in the derogation clause of the Covenant and its absence from the European Convention seemed, in principle, to be of importance. However, it gives a lot of weight, in line with the *travaux préparatoires*, to the word 'solely' as used in the Covenant. Consequently,

---

[20] *Personal Justice Denied: Report of the Commission on Wartime Relocation and Internment of Civilians* (Washington, 1982).
[21] Ibid., pp. 18–23.

it might be necessary in emergencies to take measures which involve action which affects certain groups belonging to a particular race or religion; so far as such action could be regarded as discriminatory, it would not constitute discrimination *solely* on these grounds, since the action would have been strictly required by the situation. Therefore, according to this interpretation, the non-discrimination provision of the derogation clause in the Covenant will not constitute any difficulty for the European States which at the same time are parties to the Covenant.[22] In other words, it will not restrict their powers to take measures of derogation, even against, for example, certain racial or ethnic groups, in situations of emergency.

Interestingly enough, the American Convention contains in its derogation clause the non-discrimination requirement, but without the word 'solely'; and in fact this is the only difference between it and the text of the Covenant. The absence of this word in the final text of the American Convention is not easily explained, in view of the fact that almost all the drafts of the Convention contained the word, including the Draft of the IACHR which was the working paper of the Conference.[23] Moreover, in the official records of the San José Conference, there is no mention whatsoever of the reasons why the word was suppressed. The only possible explanation is that it was suppressed due to stylistic changes made at the last moment of the Conference. In any case, the absence of the word 'solely' in the final text of the Convention does not alter the meaning given to the provision, which, in line with the interpretation under the Covenant, allows reasonable derogating measures affecting certain groups which have distinctive characteristics.[24] This interpretation has been confirmed by the IACHR in the *Nicaragua–Miskitos case*.

In conclusion, each of the three treaties allows derogating measures against certain distinctive groups, as long as those measures are not based exclusively on discriminatory grounds, but are strictly required by the situation, e.g. for military necessity, and in order to overcome a threat.

In the light of these reflections, one may wonder what the effect would be of the absence of the non-discrimination provision in the derogation clause. In fact, a derogating measure which unreasonably discriminates against a particular group on racial or ethnic grounds will be in violation of the principle of proportionality and reasonableness, and therefore international monitoring bodies would find this measure to violate the

---

[22] Council of Europe, Committee of Experts, 'Problems', p. 19, paras. 68–9. See also the doctrine of the European organs, pp. 177 ff below.

[23] See the IACHR Draft (art. 24), the Chilean and Uruguayan Drafts (art. 22), and the 1959 Draft of the Inter-American Council of Jurists. However, the 1968 IACHR Resolution on States of Emergency does not contain the word 'solely'.

[24] The same interpretation was given by Norris and Desio, 'The Suspension of Guarantees', pp. 203–7.

principles of the derogation clause. The jurisprudence of the European organs also suggests this construction.

In any case, the inclusion of a non-discrimination provision in the derogation clause of the Covenant and of the American Convention is commendable insofar as it aims to prevent the adoption of derogating measures which are really discriminatory against certain racial, ethnic, or religious groups in states of emergency. History shows that it is precisely in situations of war or public emergencies that minority groups, singled out by special characteristics, are the object of persecution and violation of fundamental rights, and, therefore need special protection against discriminatory measures. Thus, the World Jewish Congress in its proposal for a derogation clause in the Covenant pointed out that the practices of the Nazi and Fascist regimes illustrated how in time of war and emergency the rights of individuals and especially those of minorities are most likely to be infringed.[25] At the same time, Mr Jevremovic (Yugoslavia) considered that there was a danger that in emergencies 'States might choose to act with unwarranted harshness toward certain minorities which they suspect of having close ties with the enemy.'[26] The measures of the US Government against Americans of Japanese origin and of some Latin-American States against certain enemy aliens during the Second War World confirm this danger. Thus, the Special Commission set up by the US Congress concluded that the measures of detention, exclusion, and internment were not justified by military necessity but were the consequence of racial prejudice, war hysteria, and a failure of political leadership. And it added:

Widespread ignorance of Japanese Americans contributed to a policy conceived in haste and executed in an atmosphere of fear and anger at Japan. A grave injustice was done to American citizens and resident aliens of Japanese ancestry who, without individual review or any probative evidence against them, were excluded, removed and detained by the US during War World II.[27]

2. THE APPLICATION OF THE PRINCIPLE BY THE MONITORING ORGANS UNDER THE THREE TREATIES

(a) The European Organs: The Ireland v. UK Case

Although the European Convention, as has been pointed out, does not contain an explicit prohibition of non-discrimination in the derogation

---

[25] E/CN. 4/NGO/7 (6 Apr. 1950). See the strong protest of the World Jewish Congress because the principle of non-discrimination was not explicitly mentioned in the list of non-derogable rights, E/C. 2/259/Add. 1 (19 July 1950).

[26] E/CN. 4/SR. 196, p. 5, para. 17.

[27] *Personal Justice Denied*, p. 18. For the attitude of some Latin-American States, especially Peru, see ibid., pp. 305–14.

clause, one of the most important issues in the *Ireland* v. *UK case* was precisely the contention by the Irish Government that the measures derogating from articles 5 and 6 of the Convention (extra-judicial deprivation of liberty) were applied by the UK in a discriminatory manner on grounds of political opinion, and therefore in breach of article 14 of the Convention.[28]

As a matter of principle, the European Commission accepted that a measure of derogation from a provision of the Convention, even if it has been found to be justified under article 15, could have been applied in a discriminatory manner, thereby constituting a violation of article 14 of the Convention. This position of the Commission seems to imply that the general prohibition of non-discrimination of article 14 will have a similar effect in cases of derogation to that of the particular non-discrimination prohibition of the derogation clause of the Covenant. This would be so, insofar as derogating measures prima facie justified under the derogation clause could be found to violate the European Convention if applied in a discriminatory way.[29]

The main submission of the Irish Government was that 'the exercise by the respondent Government of their powers to detain and intern persons was carried out with discrimination on the ground of political opinions and this constituted a breach of article 14.'[30] In particular, the Irish Government accused the UK of failing to detain and intern members of Unionist (Protestant) terrorist organizations in the same way that they interned members of terrorist organizations on the non-Unionist (Catholic) side.[31] Moreover, in its submission to the European Court, it contended that there was no objective and reasonable justification for the difference in treatment of the terrorist organizations[32]

The respondent Government denied this accusation and pointed out that there was no discrimination in the application of the laws, and that the fact that more persons on the non-Unionist side had been detained and interned was explained by the evidence that the terrorist threat to security came mainly from them.[33]

From the evidence of the case, it seems clear that many more members of the IRA than those belonging to 'Loyalist' organizations were detained. However, this mere fact is not conclusive proof of a discriminatory treatment contrary to article 14. Several explanations were offered by the

---

[28] *Ireland* v. *UK*, Ser. B: Report of the Commission, p. 28. See also pp. 126–7. For the measures derogating from arts. 5 and 6 of the Convention, see above, Ch. 5, pp. 144 ff.

[29] Similar opinion by Van Dijk and Van Hoof, *Theory and Practice of the ECHR*, pp. 405, 407–8.

[30] *Ireland* v. *UK*, Ser. B: Report of the Commission, p. 28.

[31] Ibid., Ser. B, p. 126.

[32] Ibid., Ser. A: Judgment of the Court, para. 227.

[33] Ibid., Ser. B: Report of the Commission, p. 127.

Commission and the Court which justified this different treatment. After a detailed examination of the different phases of the crisis in Northern Ireland as well as the changes in policy by the UK authorities, the European organs found that terrorist attacks by the IRA were far more numerous than those by the Loyalists; that the IRA was a far more structured organization and therefore constituted a more serious menace than Loyalist terrorism; and finally, that it was easier to institute criminal proceedings against Loyalists than against Republicans (in fact, Loyalists were frequently brought to Court during this period).

Although there were some shortcomings in the application of the measures in the different phases, the Court considered that it was unrealistic to make very sharp distinctions between the different phases because the situation was constantly changing and therefore the authorities needed a certain amount of time to adapt themselves to the new situation. Moreover, the initial different treatment of both organizations did not continue during the last phase. In short, the Court concluded that 'the aim pursued until 5 Feb. 1973, the elimination of the most formidable organisation first of all, could be regarded as legitimate and the means employed do not appear disproportionate'.[34] Consequently, the Court reached the same conclusion as the Commission: there had not been, at any time, any discrimination contrary to article 14 in the application of these measures derogating from article 5.[35]

The *Ireland* v. *UK case* raises the interesting question about the relation between article 14 and article 15, or, in other words, about the possibility of derogation from article 14 (article 14 does not appear in the list of non-derogable rights in article 15(2)). As was pointed out before, the European Commission accepted that a measure of derogation from an article of the Convention, even if it has been found to be justified under article 15, could have been applied in a discriminatory manner, thereby constituting a violation of article 14 of the Convention. This position would give to this general non-discrimination provision (art. 14) a similar effect to that of the specific non-discrimination condition contained in the derogation clause of the Covenant. However, there is a possible obstacle to this position, which seems to be very reasonable and in line with the spirit of the Convention. In the case of *Ireland* v. *UK*, the Commission left open the question of whether a discriminatory measure could be deemed lawful for the further operation of the derogation clause on article 14, that is, it left open the question of

---

[34] Ibid., Ser. A: Judgment of the Court, p. 87, para. 230; see also, for the whole analysis of the different phases, paras. 228–30.

[35] The Commission was unanimous in its decision on this point. The Court came to the same decision by a vote of 15 votes to 2 (dissenting, Judges O'Donoghue and Matscher).

the possibility of derogation from article 14. Due to the fact that the Commission and the Court found that the derogating measures taken by the UK Government had not been applied in a discriminatory manner, they did not need to answer this question. However, one may speculate over the possible solution to this important question.

On the one hand, the interpretation given by the UK and the Irish Governments seems to support the possibility of derogation from article 14 when necessary in the circumstances. Thus, the UK Government, in its final submission to the Court, requested in the alternative that 'the measures taken during each phase were founded on the right of derogation duly exercised by the UK Government in pursuance of article 15 of the Convention'.[36] The Government maintained that the measures were strictly required by the situation.[37] The same construction, as a matter of principle, was adopted by the Attorney-General of the Irish Government.[38]

Moreover, in a separate opinion, Mr Triantafyllides could not agree with the majority of the Commission that there had been no discrimination at any time in the manner in which the derogating measures were applied, but nevertheless, he maintained that any breach of article 14 was within the limits of the right of derogation of article 15.[39]

Even if this construction seems to be logical in theory, it does not seem to make much sense when one looks at how it would be applied in practice. In other words, in the hypothetical case in which the European organs would find that a particular measure of derogation had been applied in a discriminatory manner contrary to article 14 of the Convention, it is hard to believe that the effect of the derogation clause would be to make that discriminatory application justified under the Convention. The reason for this is as follows.

As is well known, according to the jurisprudence of the European organs, discrimination is established where three elements are found to exist, that is, where (1) the facts of the case disclose a differential treatment; (2) the distinction has no legitimate aim, or, in other words, has no objective and reasonable justification; or (3) there is no proportionality between the measures employed and the aim to be achieved. This test was also applied in the *Ireland* v. *UK case*.[40]

Therefore, in order to find discrimination in the way in which a derogating measure has been applied, the monitoring organs have to

---

[36] Ibid., Ser. B: Counter-Memorial of the UK Government, ii. 238.

[37] Ibid., pp. 231–2.

[38] Ibid., p. 231.

[39] *Ireland* v. *UK*, Ser. B: Report of the Commission, p. 498, separate opinion of Mr Triantafyllides on art. 14 of the Convention.

[40] Ibid., pp. 127 ff.

reach the conclusion that a different treatment has no objective and reasonable justification in the circumstances, or that it is not proportionate. In a situation of emergency, the legitimate aim of any derogating measure must be to overcome the threat and to restore public order, and the issue of proportionality must be aproached taking into account the relationship between the means employed and the aim mentioned. Now it is difficult to see how the monitoring organ, once it has found that the application of the derogating measure has been discriminatory, either because it has no legitimate aim (i.e. it has no relation to the emergency), or because it is unreasonable or disproportionate to the threat, well nevertheless say that it is strictly necessary according to the exigencies of the situation when applying the derogation clause. This is because the test of the principle of proportionality (the main condition of derogation of article 15) is very similar to the test used by the European organs in the application of article 14. Thus, for a measure of derogation to be lawful, it must be necessary, reasonable, and proportionate to the threat, but these criteria are precisely those checked by the European organs, in the early phase, in order to assess the justification for a different treatment in the emergency. Therefore the application of the same test to the same facts, although under two different headings (first under the principle of non-discrimination and later under the derogation clause), cannot give different results. In conclusion, it is hard to imagine a situation in which a derogation measure found to be applied in a discriminatory manner could be considered to be strictly required by the situation and therefore lawful for the operation of the derogation clause.

Applying this reasoning to the *Ireland* v. *UK case*, it is hard to believe that, had the Commission and the Court found that the derogating measures had been applied in a discriminatory manner because there was no 'objective and reasonable justification' for a different treatment of both terrorist organizations (e.g., as the Irish Government seemed to imply, because there was a bias and a clear reluctance to apply these measures to the Unionist side), they would have found that discrimination to have been required by the situation. The test of 'objective and reasonable justification' in the circumstances of the emergency had already been applied in order to decide whether the different treatment was justified or not.

In two other cases relating to emergencies, the European Commission has found violations of the provision on non-discrimination (art. 14). However, in these two cases, there was no consideration of the possible relation between articles 14 and 15, because in both cases the Commission found that article 15 (the right of derogation) could not be applied as a general excuse. In the *Greek case*, the Commission found

that the laws on press censorship, and specifically the general prohibition
of notices of left-wing organizations, without any explanation of the aim
of these laws and regulations, involved a discrimination on the grounds of
political opinion contrary to article 14 of the Convention, read together
with articles 9 and 10.[41] In the *Cyprus case*, the Commission also found a
violation of the non-discrimination provision, because the acts against the
Convention committed by the Turkish Government in the north of the
island were exclusively directed against members of one of the two
communities living in Cyprus, namely, the Greek-Cypriot community.[42]

(b)   The Practice of the HR Committee

As in the case of the principle of consistency, there is little reference in
the practice of the HR Committee to the application of the principle of
non-discrimination of the derogation clause in particular cases. In fact,
the provision of non-discrimination has not raised any major problem
directly relating to the derogation clause (art. 4). However, some remarks
made by the Committee in the context of emergencies, and relating to the
general principle of non-discrimination contained in articles 2(1) and 26
of the Covenant, should be underlined.

The only case in which a State has formally derogated from the discrimination provision of article 26 has been the case of Nicaragua. Thus,
in its notices of derogation of 14 April 1982 and 1 August 1984, the
Government of Nicaragua included article 26 among those articles which
had to be derogated from in order to overcome the emergency.[43] In this
case, however, the inclusion of article 26 seems to have been a mistake
attributable to the inexperience of the Government; the Nicaraguan
Government included all the provisions of the Covenant in the relevant
decree of derogation except those considered non-derogable by article
4(2), instead of including only those articles actually suspended in the
emergency.[44]

The Review of States' reports. On reviewing States' reports, members of
the Committee have called attention to some discriminatory measures
taken by some Governments in emergencies. Perhaps the clearest
example can be found in the case of Chile in relation to discrimination on

---

[41] *Greek case* (1969), p. 164, para. 369.

[42] *Cyprus case* (1970), p. 551, para. 503.

[43] See *Human Rights: Status of International Instruments* (up to 1 Sept. 1987), UN (New York, 1988), pp. 64–5.

[44] This mistake was pointed out by the Nicaraguan representative before the Committee in the presentation of the report. A/38/40 (1983), p. 55, para. 244. (See also CCPR/C/SR. 424.) Thus, in the following notices of derogation art. 26 has disappeared from the list (n. 1), pp. 65 ff.

political grounds. On reviewing the *First Report of Chile* in 1979, Mr Graefrath considered that article 11(2) of Constitutional Act No. 3 made political discrimination a constitutional principle; at the same time, Decree-Law No. 2376 disqualified certain people from being elected as trade union leaders on political grounds; furthermore, the then new draft constitution provided for the continuation of such political discrimination. Mr Graefrath also mentioned an incredible application of the principle of non-discrimination, which was used in order to suppress all political rights. The Chilean Government, by Decree No. 77, dissolved all the Marxist political parties and then, alleging reasons of justice and non-discrimination against such parties, caused all the other political parties to be suspended by Decree No. 78.[45]

Mr Tomuschat, on the other hand, pointed out that the Covenant prohibits political discrimination and that no restriction could be placed on this principle; he found it to be significant that Chile had not included article 2 of the Covenant (a discrimination provision) among those pro-visions temporarily suspended.[46] This position of Mr Tomuschat seems surprising at first sight, if one takes into account that derogating measures against certain political groups could in theory be taken if these groups constitute a threat to the life of the nation. These measures would not be *per se* against the derogation clause, provided that they are reasonable and proportionate for overcoming the emergency; in fact, the derogation clause does not mention discrimination on political grounds to be forbidden (art. 4(1)). Nevertheless, the position of Mr Tomuschat can be explained in the particular case of Chile, because the Committee had found that the declaration of emergency and the derogation of almost all human rights were not justified because there was no 'public emergency threatening the life of the nation'. Consequently, the restrictions by the Chilean Government had to be assessed according to normal standards and, obviously, these discriminatory measures could not be held to be necessary in a non-emergency situation.[47] On reviewing the *Second Report of Chile* in 1984, the HR Committee also pointed out that 'the so-called emergency was used to justify the discriminatory measures provided for in article 8 of the 1980 Constitution'.[48]

On reviewing the *Second Report of the UK* in 1985, members of the Committee asked for clarification of several aspects relating to the

---

[45] *Year Book of the Human Rights Committee, 1979–80*, i. 17–18. See also E/CN. 4/1221, p. 27.

[46] Ibid., p. 19.

[47] See the opinion of the HR Committee, in A/34/40, p. 78.

[48] A/39/40 (1984), p. 82, para. 450. On two other occasions, members of the Committee considered that some Constitutional provisions were against the derogation clause because it seemed to allow discriminatory measures in emergencies: A/37/40, p. 56 (Guyana), A/39/40, p. 61 (Gambia).

emergency, including the operation of the Diplock Courts in Northern Ireland, the procedures of which seemed inconsistent with articles 2(1), 14, and 26 of the Covenant.[49] It seems that some members of the Committee suspected that the operation of the Diplock Courts had some elements of discrimination against the Catholic community. The UK Government answered that these courts were not in breach of articles 2(1) and 26 of the Covenant, and that there was therefore no need for derogation; moreover, the special procedures of the Diplock Courts were necessary because intimidation made a fair trial by jury impossible.[50]

The position of Mr Hammond (UK) before the Committee would seem to be correct. In fact, the Diplock Courts, in establishing trial without jury, do not contravene article 14 of the Covenant, because there is no such right (to be tried by jury) in article 14. Moreover, they do not contravene the prohibition on discrimination, because a derogating measure is discriminatory only if it has no 'objective and reasonable justification'. There is no doubt that the situation in Northern Ireland, in which a *de facto* state of emergency existed, required special judicial procedures, because intimidation of jurors made it impossible to guarantee a fair trial by jury.[51] The measures were necessary because they had an objective and reasonable justification, and therefore there was no need to enter into derogation from articles 2(1) and 26 of the Covenant.[52]

The practice of the Committee under the Optional Protocol. Under the Optional Protocol, there is only one case in which a reference was made by the Committee to a discriminatory measure in emergencies; this was the *Weinberger case.*[53] In this case, a Uruguayan citizen was deprived *inter alia* of all political rights for a period of 15 years following his sentencing to imprisonment for a period of 8 years for 'subversive association' with aggravating circumstances of conspiracy against the Uruguayan Constitution. Although article 25(2) of the Covenant (the article which recognizes political rights) allows States in certain circumstances to place restrictions on political rights, in no case can such

[49] A/40/40 (1985), p. 100, para. 528. (See also CCPR/C/SR. 594, p. 3.)

[50] Ibid., p. 101, para. 531. It seems that in practice these courts have operated fairly and without discrimination. See Bonner, *Emergency Powers*, p. 142, and K. Boyle, 'Human Rights and the Northern Ireland Emergency', in J. A. Andrews (ed.), *Human Rights in Criminal Procedure* (The Hague, 1982), p. 159.

[51] See Mr Hammond's explanation, in CCPR/C/SR. 594, p. 5. See also Boyle, 'Human Rights', p. 159.

[52] On 22 Aug. 1984, the UK Government sent a notice of termination of the derogations from arts. 9, 10(2), 10(3), 12(1), 14, 17, 19(2), 21, and 22, which had been suspended since May 1976.

[53] *Weinberger* (v. *Uruguay*), Comm. No. 26/1978, Adoption of Views (29 Oct. 1980; 11th Session).

a restriction be enforced *solely* on the grounds of political opinion; this would amount to a discrimination contrary to article 2(1) and 26 of the Covenant.[54] The main contention of the applicant on this point was that his brother's conviction was in reality due to his political opinions, which were contrary to the official ideology of the then Military Government in Uruguay. In fact, Ismael Weinberger had given information on trade union activities to a newspaper opposed to the Government, and, at the same time, he was a member of a political party which was then perfectly lawful, but which was later banned by the Military Government.[55] The Uruguayan Government did not provide the Committee with the documents of the trial which they had requested and which could have proved that the allegations of the applicant were false. Although the Government alleged that there was a situation of emergency, it did not make any submission of fact or law to justify any derogation from the standards of the Covenant.[56] Consequently, the Committee did not take the derogation clause into consideration, and had to apply the normal standards for peacetime. The Committee found *inter alia* a violation of article 25, since Weinberger had been banned from taking part in public affairs and from being elected for 15 years.[57] One wonders why the Committee, after having found that the conviction was due to political discrimination, did not include the violation of articles 2(1) and 26 in the dispositive part of its views.[58]

(c)   The Practice of the IACHR

The study of the principle of non-discrimination in the Inter-American system shows how little importance it has had in the doctrine of the IACHR not just in relation to emergencies, but also in more general terms, as a human right principle.

The principle of non-discrimination in general. If one looks at the doctrine of the Inter-American Commission as systematized in one of its official publications, there is only one reference, and that quite unimportant, to the concept of racial discrimination.[59] On the other hand, in the Com-

---

[54] Ibid., para. 15.

[55] Ibid., para. 12.

[56] Interestingly enough, the Uruguayan Government sent a notice of derogation on 30 July 1979, after a great number of applications to the HR Committee under the Optional Protocol had been sent, alleging gross violations of several articles of the Covenant.

[57] Ibid., para. 16.

[58] Cf. ibid., paras. 15 and 16.

[59] OAS, *The IACHR: Ten Years of Activities*, p. 320. However, see the interesting remarks by the Commission on 'crimes of opinion', which have a certain relation with political discrimination (p. 332).

mission's country reports there is hardly any mention of the compliance with that principle by States parties to the American Convention or to the American Declaration. The only country reports in which some reference has been made to discrimination are the *Report on El Salvador*, 1978, and the *Report on Cuba*, 1983.[60] The case-law on discrimination arising from individual communications is also very scarce.[61]

The principle of non-discrimination in relation to emergencies. Although the derogation clause of the American Convention contains a non-discrimination provision (art. 27(1)) the doctrine of the IACHR provides little insight into the application of the principle in emergencies. Moreover, when the IACHR has had to apply the standards established by the derogation clause, it has not insisted on this requirement, this being clear proof that it has not presented a special problem for the Commission. This is somewhat surprising in view of the great number of states of emergency declared in the hemisphere and the all too frequent political discrimination that takes place. Perhaps the Commission has not considered it necessary to express any view on the discriminatory manner in which certain measures of derogation have been applied, because it has focused more on the gross violations of the most fundamental rights like the rights to life and to freedom from torture and ill-treatment. It is worth noting that in the cases in which the IACHR referred to the 'most accepted doctrine' on human rights in emergencies within the Inter-American system, the principle of non-discrimination was not mentioned.[62]

---

[60] IACHR, *Report on El Salvador*, 1978, pp. 159–61 (discrimination on the grounds, of sex); IACHR, *Report on Cuba*, 1983, pp. 48–9; p. 179, para. 11; and pp. 180, paras. 12 and 15, and 182, para. 22.

[61] This principle has not had great importance in the case-law of the Inter-American system; thus, in the exhaustive index by special topic, the term 'discrimination', or any other similar term, does not appear. See Buergenthal and Norris (eds.), *The Inter-American System*, vol. iv, annexe F. However, some reference had been made to this principle in the following cases: IACHR, Comm. No. 9635 (v. Argentina); case of Mr Osvaldo Lopez (a prisoner who was denied the right of appeal for being a member of the armed forces) in IACHR, *Annual Report, 1986–7*, pp. 32–63. See also IACHR, Case No. 7465 (v. USA) (discriminatory application of the death penalty in the USA against black people), Resolution No. 10/84 (4 Oct. 1984), in Buergenthal and Norris (eds.), *The Inter-American System*, vol. iii, booklet 21. 1, pp. 157 ff.; Amnesty International, *USA: The Death Penalty* (London, 1987), pp. 54–64; IACHR, Case No. 9647 (v. USA) (death penalty of juveniles), Resolution No. 3/87 (27 Mar. 1987), OEA/Ser. L/V/II. 69, doc. 17, pp. 37–40. See the observations of the US Government on this problem, 'Request for Reconsideration of the Resolution 3/87 Made by the US', in Buergenthal (ed.), *The Inter-American System*, vol. iii, booklet 21. 3, pp. 155 ff. See also the dissenting opinion of Monroy Cabra on the resolution.

[62] IACHR, *Report on Chile*, 1985, p. 44, refers to the principle of exceptional threat and to the principle of proportionality. IACHR, *Report on Argentina*, 1980, pp. 22 ff. and *Report on Paraguay*, 1987 refer, in addition to these principles, to the non-derogability. See, however, a reference to the principle in the IACHR *Report on Suriname*, 1983, pp. 36–7.

The clearest illustration of this lack of relevance of the principle is found in the 1981 *Report on Bolivia*. In this report, the Commission reproduced the text almost verbatim, and insisted on all the requirements contained in the derogation clause, except the provision of non-discrimination.[63] Nevertheless, on some occasions the Commission has made an indirect reference to this principle.

In the *Report on Suriname*, 1983, the Commission considered that the regulation of the state of emergency in Decree A-11 was in flagrant violation of article 4 of the Covenant on a number of grounds, among them, because the decree did not prohibit discrimination in the suspension of rights.[64] In the *Report on Nicaragua*, 1981, the Commission found that the operation of Special Tribunals set up by the Nicaraguan Government in order to try the Somocistas gave rise to irregularities incompatible with the American Convention, in so far as they caused discrimination against all the accused Somocistas which consisted in denying them certain minimum guarantees of fair trial which, by their very nature, ought to be applicable to all inhabitants of the country.[65]

On the other hand, in the *Nicaragua–Miskitos case*, the IACHR has carefully analysed the principle of non-discrimination of the derogation clause (article 27(1)). As has been pointed out on several occasions, the problem examined by the Commission in this case was the compatibility of the compulsory relocation of the Miskitos with the right to residence and movement contained in the American Convention. The Commission, after accepting that there was a genuine public emergency threatening the life of the nation, and that the measure was strictly required by the situation, analysed whether the measure was based 'on one of the proscribed forms of discrimination'.[66] In this case, due to the fact that the measure of derogation was directed against a particular ethnic group, the issue of discrimination on racial grounds had to be carefully scrutinized. The Commission examined whether the compulsory relocation was a form of punishment against an ethnic group considered to be disloyal to the Government;[67] and it pointed out that, in these cases of restrictions on ethnic groups, the Government has to declare 'strictly and explicitly' the rationale behind the restrictions. The Commission gave great weight to the fact that the population which had been relocated was an indigenous one; indigenous populations have very strong ties with their lands and homes, and consequently, every compulsory movement of

---

[63] IACHR, *Report on Bolivia*, 1981, pp. 22–3.
[64] IACHR, *Report on Suriname*, 1983, pp. 36–7.
[65] IACHR, *Report on Nicaragua*, 1981, p. 93.
[66] IACHR, *Nicaragua–Miskitos case*, p. 120.
[67] For the possible reasons for this disloyalty to the Nicaraguan Government, see Mendez, 'La Participacion', pp. 306–17.

these groups is a traumatic experience affecting their way of life and even their survival.[68]

The Commission considered that the relocation, for reasons of military necessity, was not carried out in a discriminatory manner, but at the same time it held that once the emergency had ended, the Miskitos should be helped to return to their lands if they wished to do so; otherwise, their stay in the relocation camps would become a discriminatory punishment.[69]

An interesting point which emerges from this reasoning of the IACHR is that derogating measures affecting indigenous populations, e.g. compulsory relocation, have to be carefully considered because of the grave effects on the lives of the people due to their particular way of life and their connection with the land. This would mean that the test of necessity for certain derogating measures affecting indigenous populations should be stricter than in the case of other populations which do not have such strong ties with their land and whose way of life does not depend on the land as much as the indigenous population's.

It is interesting to note that the same consideration was given great weight in the case of the relocation of the Aleuts in the Second World War. Before the evacuation, in anticipation of a Japanese attack, there was a discussion on the course of action that should be taken with regard to the indigenous population: either evacuate and relocate them in camps, in order to allow them to avoid the risks of war, or leave them on their original islands because the relocation would in fact be more disruptive to their lives.[70]

3. CONCLUSION

1. The principle of non-discrimination constitutes the second condition for the derogation of rights according to the legal regime established by the derogation clause. This condition contained in the UN Covenant and in the American Convention does not appear, however, in the European Convention. The absence of this condition, which in principle seems to be of importance, has no major consequences insofar as the discriminatory application of derogating measures is also forbidden under the European Convention by the operation of the general non-discriminatory provision of article 14; this position has been confirmed by the European organs.

2. The extent of the non-discrimination provision of the derogation clause is limited by two facts: first, what it prohibits is discrimination on

---

[68] IACHR, *Nicaragua–Miskitos case*. See p. 120 n. 33, in which the Commission refers to the negative effects of the relocation of 10,000 Navajo Indians.

[69] Ibid., p. 121, para. 31.

[70] *Personal Justice Denied*, p. 19.

five grounds (race, colour, sex, language, religion or social origin), but not on the others contained in the general provision on non-discrimination in the treaties. Secondly, even on those grounds mentioned, what it prohibits is derogating measures *solely* based on those discriminatory grounds, but it allows derogating measures establishing certain differences of treatment even on racial, ethnic, or religious grounds, when reasonably justified and deemed to be necessary to overcome the emergency.

3. Although the extent of the non-discrimination provision is limited, its inclusion in the derogation clause is commendable because the risk of taking discriminatory measures based only on racial prejudice and hatred towards minorities is greater in situations of emergency. Therefore, derogating measures against certain racial, ethnic, or religious groups should be carefully scrutinized by international monitoring bodies in order to assess whether these measures have an objective justification, e.g. for reasons of military necessity, or are based exclusively on prejudice against these minorities.

4. The application of the principle of non-discrimination in states of emergency has not generally played a major role in the case-law of the international monitoring bodies under consideration. However, two cases were of some interest.

In *Ireland* v. *UK*, the applicant Government contended that the derogating measures (detention and internment without trial) had been applied in a discriminatory manner on grounds of political opinion. However, the different treatment accorded to terrorist members of non-Unionist organizations (IRA) as compared with those of a Unionist character was found by the European organs to have an objective and reasonable justification.

In the *Nicaragua–Miskitos case*, the Inter-American Commission found that the compulsory relocation of the Miskito Indians had not been discriminatory because it was due to military necessity. The Commission carefully considered the possibility of discrimination, because the measure was directed against an ethnic group which was considered to be disloyal to the Government. At the same time, it gave a lot of weight to the fact that the Miskitos constituted an indigenous population with especially strong ties with their land.

7

The Principle of Consistency

I. INTRODUCTION

The text of paragraph I of the derogation clause establishes that the right of a State to take measures of derogation in emergencies is limited by the condition that the measures must not be inconsistent with the State's other obligations under international law. The operation of the this legal criterion of the validity of the derogating measures can only come after all the other conditions of the derogation clause have been satisfied; in other words, after the monitoring organ under the treaty has accepted that (1) there exists a public emergency threatening the life of the nation, (2) that the emergency has been officially proclaimed, (3) that there has been a valid notice of derogation, (4) that it has not affected a non-derogable right, (5) that the measure is strictly required by the exigencies of the situation, and (6) that it does not involve discrimination. Once the monitoring organ has been satisfied that these conditions have been met, it should check, in order to accept the validity of the derogating measure, that the latter does not conflict with the State's other obligations under international law.[1]

In principle, this requirement seems to be of great importance because it could be interpreted as requiring the implementation through the machinery of these three human rights treaties, which are better than most of the other human rights treaties, or other obligations that States have assumed in the field of human rights. In other words, according to this interpretation, monitoring organs under the three treaties should declare null and void all measures of derogation taken by States which conflict with other human rights obligations that States have assumed under other treaties, but which are not covered by the three treaties.

However, this principle of consistency obviously has a limited applicability, and it cannot be applied in order to implement through the machinery of the three treaties all the other human rights obligations that the State has undertaken. The condition *since qua non* for the application of the principle of consistency is precisely that there must be a derogation

---

[1] See the application of the principle of consistency by the IACHR in the *Nicaragua–Miskitos case*. See below pp. 197–9.

from the State's obligations under one of the three human rights treaties under consideration. Therefore, there is here an important limitation *ratione materiae* upon the application of the principle: those other international obligations that have arisen from treaties or from general international law, and which contain obligations not covered by the three treaties, cannot be implemented by the machinery of implementation of the latter treaties.

In order to understand the possible content of the principle of consistency, the following aspects will be studied in this chapter: first of all, the origin of the principle in the three treaties seen through its legislative history; secondly, the application of the principle by the monitoring organs of the treaties, namely the European Court and Commission, the Human Rights Committee, and the Inter-American Commission; finally, the possible extent of the principle will be ascertained in the light of the two previous sections.

2. LEGISLATIVE HISTORY OF THE PRINCIPLE IN THE THREE TREATIES

As in the case of other principles governing the derogation clause, the first reference to the principle of consistency is found in the *travaux préparatoires* of the UN Covenant on Civil and Political Rights. However, the principle of consistency, unlike the principle of proportionality, was not included either in the first UK proposal of a derogation clause, or in the text approved at the 5th Session of the UN Commission on Human Rights.[2] It was sometime later, in 1950, when the US delegation proposed an alternative text to paragraph 2 of the derogation clause in the context of the debate on which rights should be declared non-derogable in public emergencies. The aim of this proposal was to substitute for the principle of non-derogability of certain human rights, a provision which simply stated that 'no derogation may be made by any State under this provision which is inconsistent with international law or with international agreements to which such State is a party'.[3] Mrs Roosevelt, head of the US delegation, explained the rationale behind this proposal as follows: due to the fact that the conduct of States in time of war had been regulated in international conventions after long debates in large international conferences, in particular the 1949 Geneva Conventions, the UN Commission should take full advantage of this legislation, instead of trying to work out which rights should be entrenched from derogation.[4] Therefore what the US delegation proposed was to include a reference to the laws of war in

[2] Doc. E/1371, art. 4, para. 1.   [3] E/CN. 4/365, p. 19.
[4] E/CN. 4/SR. 195, para. 45 (29 May 1950) (Mrs Roosevelt).

order to assess which of those human rights provisions could not be derogated from in emergencies. However, what the US delegation did not seem to realize was that the field of application of the 1949 Geneva Conventions is not the same as that of the derogation clause, the latter being much wider, including as it does public emergencies threatening the life of the nation which do not amount to war. In any case, the US proposal, as amended by Belgium, was approved as an addition to paragraph 2 of the derogation clause, which already contained a list of non-derogable rights, rather than as a substitution of the principle of non-derogability of fundamental human rights.[5] As was seen above, the principle of non-derogability was considered a fundamental principle of the derogation clause.

The reference to 'other obligations inconsistent with international law' was considered by some delegations to be 'vague and unsatisfactory', insofar as in no part of the Covenant, and *a fortiori* outside the Covenant, were the international law standards clearly stated; this could create problems because of the different views among States on some issues, for example, property. Therefore it was proposed that the reference to international law be clarified.[6] In the same context, and in order to avoid misinterpretation of the term 'other obligations under international law', the Yugoslav delegation forwarded a proposal making an express reference to the UN Charter and the Universal Declaration of Human Rights.[7] Unfortunately, they did not explain the meaning of this reference. Although some delegations supported the inclusion, most of them rejected it, mainly because the reference to the UN Charter was considered redundant, insofar as it is part of international law, and the reference to the UDHR was considered inadmissible because it was not part of international law.[8] Consequently, the principle of consistency continued to be expressed in that general reference to 'other obligations under international law.'

The introduction of the principle of consistency in the derogation clause of the European Convention on the basis of a UK proposal raised no discussion and was accepted.[9] Under the American Convention, although the different drafts submitted to the San José Conference by the

---

[5] Ibid., p. 10, paras. 48 and 156.

[6] E/CN. 4/528, p. 31, comments by the representative of Czechoslovakia before the General Assembly.

[7] E/CN. 4/L. 212. At this stage of the *travaux préparatoires*, a UK proposal to transfer this principle of consistency to para. 1 of the derogation clause succeeded. The aim was to adopt the same wording as the recently approved para. 1 of art. 15 of the ECHR (see E/CN. 4/528, Add. 1). However, this transfer did not involve any substantial change.

[8] See A/2929, para. 43. See also in more detail the reasons given by Mrs Roosevelt, E/CN. 4/SR. 330, pp. 6 and 9; Mr Bracco (Uruguay), pp. 5 and 11; Mr Juvigny (France), p. 7; Mr Nisot (Belgium), p. 10, and Mr Cassin (France), E/CN. 4/331, p. 8.

[9] ECHR, *Collected Edition of the* Travaux préparatoires, iii. 280 (art. 3, para. 1). Meeting of the Committee of Experts, Strasbourg, 6 Mar. 1950.

Inter-American Council of Jurists, the Government of Chile and the Government of Uruguay did not contain this principle, it was included in the draft of the American Convention, which was the working paper of the Conference, as a result of the influence of article 4 of the UN Covenant.[10]

3. THE APPLICATION OF THE PRINCIPLE BY THE MONITORING ORGANS UNDER THE THREE TREATIES

The application of the principle of consistency by the monitoring organs has been very rare. No organ has had the opportunity of developing and defining what kind of other international obligations could be applicable in cases of derogation in states of emergency. However, there are some interesting elements in the case-law of the three treaties which merit a brief examination.

(a) The European Convention

In the first case on derogations under the European Convention, although the European Commission declared that it had competence to decide on the existence of a public emergency threatening the life of the nation and on the proportionality of the derogating measures taken, there was no reference to the principle of consistency.[11] However, in the *Lawless case*, the European Court of Human Rights, in its judgment on the merits, recognized its competence to determine *proprio motu* whether the principle of consistency had been fulfilled, although neither the Commission nor the Irish Government had referred to the provision in the proceedings. In any case, the Court found that no facts had come to its knowledge which could invalidate the derogating measures taken by the Irish Government for being inconsistent with its other obligations under international law.[12]

The Court reached the same conclusion in the *Ireland* v. *UK* case in 1978. In this case, the Irish Government in its application of 16 December 1971 contended that the derogating measures taken by the UK Government in respect to articles 5 and 6 of the Convention were contrary *inter alia* to the principle of consistency; however, the applicant government did not provide any details of this claim in the proceedings of the case.

---

[10] 'Annotations of the IACHR to the Draft of the Convention Prepared by the Commission', in Buergenthal and Norris (eds.), *The Inter-American System*, vol. ii, booklet 13, p. 58.

[11] *First Cyprus case*, Application No. 176/56, Report of the Commission, *YBECHR* 2 (1958–9), 174–6.

[12] *Lawless case*, Ser. A: Judgment on the merits (1 July 1961), paras. 39–41.

The Court found no violation of the principle, because no facts came to its knowledge proving the inconsistency of the UK derogating measures.[13]

In the *Greek case*, the Commission also analysed as a matter of principle the fulfilment of this condition; the Greek Government had pleaded that its measures of derogation 'related to a matter in which Greece is not bound by any contrary obligation under international, contractual or customary, law'. Due to the fact that the main condition for the application of the derogation clause, the existence of a public emergency, was not satisfied in the case, the Commission considered that it was not called upon to express a view on this question.[14]

In the *Cyprus* v. *Turkey case*, a Government, for the first time in a case of derogation, tried to define the content of the principle of consistency. The applicant Government submitted two different interpretations of this principle. The first contention was that the expression 'other obligations under international law' in article 15(1) excluded wars violating such obligations as those under the UN Charter. Therefore Turkey, which was waging 'an aggressive war' against Cyprus, was not allowed to invoke the derogation clause.[15] This is an interesting construction of the principle of consistency. In the *travaux préparatoires* of the UN Covenant, from which the European Convention borrowed the derogation clause, Mr Bracco (Uruguay) gave a similar interpretation to this principle. In fact, in the discussion on whether or not to include the Yugoslavian amendment in which an express reference to the Charter was made after the words 'international law', Mr Bracco supported it on the grounds that it would make clear that the only case in which war could be recognized as giving rise to the right of derogation was the case of self-defence 'or for other reasons recognised in the Charter'.[16]

In this question, two aspects should be distinguished: one is the right to use the derogation clause as such, and the other the validity of concrete measures of derogation from the obligations under the treaty. As far as the first aspect is concerned, the question of whether States which wage an aggressive war have the right of derogation according to the human rights treaties was not really addressed in the *travaux préparatoires* of the Covenant, and the opinion of Mr Bracco was in fact no more than a personal remark. One wonders about the suitability for the human rights monitoring organs in cases of armed conflict of engaging in the delicate and highly political task of ascertaining which side was waging an aggress-

[13] *Ireland* v. *UK*, Ser. A: Judgment of 18 Jan. 1978, vol. 23, para. 222. See also the application of the Irish Government of 16 Dec. 1971. in ibid., Ser. B: Pleadings, p. 597, No. 5.
[14] *The Greek case*, Report of the Commission (1969), *YBECHR* 12 (1969), 112–13, paras. 226–9.
[15] *Cyprus* v. *Turkey*, Report of the Commission (10 July 1976), *EHRR* 4 (1982), 552–3, paras. 510, 512.
[16] E/CN. 4/330 (1 July 1952), pp. 5, 11.

ive war in order to deprive it of the right of derogation. In practice, the monitoring organs have never used this reason to deprive a State of this right. The approach of the UN Human Rights Committee seems to confirm this pragmatic position, leaving aside the question of the aggressor; thus, in reviewing the reports from Iran and Iraq and trying to assess according to the derogation clause the situation concerning human rights in these two countries at war, there is no mention at all of this issue; no country was deprived of the use of the derogation clause because it was waging an aggressive war.[17]

The second aspect, at which, according to the *travaux préparatoires*, the principle of consistency was directed, concerns the concrete measures of derogation from human rights obligations and makes clear that the condition required is that those measures should not be inconsistent with other international obligations. This requirement refers concretely, as will be seen below, to those obligations of other treaties on humanitarian law covering the same field as the three human rights treaties under consideration. The principle of consistency does not refer to the right of derogation as such. Thus, it does not say that the right of derogation has to be consistent with the State's other obligations under international law, but it refers to the *concrete measures* of derogation from the standards of the treaty which have to fulfil this condition.

The confusion between these two aspects can be explained by examining the *travaux préparatoires* of the Covenant. As we have pointed out, the first proposal by the US was to substitute paragraph 2 (the principle of non-derogability) for the principle of consistency. However, the UK delegation later proposed to transfer this requirement of consistency to the first paragraph of the derogation clause.[18] At that time, there was another discussion on this first paragraph, as to the convenience of introducing the word 'war' into the Covenant. Some of the drafters were ready to accept the reference to war, provided that the principle of consistency, which had a different purpose, contained an express reference to the UN Charter, on the understanding that it would exclude the right to derogate for a State waging an aggressive war. Thus, the two questions became confused.

The second interpretation of the principle of consistency given by the Cyprus Government in its submissions is very much in line with the right construction of the principle as just explained. The Government contended that even if the acts denounced as violations of the Convention by Turkey were considered to be in accordance with article 15 of the European Convention, 'they were still inconsistent with Turkey's other obligations under international law, especially the Geneva Conventions

---

[17] See Iraq, A/35/40 (1980), pp. 22 ff.; A/42/40 (1987), pp. 86 ff. For Iran see A/37/40 (1982), pp. 66 ff.
[18] E/CN. 4/SR. 330 (1952), p. 4 (see Mr Howe's remarks).

and the Hague Regulations, and therefore, could not stand under article 15'.[19] In this submission, the applicant Government referred to the concrete measures of derogation by Turkey and to the standards set up in the other treaties of a humanitarian nature. Unfortunately, the applicant Government did not give any details about which concrete acts by the respondent Government and which standards of the Geneva Conventions it was referring to. The Commission, as in the *Greek case*, did not express any view on this question because it found that article 15 could not be applied 'in the absence of some formal and public act of derogation by Turkey'.[20]

(b)   The UN Human Rights Committee

It is interesting to note, as a confirmation of the rarity of the application of the principle of consistency by monitoring bodies, that even in the Committee's 'general comment' on the derogation clause, the only requirement which is not mentioned by the Committee is the principle of consistency.[21] Nevertheless, in the discussion on the role of the Committee in states of emergency, one of its members expressly referred to the competence of the Committee to control the fulfilment of this requirement. Moreover, he pointed out, perhaps reflecting the practice of the Greek Government before the European Commission, that 'States parties should make a clear declaration that such (derogating) measures were not inconsistent with other international obligations'.[22] In the response of the Polish Government to the UN Human Rights Commission's Resolution which established a special procedure to investigate the situation of human rights in that country after the proclamation of martial law in 1981, the Government expressly referred to the consistency of its derogating measures with its other international obligations.[23]

Neither in the review of States' reports nor in the views adopted under the Optional Protocol has the Committee made any reference to this principle. Although it cannot be considered a strict application of this principle, in reviewing the report of Afghanistan in 1985, one of the members of the Committee referred to some of the standards laid down in Protocol II to the 1949 Geneva Conventions as reflecting the substance of article 6 of the Covenant (the right to life) in situations of armed conflict, and therefore in principle applicable to the case. The Afghan Government,

---

[19] *Cyprus case*, ibid., para. 514.

[20] Ibid., para. 528. See, however, the interesting dissenting opinion of Mr Sperdutti and his position on using the standards set up in the Geneva Convention IV as a guide-line to what would be required in an occupied territory. See below, p. 204.

[21] HR Committee, General Comment 5/3 (on art. 4), in CCPR/C/21, pp. 4–5.

[22] CCPR/C/SR. 349, para. 18 (remarks by Mr Ermacora). See the *Greek case*, para. 227.

[23] E/CN. 4/1983/18 (21 Feb. 1983), 'Report on the Situation in Poland Presented by the Under-Secretary General Hugo Gobbi,' pp. 2 ff.

however, rejected the contention that the situation in Afghanistan amounted to an armed conflict, and therefore denied the application not only of the Four Geneva Conventions of 1949, but also of common article 3, which is applicable to armed conflicts of a non-international character. *A fortiori*, it would be more difficult in this case to use the standards of Protocol II to interpret the content of some of the articles of the Covenant, if one considers that Afghanistan was not party to this international treaty.[24]

(c) The IACHR

The IACHR has taken a similar approach to that of the UN Human Rights Committee. The principle of consistency is rarely mentioned even when it refers to the requirements of the derogation clause in states of emergency.[25] The only case in which it has attempted to apply the principle was in the *Nicaragua–Miskitos case*.

In this case, the IACHR examined whether the compulsory relocation of 8,500 Miskitos in five camps far away from their lands in January 1982 was a lawful measure of derogation from the right to residence and movement of article 22 of the American Convention.

In order to assess the lawfulness of the measure, the Commission analysed all the requirements of article 27. Once the Commission was satisfied that there was a public emergency threatening the life of the nation and that the compulsory relocation was proportionate to the danger produced by the attacks of the 'contras' from Honduras, it dealt with the principle of consistency and that of non-discrimination; it is clear that these are two different requirements and not one, as the Commission appears to think.[26] When referring to the doctrine of international law which should be taken into consideration on this matter, as forming part of the other international obligations, the Commission said:

the preponderant doctrine is that massive relocation of population groups may be juridically valid if done with the consent of the population involved. In fact, with the exception of some cases of relocation of Indians, which are subject to criticism, the large majority of population relocations for reasons of economic development have taken place after negotiations with the populace concerned, and with assurances of adequate compensation.[27]

In order to support this position, the Commission quoted the view of the Institute of International Law, which in 1952 adopted the position that

---

[24] CCPR/C/SR. 604 (1985), p. 6, para. 36, opinion of Mr Tomuschat. See also p. 12, para. 50.

[25] See *inter alia* IACHR, *Report on Bolivia*, 1981, p. 22; IACHR, *Annual Report, 1986–7* (Nicaragua), pp. 258 ff. However, see IACHR, *Report on Suriname*, 1983, para. 82.

[26] IACHR, *Nicaragua–Miskitos case* (1983), p. 119, para. 26.

[27] Ibid., p. 120, para. 27.

population relocations may be legal only if they are voluntary.[28]

The interpretation of this opinion of the Commission presents several problems. First of all, it is not clear which of the two situations considered by the Commission in the report, namely, relocation for military necessity and relocation for economic reasons, is involved in the general statement in the first sentence. Secondly, even accepting that it only refers to the situation of relocation for economic reasons, the existence of a rule of international law by which a relocation is valid only if done with the consent of the population involved is not completely certain. These two questions will now be examined.

In this case, there are two situations which present two different problems. One is the compulsory relocation of the Miskitos for military reasons; the second their compulsory relocation for economic reasons. According to the facts of the case, the Nicaraguan Government had planned to relocate part of the Miskito population for economic reasons long before the emergency arose. However, when the military situation worsened, it had to change that economic plan and evacuate the Indian population from that area in order to save their lives and to answer effectively the attacks of the 'contras' from Honduras. In the previous paragraphs, the Commission made a distinction between these two situations, but in the paragraph concerning the principle of consistency, it does not. Therefore, the first sentence of the above-quoted paragraph could be interpreted as referring to both situations: relocation for military and economic reasons. According to this interpretation, the relocation for urgent military reasons, as was the case according to the Commission, would require under international law the consent of the population. This conclusion seems neither reasonable nor in line with the regime established in the provisions of international law of war in cases of transfer of population.[29] Therefore, one has to conclude that the entire paragraph refers exclusively to the compulsory relocation of population for economic reasons, as appears in the second sentence. If this is the case, it has nothing to do with the state of emergency in Nicaragua at the time, but with an economic plan previously established for that population. Consequently, it seems odd to apply this criterion of the principle of consistency, which corresponds to the derogation clause for emergencies, to an alien situation, a situation which arises once the emergency has ended. Of course, the aim of the Commission was to avoid the use by the Nicaraguan Government of the state of emergency for military reasons to perpetuate a *de facto* compulsory relocation for economic reasons. The Commission stressed the fact that once the reasons for the emergency had

---

[28] Ibid., p. 120 n. 32. See also p. 17 n. 1.

[29] See art. 49 of the Geneva Convention IV, 1949 in cases of forcible transfer of population for military reasons in occupied territories. Also art. 17 of Protocol II accepts the forcible transfer of populations for military reasons.

disappeared, the Nicaraguan Government had the obligation of allowing the Miskitos the freedom to return to their homes if they wished to do so. In this latter case, already in peacetime and without any reason for limiting the right to residence and movement, the relocation for reasons of economic development should be negotiated and accepted by the Miskitos. In short, this is not a strict reference to the principle of consistency in the application of the derogation clause, because it does not relate to the case of compulsory relocation for military reasons in emergency, but to the relocation for economic reasons which would occur after the emergency had ended.

On the other hand, even in the case of relocation for economic reasons, it is not clear that a norm of general international law exists by which States should relocate population with their consent; here, the Commission seems to refer to the existence of a norm based on the practice of States and expressed in the position of the Institute of International Law in 1952. It is true that both the practice of States (although scarce) and elementary considerations of humanity support the view that the decision of relocation should be taken after holding negotiations and consultations with the indigenous populations concerned. However, this is not the same as saying that a norm of international law exists by which there is an international obligation for States to require the consent of the population for any relocation. Moreover, the reference to the position of the International Law Institute adopted at its meeting in Siena in 1952 is not entirely pertinent, insofar as it mainly treated the question of the *international* transfer of populations through agreements between States, and usually linked to situations of war.[30] In the *Nicaragua–Miskitos case*, on the other hand, the main question was the *internal* transfer of population within the boundaries of the State for economic reasons in time of peace.

In the similar *Cyprus* v. *Turkey case*, a forcible transfer of population due to military necessity can be found. In this case, the occupation by the Turkish Army of the northern part of the island provoked the flight of almost all the Greek-Cypriot population to the south of the island. Many of the allegations by the applicant Government of violation of human rights referred to the displacement on a massive scale of that population, the expulsion of some 200,000 Greek-Cypriots, and the refusal to allow them to return to their homes.[31] The Commission, in its analysis of the case, distinguishes between the movement of persons provoked by military action and that not directly connected with that action.[32] However, the Commission did not find it necessary to pronounce upon the imputability to Turkey, and therefore upon the lawfulness under the

[30] 'Les Transferts internationaux de populations (Rapport présentée par M. G. Balladore Pallieri)', 44 *Annuaire de l'Institut de droit International*, 44 (1952), 138–99.
[31] *Cyprus* v. *Turkey case* (1976), para. 89.
[32] Ibid., para. 91.

European Convention, of the displacement of persons provoked by military reasons in the first phase of the invasion of Northern Cyprus. Its finding rather refers to the continuing denial to that population, once the military operations had ended, of the possibility of returning to their homes. It stated that:

the prevention of the physical possibility of the return of Greek Cypriot refugees to their homes in the north of Cyprus amounts to an infringement, imputable to Turkey, of their right to respect for their homes as guaranteed in article 8(1) of the Convention.[33]

The Commission rightly considered that the infringement could not be justified on any of the grounds established in paragraph 2 of the same article.

4. THE DELIMITATION OF THE CONTENT OF THE EXPRESSION 'OTHER OBLIGATIONS UNDER INTERNATIONAL LAW'

The study of the legislative history and the application of the principle of consistency by the monitoring organs of the three treaties has offered little guidance on the possible substantive content and implications of the principle in emergencies. The task of ascertaining this is difficult; Professor Fawcett, commenting on this principle under the European Convention, pointed out that 'it is not easy to see in what circumstances this clause could be applied so as to invalidate a derogation of rights under article 15'.[34] However, in the Parliamentary Conference on Human Rights in Vienna in 1971, the International Institute of Human Rights in a written communication stated that

the Convention does not state what other obligations under international law are meant, nor does the Commission's case law provide an answer to the question. It might well be a question in the first place, in time of war or armed conflict not of an international nature, of Geneva law and in particular the Geneva Conventions of 1949. The fact that these 'other obligations under international law' are embodied in the law of the European Convention on Human Rights is indisputably an *enriching factor*, and for this reason it would be interesting to know their precise extent, particularly because, through these 'other obligations under international law' (and international law is founded on custom as well as treaty), the Commission and the Court will have to ensure that they are observed by the Contracting States.[35]

---

[33] Ibid., para. 208.

[34] Fawcett, *Application*, p. 312.

[35] Council of Europe, Consultative Assembly, Parliamentary Conference on Human Rights, Vienna, 18–20 Oct. 1971, written communication submitted by the International Institute of Human Rights (Strasbourg), p. 83.

In line with this position an attempt will be made in the rest of this chapter to identify possible international obligations related to human rights'which could be applicable through the principle of consistency.

In principle, international obligations can arise from general international law and from treaties. However, it is difficult to see how there could exist additional obligations of respecting human rights in states of emergency arising from customary international law over and above those established by the legal regime of the derogation clause. As is well known, the derogation clause in the three treaties establishes clear guidelines on the legal regime applicable in states of emergency, whereas so far there have been no clear norms of general international law on this point; what seems more likely is that some of the principles and norms established in the derogation clause, but not all, could be deemed to constitute emerging norms of customary international law. This will be considered further in the last part of the book. In any case, it is foreseeable that in the future the application of the principle of consistency could have more relevance, due to the growth both of customary international law and of general principles of law concerning human rights.[36]

On the other hand, some possible additional norms of customary international law in time of war may arise in fact from the standards set up in the 1949 Geneva Conventions and possibly from the two additional Protocols. However, these two Protocols of 1977 not only supplement and restate more clearly what could be said to constitute customary international law, but also develop it, and this is particularly so in the case of the Second Protocol. In view of the number of ratifications they have achieved so far, it can hardly be said that the whole set of rules of the Protocols are declaratory of customary international law.[37]

In any case, the fact remains that the most likely types of rules applicable through the principle of consistency are those derived from conventional law. In particular, the most important set of rules which refer to those human rights covered in the three main treaties under consideration are the laws of war. The implications of this set of rules for emergencies will now be examined.

(a) The Possible Additional Obligations Arising from the Laws of War

The most relevant of the laws of war, as far as human rights standards are concerned, are the Four Geneva Conventions of 1949 and the two Protocols. However, the principle of consistency could only take into account some of these additional standards in those cases in which the

---

[36] See the remarks of the ILA Paris Report (1984), p. 70 and Montreal Report (1982), p. 94.

[37] As of 30 Oct. 1990, 98 States were parties to Protocol I and 82 to Protocol II.

above-mentioned laws are applicable. The field of application of the Geneva Conventions is armed conflicts of an international character and, in the case of common article 3 and the second Protocol, armed conflicts of a non-international character. This means that a less grave public emergency, which does not amount to an armed conflict, does not justify the application of the Geneva Conventions, and therefore the additional obligations cannot arise. At the same time, some differences *ratione personae* in the application of the laws of war and the human rights treaties should be taken into account in order to check the possible applicability of the principle of consistency in each particular case.[38]

Although a detailed exposition of the whole system of the laws of war relating to human rights, in particular of the Geneva Conventions and Protocols, is outside the scope of this study, a brief discussion seems desirable. There is no doubt that there is a certain overlap between these two branches of international law in the protection of human beings in armed conflicts. The Geneva Conventions and Protocols tried specifically to protect all human beings affected by armed conflict, and especially those *hors de combat*, the prisoners-of-war and civilian population. The concrete rights that the Geneva Conventions and the Protocols give to these protected persons can be put under the heading of the right to humane treatment; in fact, the objective of humanitarian law is to safeguard the lives and dignity of human beings caught up in armed conflicts. The rights protected in the Geneva Conventions and Protocols overlap to some extent with those included in the human rights treaties under consideration. The Geneva Conventions contain specific provisions covering the four common non-derogable rights of the human rights treaties. There are two areas, however, in which the principle of consistency of the derogation clause could in principle be applied: the right to humane treatment and the set of rights relating to fair trial and due process of law.

(*i*) *Humane Treatment.* Human rights treaties contain a list of non-derogable rights which protect human beings *inter alia* from torture and inhuman or degrading treatment or punishment. The human rights treaties do not specify those standards which have been applied in particular cases by the monitoring organs. The Geneva Conventions, on the other hand, contain not only provisions forbidding certain treatment by States in respect of protected persons, but also very concrete and detailed rules regarding positive obligations of States, for instance, in respect of those wounded and sick. For example, common article 3 of the Four Geneva Conventions established that the sick and wounded 'shall be

---

[38] See *inter alia*. M. El Kouhene, *Les Garanties fondamentales de la personne en droit humanitaire et droits de l'homme* (The Hague, 1985), pp. 13 ff.

collected and cared for'. In order to fulfil this objective, the First and Second Geneva Conventions set up a legal regime granting protection e.g. to medical units and the personnel involved.[39] The Third Geneva Convention on the Treatment of Prisoners-of-War prohibits any 'moral or physical coercion' to secure from them information;[40] this provision in principle seems to go beyond what is called 'inhuman or degrading treatment' in the human rights treaties. At the same time, the Convention prohibits holding prisoners in cells without daylight,[41] or having them engaged in humiliating acts of labour.[42] It also contains a number of rules protecting the health, security, safety, and honour of the prisoners-of-war. The Fourth Geneva Convention establishes very detailed regulations for civilian internees, even to the extent of establishing provisions on food and clothing, hygiene and medical attention, and religious, intellectual and physical activities.[43]

Of course, human rights treaties do not contain all of these numerous and detailed obligations; therefore, it would be *ratione materie* outside the scope of the human rights treaties to try to implement them through the principle of consistency. However, some of the human rights treaties, e.g. the UN Covenant and the American Convention, also contain a positive right to 'humane treatment'.[44] Under the American Convention the right to humane treatment is a non-derogable right, with the result that no valid derogation can be made and consequently no application of the principle of consistency is needed. However, under the UN Covenant, the right to humane treatment (art. 10(1)) is not listed among the non-derogable rights, so in principle it could be derogated from. In these instances, the principle of consistency of the derogation clause could be deemed applicable, as it establishes as a condition for derogating the consistency with the State's other obligations under international law. Therefore, through the machinery of implementation of the Covenant, those provisions of the Geneva Conventions which specified the humane treatment due to protected persons could be applied, and could therefore nullify a prima-facie valid derogation under the Covenant.

In any case, this construction seems to be a little artificial and over-sophisticated. It seems more plausible that the monitoring organs of the treaties could take some of the provisions established in the Geneva Conventions as an indication of the humane treatment due to human beings in an armed conflict and therefore also applicable under the

---

[39] See ch. 3 of the First and Second 1949 Geneva Conventions.

[40] Art. 17 of Convention III; see also the same provision in art. 31 of Convention IV in respect of the civilian population.

[41] Art. 87 of Convention III.

[42] Art. 52 of Convention III.

[43] See arts. 89–96 of Convention IV.

[44] ICCPR, art. 10(1) and ACHR, art. 5.

human rights treaties. Thus, in his dissenting opinion in the *Cyprus* v. *Turkey case*, Mr Sperduti defended the reference to the Geneva Convention IV, in order to discover the acceptable standards of behaviour for States in occupied territories under the European Convention; in this case, the standards referred to were those corresponding to the right to respect for private life and family recognised in article 8 of the European Convention.[45]

(ii) *Fair Trial and Due Process of Law*. The other area in which the principle of consistency could enlarge the protection of human beings in armed conflict when applied to the derogation clause is in respect of the right to fair trial and due process of law of those charged with criminal offences. As we have seen above, in the human rights treaties, these rights in emergencies are in principle derogable. However, according to the Geneva Conventions and Protocols, these rights can never be suspended for any reason whatsoever; the violation of these provisions constitutes a grave breach of the Conventions.[46] The due process rights covered by Conventios III and VI and the two Protocols are almost the same as those covered by article 6 of the European Convention, article 14 of the UN Covenant, and article 8 of the American Convention.[47] So, at least in theory, even if the derogating State could find that the derogation of one of the rights to fair trial could be strictly required by the exigencies of the situation, the derogation clause, which forbid the taking of measures inconsistent with the State's other obligations under international law, would preclude the adoption of that measure.

(iii) *Safeguards for Internees in Convention IV*. Detention without trial of civilians in armed conflict is subjected to certain guarantees in Convention IV. It requires that this extreme measure be taken only in the case of absolute necessity and for the security of the State.[48] In the decision-making process, protected persons are entitled to 'a regular procedure', which includes the right of appeal with the least possible delay.[49] Furthermore, detention without trial, referred in the Convention as 'internment', should be submitted at least twice a year to a periodical review by a court or administrative board, 'with a view of favourable amendment if the initial circumstances permit'.[50] This court or administrative board must offer the necessary guarantees of independence and impartiality.[51] At the

---

[45] *Cyprus* v. *Turkey case* (1976), p. 564.
[46] Art. 130 of Convention III and art. 147 of Convention IV.
[47] See above, Ch. 4, for a detailed study of these provisions.
[48] Arts. 42(1) and 78(1) of Convention IV.
[49] Art. 78(2) of Convention IV.
[50] Arts. 43 and 78(2) of Convention IV.
[51] ICRC Commentary to Convention IV, p. 260.

same time, the Detaining Power should send the names of those detained and those released to the Protecting Power, and notify the decisions of the board according to the procedures described above. It is interesting to note the similarities of some of these guarantees for those in administrative detention with the ones formulated by the monitoring organs of the human rights treaties.[52]

Finally, many of these guarantees included in the Geneva Conventions and Protocols are prescribed for 'protected persons'. Therefore it might well be that when the human rights treaties are applied in conjunction with the laws of war, the operation of the principle of consistency will oblige the derogating State to abstain from taking measures inconsistent with the Geneva Conventions. This would be applicable only to 'protected persons' in areas like humane treatment, fair trial, and guarantees for those detained without trial. However, one of the main principles of international law in the area of human rights is that the measures taken should not be discriminatory. Therefore, by the operation of the principle of non-discrimination, the State should have to extend those additional protective measures to all persons under its jurisdiction. This will add another important safeguard to all human beings in armed conflicts.

Other possible obligations under international law. Other possible additional obligations applicable through the principle of consistency of the derogation clause could arise from other specific human rights treaties covering some of the rights contained in the three general treaties under consideration. In the case of treaties with derogation clauses, the possible operation of the principle of consistency should be examined in each case, taking into account the terms of the treaties involved; an illustration of how this principle operates will follow.

As is well known, the list of non-derogable rights under the American Convention is larger that the one included in the UN Covenant. Therefore, a State party to the Covenant and at the same time party to the American Convention will have its right of derogation further limited. In fact, according to the principle of consistency of the Covenant, States cannot take measures of derogation incompatible with their other obligations under international law; in this case, the other international obligation involved would be to respect those additional non-derogable rights recognized in the American Convention.

5. CONCLUSION

1. The principle of consistency in the derogation clause prohibits States from taking derogating measures which are inconsistent with other inter-

---

[52] See above, Ch. 4.

national obligations of the States, even if prima facie those measures are lawful under the three main treaties. The consequence of this principle of consistency is in theory very important, because it would imply to give some effect through the machinery of implementation of the three treaties, which is more effective than that of any other treaty, to other international obligations in the field of human rights. However, in order to apply the principle, a valid derogation from an obligation contained in these three treaties should take place; this condition will limit its application.

2. The legislative history of the principle and its application by the monitoring bodies under the three treaties offers little guidance in the ascertaining of the practical implications of the operation of the principle in emergencies. However, due to the growth of the international obligations of States in the field of human rights, not only through treaties, but also through customary norms and general principles, a greater application of this principle could be foreseen in the future. The body of law which now seems more relevant for the application of the principle is that of the laws of war, especially the 1949 Geneva Conventions and the two Protocols of 1977, in the areas of human treatment, fair trial, and due process of law, and the guarantees against arbitrary detention. However, for the application of the principle, due account must be taken of the field of application of these laws which does not always coincide with the concept of 'public emergency threatening the life of the nation' of the derogation clause.

PART II

Human Rights in States of Emergency in General International Law

Introduction

Once the question of the legal regime governing human rights in states of emergency according to treaty law has been dealt with, the next question that must be looked at is the identification of the principles governing human rights in emergencies according to general international law. This question is of great importance, first of all because there are some international treaties on human rights which have no derogation clauses and therefore do not explicitly indicate the legal regime applicable to them in situations of emergency. Two examples could illustrate this point, one being the African Charter and the other some ILO Conventions dealing with human rights issues.

The African Charter is the only regional treaty dealing with the whole area of civil and political rights which does not contain a derogation clause. Unfortunately, the *travaux préparatoires* of the African Charter have not yet been published, so the reasons behind the decision not to include a derogation clause in the Charter are not public. However, it seems that this decision not to include a derogation clause was due to the different positions on this issue and to the fact that there was not unanimous agreement on the need to adopt one. This lack of agreement could be explained by the very nature of the African Charter. In fact, the African Charter, despite having certain strong similarities to the Covenant and the European and American Conventions, none the less has notable differences from these Conventions. The Charter includes not only civil and political rights but also economic, social, and cultural rights, as well as some of the new 'third generation rights'.[1] As can easily be understood, the task of deciding which of the rights in the Charter could be deemed non-derogable in emergencies is a very difficult one, especially so because of the lack of precedents found in other treaties dealing with economic and social rights. It is not surprising therefore that, due to the complex nature of the African Charter, no agreement was reached over the inclusion of a derogation clause. Its absence, however, leaves the African Charter in an uncertain situation regarding derogations in emergencies.[2] Could this absence be interpreted in the sense that no

---

[1] For an analysis of the characteristics of the Charter, see E. Bello, 'The African Charter on Human and Peoples' Rights', *Recueil*, 194. 5 (1985), 9–268. See also Int. Commiss. of Jurists, *Human and Peoples' Rights in Africa and the African Charter*, Nairobi Conference, 1985 (Geneva, 1986), keynote address by Keba Mbaye. See also U. O. Umozurike, 'The African Charter on Human and Peoples' Rights', *AJIL* 77 (1983), 902–12.

[2] See Bello, 'African Charter', pp. 70–2.

right could be derogated from in emergencies? Or could the fact of this absence mean that all rights can be derogated from in emergencies?

The first position is too rigid, and no State could accept it in practice. It is obvious that in emergencies States need to take measures derogating from some of the standards which obtain in peacetime in order to overcome the threat to the nation. This logical position seems to have been adopted by some African States. Thus, Zimbabwe, which is party to the African Charter but not to the Covenant, declared a state of emergency on 22 July 1987 and took measures derogating from the right to trial without undue delay (African Charter, article 7 (*d*)(1)). The second position, according to which all rights can be derogated from, is also unacceptable, because it would lead to a widespread disregard for human rights in the Continent. Therefore this uncertain situation has to be resolved in the future by the new African Commission of Human Rights according to the principles of general international law. Similarly, several ILO Conventions dealing with human rights have no derogation clause; hence, a problem arises when States use the general defence of emergency as a justification for non-fulfilment of the treaty obligations. The ILO organs have to find some principles of general international law in order to solve this problem.

The task of finding principles governing human rights in emergencies in general international law is important for a second reason: because there are States which are not parties to any general treaty on human rights despite being members of various international organizations. In this case, the great problem of international monitoring organs when assessing their international obligations on human rights in emergencies outside treaty-law is to find general principles of customary law. For instance, the IACHR often has to face this problem because states of emergency have been a common phenomenon in Latin-America in recent times. However, the situation in which the IACHR found itself differed from that in which the ILO organs found themselves in the above-mentioned cases. The difference is that the ILO organs have to find general principles of law which could justify the non-performance of international obligations arising from treaties without derogation clauses; these international obligations are clearly defined in the ILO Conventions. However, the IACHR has to find general principles of law to apply to those State members of the OAS but not parties to a treaty (the American Convention); in this case, their international obligations (which are not clearly defined) arise from the OAS Charter and from the American Declaration of the Rights and Duties of Man (1948). However, at the time it was signed, the American Declaration, which is not a treaty, was not intended to create legal obligations for States; but with the passage of time, it has gained legal force.[3] In fact, when the New Statute of the Inter-American Commission was adopted by the Ninth Regular General Assembly in 1979, it specified that for States non-parties to the American Convention, human rights are to be understood as being those embodied in the American Declaration. The IACHR has in practice consistently applied those standards of the Declaration in its case-law. However, the American Declaration has no derogation clause; the Commission has therefore to find some general principles regulating this matter.

The jurisprudence of the IACHR in respect to those States non-parties to the American Convention could be of special interest for the UN system because of the similarity between the two systems. Thus, within the UN framework, there are States non-parties to the International Covenant on Civil and Political Rights but which are parties to the UN Charter. Therefore the international obligations concerning human rights for those States non-parties to the Covenant arise from the UN Charter and from the UDHR, as authoritative expression of the obligations of the Charter, in so far as they constitute general international law. At this stage in the evolution of international law, it is clear that customary law imposes international obligations in the area of human rights which are binding even for States which are non-parties of human rights treaties. However, once the content of the customary international law applicable to normal situations has been ascertained, the problem is to determine the principles regulating the human rights obligations of States in situations of emergency. For example, a serious legal analysis of the present situation in South Africa, a State which is non-party to any general treaty on human rights[4] but which is in a state of emergency, would require a clear understanding of the principles of general international law regarding human rights in states of emergency. The international community and the UN specialized bodies dealing with human rights issues need a legal doctrine to assess States' obligations in these situations. In fact, there is a certain hesitation within these organs when

---

[3] See Buergenthal and Norris (eds.), *The Inter-American System*, vol. i. For the position reaffirming the legally binding force of the Declaration, see Buergenthal *et al.* (eds.), *Protecting Human Rights in the Americas*, 3rd edn., pp. 39 ff. See also the interesting analysis on this point by a former President of the IA Court HR, P. Nikken, *La Proteccion internacional de los Derechos Humanos: Su desarrollo progresivo* (Madrid, 1987), pp. 261 ff. See also, IACHR, advisory opinion of 14 July 1989, 'Interpretation of the American Declaration of the Rights and Duties of Man within the Framework of Art. 64 of the ACHR'.

[4] Here, a general treaty is understood to be the opposite of a specialized treaty on a particular subject (e.g. on racial discrimination), a treaty such as the ICCPR which establishes a general regime of obligations on civil and political rights, with a derogation clause for emergencies. For the international treaty obligations of South Africa, see Marie, 'International Instruments', p. 139.

they have to assess States' obligations on human rights in emergencies according to general international law, precisely because of the lack of a clear understanding of the legal principles applicable.

The plan of this last part of the study, which aims to identify some principles governing human rights in states of emergency according to general international law, will be the following. First of all, two pre-liminary questions will be analysed; the first refers to the general issue of whether there exist obligations to respect human rights standards in customary international law; the second deals with the special types of evidence required to prove the existence of customary norms in the field of human rights. In fact, it has already been pointed out that just as special rules concerning reciprocity, breach, and interpretation of treaties often apply to human rights instruments, different types of evidence may be relevant to the creation of customary human rights law.[5] This con-tention needs careful examination.

In order to ascertain the existence of general principles governing human rights in emergencies, two main lines of inquiry will be followed. The first line of inquiry (which will be studied in Chapter 9) will look at the main doctrines in general international law which justify non-compliance with international obligations. In fact, once it has been estab-lished that there exist obligations to respect human rights standards in customary international law, the only way for States to justify their non-compliance with these obligations in emergencies would be to adduce some appropriate general doctrine recognized in international law. In the case of states of emergency, it seems that the doctrine which prima facie corresponds best with that situation is the *doctrine of State necessity*. Therefore, this doctrine and the main principles regulating its application call for examination. An illustration of the application of this general doctrine in cases relating to human rights in emergencies will be provided by the decisions of the ILO organs in the *Greek* and *Polish cases*.

The second line of inquiry (which will be studied in Chapter 10) will analyse the contention found in the literature and offered by some monitoring organs that some of the main principles of the derogation clause are emerging as principles of general international law. Two pre-liminary issues will be addressed. First, the contention that the derogation clause is a particular application and adaptation of the doctrine of necessity to the subject of human rights in emergencies. Second, the process by which treaty norms could become customary law; in this respect, the 'norm-creating character' of the ICCPR and of some of the principles of the derogation clause will be examined.

The main part of Chapter 10 will be dedicated to the study of the

---

[5] See e.g. Meron, *Human Rights and Humanitarian Norms*, p. 100.

concrete evidence concerning the emergence of some of the principles of the derogation clause as general international law. As it is perhaps the most important part of this evidence, it will be necessary to study in detail the application of these principles by international organs as a matter of general international law; special attention will be paid to the jurisprudence of the IACHR. Furthermore, special reference will be made to the work of the UN organs when dealing with this question, especially the work of the UN Special Rapporteur on States of Emergency, and that of the UN Commission on Human Rights.

Another interesting aspect that will be studied is whether some of the principles of the derogation clause have been deemed to be applicable to other areas of international law. If some of these principles are found to be applicable to other areas, this will underline their significance as principles of general international law. In particular, the area of state responsibility for injury to aliens (in the case of human rights of aliens in emergencies) and the law of belligerent occupation will be examined. Another piece of evidence will be explored, namely, the possible status of some of the principles of the derogation clause as 'general principles of law recognised by civilised nations', according to article 38(1)(c) of the Statute of the ICJ.

Finally, in the light of all the evidence collected through these two lines of inquiry, it will be asked which principles could be deemed to constitute emerging norms of customary international law on human rights in states of emergency.

8

Customary International Law and Human Rights: Two Preliminary Questions

Before dealing with the main question of this second part of the study, namely, the identification of the principles governing human rights in emergencies according to general international law, two preliminary questions are to be addressed. The first refers to the problem of human rights in customary law, and the second to the special types of evidence required to prove the existence of customary norms on human rights.

I. THE EXISTENCE OF HUMAN RIGHTS STANDARDS IN CUSTOMARY LAW

There is little doubt at this stage in the evolution of international law that States have legal obligations to respect human rights arising from general international law. The UN Charter and the work of the organization have greatly contributed to the generation of the customary international law of human rights. Through its different organs the UN has played an important role not only in the formulation of human rights standards through the adoption of multiple conventions, declarations, and re-solutions, but also in their implementation. As is well known, in articles 55 and 56 of the Charter all member States pledge themselves to take joint and separate action in co-operation with the UN for the achievement of universal respect for, and observance of, human rights and funda-mental freedoms. This language has been seen by the political and judicial organs of the UN as establishing legal obligations for States members.[1] However, agreeing upon the content of these obligations has proved to be problematic. The 1948 Universal Declaration of Human Rights contains a list of rights; but as a General Assembly resolution the instrument was not binding as such. None the less, it has been considered as an 'authoritative guide . . . to the interpretation of the provisions in the

---

[1] See *inter alia* E. Jimenez de Arechaga, 'International Law in the Past Third of a Century', *Recueil*, 159, 1 (1978), 172–7; Brownlie, *Principles*, pp. 570–1; G. T. Tunkin, 'General International Law in the International System', *Recueil*, 147. 4 (1975), 95 ff.; Tanaka, dissenting opinion, *South West Africa cases* (second phase), 1966, *ICS Report, 1966*, pp. 286 ff.

Charter',[2] and there is no doubt that some of its provisions can be deemed to constitute customary law or general principles of law.[3] The further approval in 1966 of the two Covenants also had the effect of rendering the definition of the rights listed in the Universal Declaration more exact.[4] However, the problem remains of ascertaining precisely which standards have already become customary law.[5]

The recent US Third Restatement of the Foreign Relations Law considers a State to be in violation of customary international law if, as a matter of state policy, it practises, encourages, or condones; genocide, slavery or slave trade, the murder or causing of the disappearance of individuals, torture or any other cruel, inhuman, or degrading treatment or punishment, prolonged arbitrary detention, systematic racial discrimination, or a consistent pattern of gross violations of internationally recognized human rights.[6] This list, which was presented in the Restatement as being cautious and conservative, has been criticized for being too restrictive. Meron, for example, believes that the due process guarantees, the right of self-determination, the right of detainees to humane treatment (ICCPR, art. 10), and the principle of non-retroactivity should be added to it.[7] Lillich considered the right recognized in articles 7 and 14 of the UDHR (the right to equality and non-discrimination) to be a customary human right, and further believed that the right to leave any country and to return to one's own country is a possible candidate.[8] Other writers have tentatively mentioned the principle of non-refoulement in the context of article 3 of the UN Convention against Torture.[9] At the same time, the IACHR held in a recent case that the prohibition of the execution of juveniles is emerging as a customary international norm.[10]

---

[2] Brownlie, *Principles*, pp. 570–1.

[3] Among the writers who consider that the UDHR has become customary law, see Waldock, *Human Rights in Contemporary International Law*, p. 15; J. P. Humphrey, 'The UDHR: Its History, Impact and Judicial Character', in Ramcharan (ed.), *Thirty Years after the UDHR*, pp. 21–40. McDougal holds the extreme view that the UDHR has become not only customary law but also norms of *ius cogens* ('Human Rights and World Public Order', p. 274).

[4] e.g. Tunkin thinks that many rules of the Covenants are general principles of law common to different legal systems, see 'General International Law', p. 102. See also N. Kaufman and S. Mosher. 'General Principles of Law and the UN Covenant on Civil and Political Rights', *ICLQ* 27 (1978), 596–613.

[5] For an interesting view of the problem of different opinions concerning which rights are the most important and possible candidates to become customary law, see Meron, 'On a Hierarchy of International Human Rights', *AJIL* 80 (1986), 1–23.

[6] Restatement of the Law Third, Restatement of the Foreign Relations Law of the USA (1988), §702, p. 161 (hereinafter 'US Restatement').

[7] Meron, *Human Rights and Humanitarian Norms*, pp. 95–6.

[8] R. Lillich, 'Civil Rights', in Meron (ed.), *Human Rights in International Law*, p. 151.

[9] See e.g. Professor Kooijmans, the UN Special Rapporteur on Torture, UN doc. E/CN. 4/1987/13, para. 10 (quoted in Meron, *Human Rights and Humanitarian Norms*, p. 97).

[10] IACHR. Resolution No. 3/87 (27 Mar. 1987), Case No. 9647 (US), p. 38, para. 60.

Despite the fact that there is no agreement on its contents, it seems to be clear that the list of customary rights cannot be a closed one but rather has a dynamic and flexible character, and that soon several candidates may also become customary norms. What is needed is a careful and detailed analysis of each particular right in order to ascertain whether it has been accepted by the practice of States and fulfils the requirement of *opinio iuris*.

However, the main object of our inquiry is not the elaboration of a list of those rights which are norms of customary international law, but to ascertain the principles regulating human rights in emergencies. In fact, it may well be that some of the rights contained in the list of customary rights could be derogated from according to general principles applicable in emergencies. Thus, even though most of the rights in the list of the US Restatement mentioned above are also non-derogable in states of emergency, at least two of them present some problems. First, prolonged detention is not considered to be arbitrary and is therefore legitimate in emergencies when it fulfils the main conditions of the derogation clause. Secondly, in a situation of emergency, a State may also need to derogate from some of the 'internationally recognised human rights'.

Therefore, before establishing any concrete list of rights which could not be derogated from in emergencies, it is important to find out what the main principles regulating human rights in states of emergency according to general international law actually are.

2. THE SPECIAL EVIDENCE REQUIRED TO PROVE THE EXISTENCE OF CUSTOMARY NORMS IN THE AREA OF HUMAN RIGHTS

Before examining the existing evidence concerning the principles governing human rights in emergencies in general international law, it seems pertinent to add some remarks on the special type of evidence used in this field of human rights to prove the existence of international custom. Of course, this is not a plea for a special status for human rights in the process of formation of customary rules, nor a defence of the idea that human rights should follow different rules from those of other areas of international law. Human rights is a branch of international law and therefore subject to its general rules, which, in the case of the formation of customary law, require the existence of the two traditional elements: a general practice and *opinio iuris*. However, as Meron has pointed out,

Just as special rules concerning reciprocity, breach and interpretation of treaties often apply to human rights instruments, different types of evidence may be relevant to the creation of customary human rights law.[11]

[11] Meron. *Human Rights and Humanitarian Norms*, p. 100.

This same position was adopted by the US Restatement when it noticed that 'the practice of States that is accepted as building customary international law of human rights included some forms of conduct different from that building customary international law generally'.[12] Schachter, who held the same view, explains the reasons for this as follows:

States do not usually make claims on other states or protest violations that do not affect their nationals. In that sense, one can find scant state practice accompanied by opinio juris. Arbitral awards and international judicial decisions are also rare except in tribunals based on treaties such as the European and Inter-American Courts of human rights. The arguments advanced in support of a finding that rights are a part of customary law rely on different kinds of evidence.[13]

Meron finds in the types of evidence mentioned in the US Restatement, which are backed by the ICJ (especially in the *Nicaragua case*[14]) and in some judicial decisions of national courts, 'an appropriate adaptation to human rights of accepted methods of building customary law'.[15]

The Restatement enumerates the following types of practice as building customary law in this area:

virtually universal adherence to the UN Charter and its human rights provisions, and virtually universal and frequently reiterated acceptance of the Universal Declaration even if only in principle; virtually universal participation of states in the preparation and adoption of international agreements recognizing human rights principles generally or particular rights; the adoption of human rights principles by states in regional organizations in Europe, Latin America, and Africa ...; general support by states for UN resolutions declaring, recognizing, invoking, and applying international human rights principles as international law; action by states to conform their national law or practice to standards or principles declared by international bodies, and the incorporation of human rights provisions, directly or by reference, in national constitutions and laws; invocations of human rights principles in national policy, in diplomatic practice, in international organisations, activities and actions; and other diplomatic communications or action by states reflecting the view that certain practices violate international human rights law, including condemnation and other adverse state reactions to violations by other states.[16]

The recent decision of the ICJ in the *Nicaragua case* supports this approach taken by the US Restatement on the evidence necessary to prove the existence of customary law. According to this view, the Court focused not so much on State practice,

---

[12] US Restatement, vol. ii, §701, p. 154 n. 2.

[13] Schachter, 'General Course', p. 440.

[14] 'Military and Paramilitary Activities in and against Nicaragua', (*Nic.* v. *US*), Merits, *ICJ Reports*, 1986 (Judgment of 27 June).

[15] Meron, *Human Rights and Humanitarian Norms*, p. 101.

[16] US Restatement, vol. ii, §702, p. 154 n. 2. Schachter also refers to the same type of evidence for building customary international law on human rights, 'General Course', pp. 332–8.

but on opinio iuris found in verbal statements by governmental representatives to international organizations, the content of resolutions, declarations and other normative instruments adopted by such organizations, and the consent of States to such instruments . . . . The Nicaragua judgment simultaneously strengthens the law-making force of General Assembly Resolutions and de-emphasizes the importance of practice as one of the two elements necessary for the formation of customary international law.[17]

The position of the Court in the *Nicaragua case* was not completely new and found important antecedents in earlier judgments, especially those involving human rights issues.[18] One consequence of this position seems to be that the burden of proof in establishing customary norms on human rights would be less onerous than in other fields of international law.

Grounds for this position are found in the nature and goals of human rights, which give expression of community values and focus on the protection of individuals rather than on the reciprocal interests of States.[19] Referring to the codification of humanitarian norms, Meron has pointed out that:

the law-making process does not merely 'photograph' or declare the current state of international practice. Far from it. Rather, the law-making process attempts to articulate and emphasize norms and values that, in the judgment of some state, deserve promotion and acceptance by all states so as to establish a code for the better conduct of nations. This applies in particular to instruments designed to humanize the behaviour of states in armed conflicts which is characterized by violence and violations . . .[20]

This position applies in a very appropriate way to the principles regulating human rights in situations of war or grave public emergency. These types of evidence mentioned by the US Restatement and other writers will be referred to in order to show the emergence of some principles relating to human rights in emergencies in the context of general international law. Special importance will be given to the practice of judicial, quasi-judicial, and monitoring bodies in the application of general principles, and the acquiescence in those standards by States non-parties to the treaties.

In order to ascertain the existence of general principles governing human rights in emergencies, two main lines of inquiry will be followed. The first line of inquiry will look at the main doctrines in general international law which justify non-compliance with international obligations. In the case of states of emergency, it seems that the doctrine

---

[17] Meron, *Human Rights and Humanitarian Norms*, p. 107.
[18] Ibid., pp. 106–14.
[19] Ibid., p. 131.
[20] T. Meron, 'The Geneva Conventions as Customary Law', *AJIL* 81 (1987), 361 and 363.

which prima facie corresponds best with that situation is the doctrine of State necessity; therefore, this doctrine and the main principles regulating its application will be carefully examined. The second line of inquiry will analyse the contention found in the literature and offered by some monitoring organs that some of the main principles of the derogation clause are emerging as principles of general international law.

9

First Line of Inquiry: Human Rights in Emergencies: The Doctrine of State Necessity

1. LEGAL DOCTRINES IN INTERNATIONAL LAW WHICH JUSTIFY THE NON-COMPLIANCE WITH INTERNATIONAL OBLIGATIONS: CASES OF *FORCE MAJEURE*, SELF-DEFENCE, AND NECESSITY

Once it has been established that States have human rights obligations arising from customary international law, the question that should be dealt with is that of finding some doctrine which could justify the non-compliance with these obligations in situations of emergency. Within the framework of general international law, it seems that the most suitable doctrine on which States could base their plea for non-compliance with international obligations in emergencies would be that of the excuses recognized in the law of State responsibility. In other words, the only way in which States can justify their non-compliance with customary norms on human rights in emergencies would be by pleading *force majeure*, self-defence, or state of necessity. These excuses have been recognized by the ILC in the body of principles of the law of State responsibility. The relation of these doctrines with states of emergency needs examination.

As was pointed out in Chapter 1, states of emergency could arise in several different situations, for instance, international war, internal armed conflicts or grave unrest, natural disasters, etc. Broadly speaking, one could say that in principle the situation of international war corresponds best with the doctrine of self-defence, since the State has to take derogating measures to defend itself against an external attack; that the situation of grave internal unrest corresponds best with the doctrine of necessity; and that natural disasters correspond best with *force majeure*. However, two remarks should be made on this point: first, these characterizations should not be exaggerated, because very often the distinctions between the three excuses are not clear.[1] Second, in any case, it is the doctrine of necessity which seems to correspond best with the most common forms of emergencies. This is for the following reasons.

---

[1] This fact is very well illustrated in the ILO doctrine in the *Greek case*, see below, pp. 223 ff.

As far as the doctrine of self-defence is concerned, the measures derogating from human rights standards in states of emergency are not exclusively directed against those who could be called 'unjust aggressors', either external or internal, but rather affect the whole bulk of the population living under the jurisdiction of the State. Therefore, one of the main characteristics of self-defence, that the reaction to the attack be directed against the aggressor, is not always present in states of emergency. As far as the doctrine of *force majeure* is concerned, two points are relevant. First, this doctrine normally applies only in cases of emergencies brought about by natural disasters; and such states of emergency are not the most common.[2] Secondly, *force majeure* has been described as 'an external and irresistible force which operates independently of the will of the agent'.[3] In other words, *force majeure* does not leave any room for a choice of means in dealing with the emergency. However, in a public emergency, States usually deliberate very carefully about the course of action they should take in order to restore public order; they take great care to balance the extent of the suspension of individual rights against the gravity of the threat. This fundamental characteristic is present in the doctrine of necessity;[4] and this is the reason why it is this latter doctrine which seems to correspond best with emergencies.

2. THE DOCTRINE OF STATE NECESSITY

(a) *Characteristics*

The doctrine of State necessity has been incorporated into the body of principles of the law of State responsibility as one of the legal justifications excluding responsibility for a breach of an international obligation.[5] The doctrine of necessity has been considered by the ILC to be 'deeply rooted in the general theory of law'.[6] Due to the abuses that recurrence to this doctrine by States has brought about, 'the excuse of necessity may con-

---

[2] In fact, in the long history of the case-law of the European Convention, a public emergency due to natural disaster has never arisen. See S. Marks, 'Principles and Norms of Human Rights Applicable in Emergency Situations: Underdevelopment, Catastrophes and Armed Conflicts', in K. Vasak (ed.), *International Dimensions of Human Rights*, (2 vols.; Westport, Conn., 1982), i. 175–213.

[3] Jimenez de Arechaga, 'International Responsibility', in Sorensen (ed.), *Manual of Public International Law* (1968), p. 544.

[4] See Report by Ago on State Responsibility, R. Ago, 'The International Wrongful Art of the State, Source of International Responsibility', art. 33, State of Necessity, *YBILC* 2. 1 (1980), 14.

[5] See ILC, Report of the Commission to the General Assembly on the work of its 32nd Session, *YBILC*, 2.2 (1980), 33 ff. (art. 33).

[6] ILC A/35/40, p. 104.

ceivably be accepted in international law only on condition that it is absolutely of an exceptional nature';[7] this is why it must be subjected to very strict conditions.

The doctrine of necessity has been described as 'a factual situation in which a State asserts the existence of an interest of such vital importance to it that the obligation it may have to respect a specific subjective right of another State must yield because respecting it would . . . be incompatible with safeguarding the interest in question'.[8] Ergec, on the other hand, thinks that

l'état de nécessité constitue la cause de justification d'un acte commis délibérément en violation d'une règle impérative en vue de sauvegarder, en présence d'un danger grave et imminent, une valeur manifestement supérieure à celle protégée par cette règle.[9]

The doctrine of necessity in its classic context therefore involves a conflict between the interest of State A to safeguard one of its essential interests, and the interest of State B to have its right respected. The interest of State A prevails because of the greater importance of this interest in a particular circumstance. The law recognizes this pre-eminence, and consequently State A chooses deliberately to breach that international obligation in respect to State B in order to safeguard an essential interest.

(b)   Conditions of Application

Among the 'particularly strict conditions' that have to be met in order to claim the application of this doctrine, Ago has indicated the following:

the interest threatened must be one of exceptional importance;
the threat to that essential interest must be extremely grave and actual or imminent;
the adoption of conduct not in conformity with an international obligation must be the only means available to avert the extremely grave danger;
the action must be strictly necessary for that purpose; any excessive action will be *ipso facto* a wrongful act.[10]

---

[7] 'Ago Report', para. 12.

[8] Ibid., para. 10.

[9] Ergec, *Les Droits de l'homme*, p. 45.

[10] 'Ago Report', pp. 19–20. For a good exposition of the doctrine of necessity in classic international law, see P. A. Pillitu, *Lo stato di necessità nel diritto internazionale* (Perugia, 1981), pp. 1–180. For recent works see also *inter alia* J. Barboza, 'Necessity (Revisited) in International Law', in *Essays in Honour of Judge Lachs* (1984), pp. 27–43; S. P. Jagota, 'State Responsibility: Circumstances Precluding Wrongfulness', *NYIL* 16 (1985), 266–71; J. A. Salmon, 'Faut-il codifier l'état de nécessité en droit international?', in *Essays in Honour of Judge Lachs*, pp. 236–70.

On close inspection, derogations from human rights standards in emergencies can be seen to have several of these characteristics. There are two conflicting interests: the interest that other States have in the derogating State respecting fundamental human rights obligations (and one can add the interest of those under the jurisdiction of the State to have their human rights respected), and the interest of the derogating State in safeguarding the life of the nation, or rather, in safeguarding the rights of the whole society.

The strict conditions for the application of the doctrine of necessity can be expressed in the language of some of the principles contained in the derogation clause of the treaties. Thus, the threat to the life of the nation must be extremely grave (the principle of exceptional threat); and the suspension of human rights obligations must be used only as a last resort once all other legitimate means have been exhausted. At the same time, derogating measures must be proportionate to the threat without exceeding it (the principle of proportionality); finally, they must also be terminated once the threat has ended (the principle of temporariness). Moreover, the ILC has established that a State cannot invoke the plea of necessity when 'the international obligation with which the act of the State is not in conformity arises out of a peremptory norm of general international law'.[11] Interestingly enough, the principle of non-derogability of fundamental rights seems to reflect a similar idea.

Another characteristic of the strict conditions under which the doctrine of necessity applies is what can be called the 'international accountability' of those States which plead necessity. In fact, States pleading necessity have to prove, if required to do so, that the conditions necessary for the application of that doctrine have been met. The State is not the sole judge in determining the existence of the danger threatening one of its essential interests, and it has to justify the need for such conduct in breach of an international obligation, i.e. the fulfilment of the principle of proportionality.[12]

As an illustration of the application of the doctrine of state necessity to human rights obligations in emergencies, the decisions of the ILO organs will be analysed. The two most important cases involving the identification of principles applicable to human rights in emergencies were the *Greek case* in 1971 and the *Polish case* in 1984.

(c)   Decisions of the ILO Organs: The Greek and Polish cases

As a consequence of the military coup, and in relation to ILO Conventions No. 87 on the Freedom of Association and Protection of the

[11] ILC, Report of the ILC to the General Assembly, p. 34 (draft art. 33(2) (*a*)).
[12] ILC, Report to the General Assembly, p. 50.

Right to Organize (1948) and No. 98 on the Right to Organize and Collective Bargaining (1949), Greece contended in 1967, *inter alia*, that 'it was relieved from the obligation of compliance with the Conventions by reason of the state of emergency which led to the proclamation of the siege law in April 1967'.[13] Greece pointed out that 'state of emergency' was as familiar a concept in public law as that of *force majeure* in private law, and therefore that according to its constitution, it was entitled not to comply with those obligations. The ILO Special Commission, appointed under article 26 of the ILO Constitution to examine the complaints, recognized that 'in neither of these Conventions is there any provision allowing the possibility of basing a plea of emergency, as an exception to the obligations arising under the Conventions, on the terms thereof'.[14] The Commission took the view that, according to an accepted principle of international law, a State cannot rely on the terms of its own national laws to justify a non-performance of an international obligation; and it added: 'any doubt concerning the extent of such obligation must be determined by exclusive reference to the relevant principles of international law, whether made express by the parties to a treaty or derived from another source of international law, in particular, international custom and general principles of law'.[15] Applying international custom and general principles of law, the Commission said:

The position of pleas of emergency or necessity in international custom may be said to correspond essentially, within the peculiar framework of the international community, to the place given to pleas of *force majeure* or legitimate self-defence in national systems of law. A plea of *force majeure* generally requires a showing of irresistible force of circumstances. A plea of legitimate self-defence requires a showing both of imminent danger and of a proportionate relationship between the danger and the measures adopted for defence. Both the general principle of law derived from national practice and international custom are based on the assumption that the non-performance of a legal duty can be justified only where there is impossibility of proceeding by any other method than the one contrary to law. It must also be shown that the action sought to be justified under the plea is limited, both in extent and in time, to what is immediately necessary.[16]

Some important conclusions can be drawn from this finding:

1. The ILO Special Commission identifies the plea of emergency with that of necessity in general international law.

2. The Commission links the doctrines of necessity, *force majeure*,

---

[13] ILO, *Report of the Commission Appointed under Article 26 of the Constitution of the ILO to Examine the Complaints concerning the Observance by Greece to the Freedom of Association Conventions*, *ILO Official Bulletin*, 54, Special Supplement, No. 2 (1971).

[14] Ibid., p. 25, para. 109.

[15] Ibid., p. 25, para. 108.

[16] Ibid., p. 26, para. 110.

and self-defence. Although these doctrines have different characteristics, some of the main principles regulating their application are similar.

3. These principles are spelled out by the Commission in its application to the case of emergency in human rights treaties:

a plea for a non-compliance with the Conventions is only justified in the case of an exceptional situation (imminent danger);

the non-compliance with the Convention related to the emergency must be temporary (limited in time);

non-compliance must also be used only as a last resort, once other means have failed;

there must be a proportionality between the danger and the measures adopted (limited to what is immediately necessary);

the State pleading the existence of an emergency has the burden of proof, and the international monitoring organs have the competence to make an independent determination of the existence of the conditions for the application of the plea; in other words, the State is not the sole judge.[17]

Interestingly enough, these principles, which the Commission has drawn from international custom and general principles of law, are similar to some of the basic principles implicit in the derogation clause, namely, the principles of exceptional threat, temporariness, proportionality, and international accountability of the State pleading non-compliance with international obligations due to emergency.

The doctrine established by the Commission in the *Greek case* was reaffirmed with even greater conviction in the *Polish case*, a case which has similar characteristics to the former one. The case arose as a consequence of the martial law established in Poland in 1981, and one of the central issues, as in the *Greek case*, was the country's international obligations relating to Conventions 87 and 98. The Commission stated:

the plea of emergency to justify restrictions on the civil liberties that are essential to the proper exercise of trade union rights can be advanced before an international authority only in circumstances of extreme gravity constituting a case of *force majeure* and provided that any measures affecting the application of the Convention are limited in scope and in duration to what is strictly necessary to deal with the situation in question.[18]

---

[17] The Commission rejected the plea of emergency by Greece because the exceptional character of the situation was not established by the Greek Government, ibid., p. 26, para. 112.

[18] ILO, *Report of the Commission Instituted under Article 26 of the Constitution of the ILO to Examine the Complaint on the Observance by Poland of the Freedom of Association and Protection of the Right to Organise Convention, 1948 (No. 96), Presented by Delegates at the 68th Session of the International Labour Conference*, ILO *Official Bulletin*, 57 (1984), Special Supplement, Ser. B, pp. 126–7.

As in the Greek case, the principles applied by the Commission were the two fundamental principles of exceptional threat (justification for restrictions can only arise in emergencies of extreme gravity), and the principle of proportionality (the extent of the restrictions should be limited to what is strictly required by the exigencies of the situation); moreover, the Commission also refers to the burden of proof by the State pleading the emergency and to the competence of the supervisory bodies to examine whether all the conditions have been met.

The position of the ILO Special Commissions in the *Greek* and *Polish cases* finds confirmation in other cases under the supervision of ILO organs.[19] Thus, in a case relating to Nicaragua, the Committee on the Freedom of Association insisted on the application of the principle of proportionality and of another principle well established by human rights bodies, that is, that the state of emergency should be limited to a certain geographical area, the one affected by the armed conflict.[20]

Another interesting case is the one relating to the Forced Labour Conventions of 1930 (No. 29) and of 1957 (No. 105).[21] The first of these Conventions does not include in its scope labour exacted in certain specific cases of emergency which it goes on to define. However, it does not say anything about the legal regime regulating that compulsory work or service. In order to fill the gap, the Committee of Experts of the Application of Conventions and Recommendations has applied three principles: (1) the principle of proportionality ('the duration and extent of compulsory services, as well as the purpose for which it is used, should be limited to what is strictly required by the exigencies of the situation'); (2) the principle of non-discrimination (general ILO standards of non-discrimination should also be applied), and (3) the principle of the competence of ILO organs to carry out an appraisal of these conditions.[22] These very same principles have also been applied by the ILO organs in respect of the second Convention mentioned above, that of the Abolition of Forced Labour of 1953, which has no derogation clause.[23]

---

[19] See the case-law concerning the Convention on Freedom of Association contained in ILO, *Freedom of Association: Digest of Decisions* (1976), p. 124 ff. (there is an acceptance of human rights restrictions in emergencies from normal standards, provided that they are subjected to the principles of temporariness, exceptional threat, and proportionality).

[20] ILO, Nicaragua Cases No. 1129 and 1351, ILO *Official Bulletin*, 70 (1987), 138.

[21] See Brownlie, *Basic Documents on Human Rights*, pp. 176–89.

[22] ILO, 'Comparative Analysis of the UN International Covenants on Human Rights and the ILO Conventions and Recommendations', in ILO *Official Bulletin* 52 (1969), 206.

[23] Ibid.

Second Line of Inquiry: The Emergence of Some of the Principles of the Derogation Clause as Principles of General International Law

Having shown that the doctrine of necessity is the doctrine which best corresponds with the plea of emergency in general international law, and having analysed how it has been applied by the ILO organs in certain cases relating to human rights in emergencies, we may move on to examine another approach to this question. This second approach mantains that the legal regime of the derogation clause of the main treaties already studied is basically a particular application of the doctrine of necessity to the subject of human rights in emergencies. Moreover, this approach also contends that the main principles of the derogation clause are emerging as principles of general international law. This approach, which has support in the doctrine and in the practice of international monitoring organs, requires close examination.

1. THE LEGAL NATURE OF THE DEROGATION CLAUSE

It is true that the legal nature of the derogation clause has not received much attention by the doctrine.[1] Nevertheless, it has been qualified in several different ways; as a clause of suspension,[2] as a clause *rebus sic stantibus*,[3] and as a clause reflecting the principle of 'supervening impossibility of performance' according to article 61 of the Vienna Convention on the Law of Treaties[4]. It has also been argued that the derogation clause has some relation with the doctrines of self-defence, *force*

---

[1] The only serious attempt to analyse its nature has been carried out by R. Ergec, *Les Droits de l'homme a l'épreuve des circonstances exceptionnelles* (Brussels, 1987), pp. 39 ff.

[2] Capotorti, 'L'Extinction et la suspension des traités', in *Recueil*, 134 (1971), 488.

[3] A. Kiss, 'Les Fonctions du secretaire-general du Conseil de l'Europe comme depositaire des conventions européennes', *AFDI* 2 (1956), p. 685. Also Mr Ermacora, in the *Cyprus case*, p. 179.

[4] E. Suy, 'Droits des traités et droits de l'homme', in *Völkerrecht als Rechtsordnung Internationale Gerichtsbarkeit Menschenrechte*, Festschrift for H. Mösler (Berlin, 1983), p. 944.

majeure and necessity. However, the theory which sees in the derogation clause a particular application of the doctrine of necessity is perhaps the one which enjoys the greatest support. Thus, Ago sees in the derogation clause of the Covenant a particular adaptation of the doctrine of necessity; this adaptation has been carried out taking into account the particular subject-matter and the nature of the legal obligations involved.[5] For his part, Ergec has seen the derogation clause as the concrete application, and adaptation, of the doctrine of necessity to the specific subject of human rights in emergencies.[6] Meanwhile, Castberg has pointed out that the drafters of the Covenant, in order to prevent arbitrary derogations taking place through the application of the dangerous doctrine of necessity in general international law, decided to establish an appropriate legal regime for those situations through the derogation clause. They did this because they realized that States may inevitably find themselves in situations in which it is impossible for them to comply with all the human rights obligations.[7] The International Law Association has also noticed that in drafting the Covenant the point of view which ultimately won through was that the doctrine of necessity permitting derogations from some rights had to be reflected in the major human rights treaties.[8] As far as the European Convention is concerned, Ganshof van der Meersch sees in the derogation clause the transposition to the international arena of the doctrine of necessity which is found in public law and which allows the suspension of human rights in emergencies.[9]

As has been pointed out before, the derogation clause can be seen to have adopted several of the main principles of the doctrine of necessity. In the first part of this study, seven main principles constituting the legal regime of the derogation clause have been analysed. These principles can be classified in three categories.

2. THREE TYPES OF PRINCIPLES

1. There are some principles which are a clear reflection of the principles of the doctrine of necessity, namely, the principle of exceptional threat, the principle of proportionality, and the principle of temporariness. Moreover, the principle of non-derogability of some fundamental rights can be deemed to constitute an application of the doctrine of necessity as formulated by the ILC. The ILC says that a State cannot

---

[5] 'Ago Report' p. 45.
[6] Ergec, *Les Droits de l'homme*, p. 53.
[7] F. Castberg, *The European Convention on Human Rights* (Leiden, 1974), p. 165.
[8] ILA Warsaw Report (1988), p. 40.
[9] W. J. Ganshof van der Meersch, *Organisations européennes* (Brussels, 1966), p. 286.

invoke the plea of necessity when 'the international obligation with which the act of the State is not in conformity arises out of a peremptory norm of general international law.'[10] In so far as the principle of non-derogability of fundamental rights refers to the four common non-derogable rights which are recognized as norms of *ius cogens*, it could be said that States cannot justify their non-compliance with these obligations using the plea of emergency.[11]

2. The derogation clause also contains at least one principle which has great importance in the human rights field, and which does not explicitly appear among the principles of the doctrine of necessity; this principle is the principle of non-discrimination.[12]

3. Finally, the derogation clause contains two principles which can be seen to be of a 'procedural' nature and which seem in principle to be more suitable for application within the framework of treaty law than in general international law. These principles are the principles of proclamation and notification. Before analysing which of the principles of the derogation clause seem to be emerging as principles of general international law, a few remarks should be made concerning the process by which treaty provisions could become customary norms.

3. THE DEROGATION CLAUSE AND CUSTOMARY INTERNATIONAL LAW: THE 'NORM-CREATING CHARACTER' OF THE ICCPR AND THE DEROGATION CLAUSE

The recent importance given to the process by which rules of law formulated in treaties could become rules of customary international law has been recognized by the ICJ. As has been pointed out,

the reason for this is that in the last 30 years the international community has been engaged, under the auspices of the UN, in the task of codification and progressive development of international law of an unprecedented scale.[13]

This has been done through general multilateral conventions covering whole branches of international law adopted in international conferences with the participation of a great number of States from all geographical

---

[10] ILC, Report to the General Assembly, *YBILC* 2.2 (1980), 34 (draft art. 33(2) (*a*)).

[11] See the interesting observation by Meron on the fact that 'peremptory norms' in general international law may possibly have a wider content than the concept of *ius cogens* in treaty law. Meron, *Human Rights and Humanitarian Norms*, pp. 220–2.

[12] However, one can say that this principle is implicit in the concept of proportionality and the purposive elements in necessity (see above, p. 169).

[13] Jimenez de Arechaga, 'International Law', p. 13.

areas. Jimenez de Arechaga, talking about the process of formation of customary international law from a treaty norm as found by the ICJ in recent times, described three modalities: the declaratory, the crystalliz-ing, and the generating effect. In the first case, the treaty norm is no more than the formal and written expression of a pre-existing rule of customary law already in force. In the second case, a rule which was emerging, or in *statu nascendi*, crystallizes as a customary rule and receives its first written expression through the dynamic of the international con-ference which concludes the treaty. Finally, in the third case, a treaty norm, not being declaratory of an existing norm or the codification of an emerging rule, may become a rule of customary international law through the subsequent practice of States.[14]

Although it is difficult in some cases to know if a treaty norm has codified or developed international law, it seems that prima facie the case of the derogation clause would better be described as a case in which, as a consequence of subsequent practice, the treaty norms (in this case some of the principles of article 4 of the ICCPR), are emerging as norms of customary law;[15] so the treaty norms would therefore have the 'generat-ing effect' of creating a customary rule. Sorensen, commenting on this generating process, says

The Convention may serve as an authoritative guide for the practice of States faced with the relevant new legal problems, and its provisions thus become the nucleus around which a new set of generally legal rules may crystallize.[16]

Following the reasoning of the Court in the *North Sea Continental Shelf cases*, the first requirement for making this process possible is that the provision invoked would have a 'norm-creating character'.[17] It seems clear that the ICCPR and some of the principles of the derogation clause in particular have this character.

The norm-creating character of the ICCPR. As is well-known, what the drafters of the ICCPR tried to do was to define with greater precision, in the context of a treaty, the international obligations of States in the field of human rights, obligations which were not defined in the UN Charter and which were couched in very general terms in the UDHR. The drafters of the Covenant tried for the first time to regulate human rights obligations in the first international instrument of a universal character

---

[14] Ibid., pp. 14–15.

[15] The derogation clause of the Covenant is singled out here because the ICCPR is a quasi-universal treaty, whereas the European and American Conventions are regional treaties.

[16] *North Sea Continental Shelf cases*, *ICJ Reports*, 1969, dissenting opinion, p. 244.

[17] Ibid., pp. 41–42, para. 72. See also Jimenez de Arechaga, 'International Law,' p. 22.

dealing with general obligations relating to civil and political rights.[18] The fact that the treaty has a law-making character seems therefore self-evident. Moreover, the ICCPR was the result of a long drafting process (from 1947 to 1966) carried out under the direction of the UN Human Rights Commission (1947–55), later discussed by the UN General Assembly, open to the participation of all States,[19] and finally adopted by the General Assembly in 1966 almost unanimously.[20] Up until 1 January 1991, it had been ratified by ninety-two States from every geographical area; this fact underlines the universal character of the treaty. The UN General Assembly has constantly recommended that those States which have not yet done so should ratify the Covenant.

The norm-creating character of the derogation clause. The fact that the derogation clause contained in article 4 of the Covenant at least in part has a norm-creating character also seems to be obvious. As has been pointed out in the detailed study of the derogation clause, its main aim was to create a legal regime regulating human rights in states of emergency. This was a new and a very important matter, because it is precisely in emergencies that the protection of human rights is most needed. The derogation clause was the result of a long drafting process in which a balance was carefully worked out between the interest of the State to defend the life of the nation when gravely threatened and the respect for individual human rights; the result has been seen to be a success.[21] Thus, the derogation clause has been considered to be a key article of the Covenant.[22]

Moreover, the derogation clause has not been the object of substantial reservations by States; these reservations could in theory have undermined its main principles. (The only substantial reservation was made by

---

[18] The European Convention which was signed in 1950 and which has a similar content to the ICCPR was a regional treaty. Other human rights treaties concluded under the auspices of the UN, e.g. the Genocide Convention of 1948 and the Slavery Convention of 1956, are specialized treaties covering one single right or aspect; the ICCPR, on the other hand, is a general treaty in the sense that it tries to regulate the whole field of civil and political rights.

[19] The US *Restatement* (p. 154) accepts the virtually universal participation of States in the preparation of international agreements recognizing human rights principles as evidence of practice building customary law. This position requires some caution. In fact it is possible to imagine a case in which a great number of States participate in the preparation of a treaty which is not then ratified by the majority of States. However, this did not happen in the case of the ICCPR.

[20] The ICCPR was adopted and open to signature on 16 Dec. 1966 by GA Resolution 2200A (XXI) with 106 in favour, and with no votes against or abstentions. For an account of the process of drafting the Covenant, see V. Pachota, 'The Development of the ICCPR,' in Henkin (ed.), *The International Bill of Rights* (New York, 1981), p. 37.

[21] Higgins, 'Derogations', p. 319.

[22] The derogation clause was described by a member of the UN HR Committee as 'the cornerstone of the human rights treaties' (see above, p. 1).

Trinidad and Tobago in 1978, but it provoked a formal protest from other States.[23]) Reservations to fundamental principles of the derogation clause seem to be against the 'purpose and the object of the treaty' and are therefore forbidden by the Vienna Convention on the Law of Treaties. This reaffirms the 'norm-creating character' of the derogation clause, in so far as the idea of making far-reaching reservations to a provision seems to go against its 'norm-creating character'.[24]

The fact that the derogation clause in article 4 of the Covenant qualifies in principle as a potential provision of a 'norm-creating character' does not necessarily mean that all the principles of the provision are emerging, or are equally qualified to become, norms of customary international law. It is a well-known doctrine that some principles within a provision may become principles of general international law, while others do not. In fact, there is no need to expect the derogation clause as such to become part of general international law. This is because the derogation clause itself establishes a mechanism which is more appropriate for treaty law; in fact, the State which wants to derogate from the treaty provisions has formally to rely on the right to derogation recognized in the derogation clause. According to this procedure, the State has officially to proclaim the state of emergency and to notify the other States parties of the provisions derogated from and the reasons therefore (principles of proclamation and notification). However, in general international law, the doctrine of necessity can always be applied by the State in exceptional circumstances without any need for the State to follow any special procedures.

Having said that, this position does not mean that in principle these provisions of a 'procedural' character cannot become norms of general international law. There is nothing in principle which prevents this process from taking place. In fact, as far as a 'procedural' principle in a treaty becoming an emergent principle of general international law is concerned, an interesting parallel can be found in the law of the sea in the delimitation of the continental shelf between adjacent or opposite States. Article 6 of the 1958 Continental Shelf Convention establishes that the delimitation shall be determined first of all by agreement between the States. In the absence of agreement, article 6 provides the main criteria for delimitation.[25] In recent cases on delimitation, the ICJ referred to

---

[23] The reservation referred to the principle of non-derogability. See UN, *Multilateral Treaties*, p. 135. The Federal Republic of Germany and the Netherlands raised a formal objection; they considered it incompatible with the object and purpose of the Covenant, ibid., pp. 137–8. See above, pp. 131–2 ff.

[24] See the reflections of the ICJ on this point in the *North Sea Continental Shelf cases, ICJ Reports*, 1969, pp. 38 ff., paras. 63 ff.

[25] See also art. 83 of the 1982 Convention on the Law of the Sea.

this procedural principle of delimitation by agreement as a principle of general international law.[26]

As far as the 'procedural' principles of the derogation clause are concerned, it could well be that an international obligation could emerge by which States have officially to proclaim a state of emergency and to make public the derogating measures taken, in order to justify their non-compliance with customary human rights norms; this practice, which is well-established in most municipal systems, can also become a principle of general international law. In fact, there is some evidence in the doctrine and in the practice of monitoring organs to support this contention.[27]

Furthermore, in respect to the notification requirement, a practice could develop sometime in the future by which all States finding themselves in emergencies would have to notify the international community (perhaps the UN General Assembly through the UN Secretary-General or the Commission on Human Rights) of those derogations from human rights standards recognized as such under customary international law.[28] Even though the present development within the UN Human Rights Commission through the Special Rapporteur on States of Emergency seems to be moving in that direction, the moment when the obligation of notification, as a matter of customary law, will be a reality seems much more remote.[29]

In conclusion, of the main principles constituting the general legal regime of the derogation clause, the principles which seem to be prima-facie clear candidates for becoming principles of general international law regulating human rights in emergencies are the following: the principle of exceptional threat, the principle of non-derogability of fundamental rights, the principle of proportionality, and the principle of non-discrimination. What should now be analysed is the existing evidence concerning the emergence of these principles as principles of general international law.

---

[26] *North Sea Continental Shelf cases*, Judgment, *ICJ Reports*, 1969, at p. 54. *Continental Shelf (Tunisia/Libya)*, Judgment, *ICJ Reports*, 1982, p. 92 and para. 37. *Delimitation of the Maritime Boundary in the Gulf of Maine Area*, Judgment, *ICJ Reports*, 1984, pp. 266, 311–2. *Continental Shelf (Libya/Malta)*, Judgment, *ICJ Reports*, 1985, pp. 31–56. See also R. R. Churchill and A. V. Lowe, *The Law of the Sea*, (2nd rev. edn., Manchester, 1988), p. 156 ('delimitation by agreement remains the primary rule of international law').

[27] In respect to the principle of proclamation in the 'Siracusa Principles', principle No. 75; see also the comments by D. O'Donnell and J. F. Hartman in *HR Quarterly*, 7 (1985), 34, 131. See also the Resolution of the UN Sub-Commission 1988/24, below, p. 246 (see also pp. 244–5).

[28] A proposal in this direction was presented in the *travaux préparatoires* of the Covenant; see A/2929, para. 47.

[29] A practice is developing under the Special Rapporteur on States of Emergency in the sense that States communicate to him the declaration of emergency, the derogated provisions, and the reasons thereof.

4. EVIDENCE OF THE EMERGENCE OF SOME OF THE PRINCIPLES OF THE DEROGATION CLAUSE AS CUSTOMARY LAW

Among the types of evidence mentioned by the US *Restatement* and other sources in order to prove the existence of customary law,[30] the following have a special importance in showing the customary character of some of the principles of the derogation clause: the probative value of law-making treaties, the repetition of the same norm in several human rights treaties, the practice of judicial, quasi-judicial, and monitoring bodies in the application of norms and principles, and the acquiescence with those standards by States non-parties to the treaties.

(a)   The Probative Value of Law-Making Treaties

As has been pointed out before, most of the international obligations relating to human rights are contained in treaties. However, some norms and principles of these main treaties may be regarded as expressing, or as generating, rules of customary law. In view of the fact that most of the human rights instruments state largely new norms, it would seem that these treaty norms, rather than being an expression of pre-existent customary rules, have in general generated new customary rules. Thus, some of these law-making treaties may also acquire a probative value in establishing the customary character of new norms. This is especially true in the case of multilateral treaties containing general standards, concluded after a long period of discussion and with the participation of a great number of States. Undoubtedly, the ICCPR qualifies as one of these treaties, and the derogation clause can be seen as a provision which contains principles of a norm-creating character.[31]

(b)   The Repetition of the Same Norm in Several Human Rights Treaties

Another element, which has been considered to be an important expression of State practice and evidence of customary law, is the repetition of the same norm in several human rights treaties, especially if these treaties are multilateral. Thus, more evidence of the general acceptance of the principles of the derogation clause of article 4 of the Covenant could be found in their being repeated in the main regional treaties, namely, in article 15 of the European Convention and article 27 of the American Convention; a derogation clause is also contained in the new Draft

---

[30] See above, Ch. 8.
[31] See above, pp. 231 ff.; see also pp. 230–1 for the universal character of the Covenant.

Convention for the Arab World.[32] Moreover, the fact that the same provision, in this case the derogation clause, is restated after eighteen years in identical terms can be seen as strengthening the principles contained within it.[33]

A recent finding of the West German Supreme Constitutional Court used the repetition of a treaty-norm in different treaties as evidence of customary law; the issue at stake was the principle *ne bis in idem*. The Court noted: first, that the principle was recognized in article 14(7) of the Covenant, which had been ratified by eighty States from all major legal systems; secondly, reservations to that article had not decreased its effect; and, thirdly, that the same principle was recognized in article 8(4) of the American Convention and article 4 of the Seventh Protocol of the European Convention. Bearing these elements in mind, the Court considered that the principle *ne bis in idem* was a general rule of international law.[34] The use of this kind of evidence of customary law has also been adopted in the judicial decisions of other national courts. Thus, Meron has found that the review of the recent US cases 'reveals significant, though uneven and uncertain, resort to international human rights instruments as evidence of customary human rights law'.[35]

(c) The Practice of the Organs of International Organizations and Similar Agencies

As has been pointed out, 'the practice of judicial, quasi-judicial and supervisory organs has a significant role in generating customary rules'.[36] It is true that many of these bodies, such as the European Court and Commission, the Human Rights Committee, and the Inter-American Court and Commission, usually apply treaty law rather than general international law. However, their decisions and jurisprudence are often invoked outside the context of treaty law and considered to be an auth-

---

[32] Art. 42 says: '1. Any country in case of actual war, imminent danger or any crisis threatening its independence and security may declare a state of emergency and may take measures derogating from its obligations under the present Charter to the extent strictly required by the exigencies of the situation.
'2. No derogation may be permitted under the preceding provision from respect for the right to life, security of person, recognition as a person before the law, the right to a nationality, the principle of supremacy of the law or the right to freedom of religion and thought.' (Para. 3 is similar to that of the other three treaties.) See *Draft Charter on Human and People's Rights in the Arab World*, 1987, Istituto superiore internazionale di scienze criminali, p. 15. For the reasons for the absence of a derogation clause in the African Charter, see pp. 209 ff.
[33] See G. Gaja, 'A "New" Vienna Convention on Treaties between States and International Organizations', *BYBIL* 58 (1987), 268.
[34] 1987, Neue Juristische Wochenschrift (NJW), 2155 (quoted in Meron, *Human Rights as Humanitarian Norms*, pp. 132–4.
[35] Ibid., pp. 130–1.
[36] Ibid., p. 100.

oritative interpretation of human rights norms in general international law. A detailed study of the jurisprudence of these bodies when applying the derogation clause as a matter of treaty law has already been carried out; what remains is to examine the evidence found in decisions of other international organs when applying the principles of the derogation clause as a matter of general international law.

(i) *The Practice of the IACHR*. The IACHR has also applied some of the fundamental principles of the derogation clause to those States non-parties to the American Convention but members of the OAS and therefore bound by the UN Charter and the American Declaration. In this section, two points will be examined: the main principles of the derogation clause applied, and the legal basis for the extension of these principles to States non-parties to the Convention.

The Principles of the Derogation Clause Applied outside the Convention. In the last decades, the IACHR has produced a certain corpus of doctrine on different important topics in the interpretation and application of the human rights instruments in the Americas; this doctrine, which comes from the Commission's reports and case-law, has been systematized to some extent by the same Commission.[37] Not surprisingly, the subject of human rights in states of emergency has been given a lot of attention. The position of the Commission on this topic and the principles applied are as follows:

The conception of a state of emergency and the legitimacy of the institution. The Commission considers the state of emergency, which is recognized in most national legal systems, to be a normal and legitimate institution when kept within certain limits. In fact, in emergencies, human rights standards cannot be the same as in normal or peacetime, because this would entail a serious risk to the maintenance of public order and state security. There is a need on these occasions to find a balance between the need for the defence of a legally institutional order and the protection of individual rights.

Principles applied by the Commission:

1. *Principle of exceptional threat*. The declaration of a state of emergency and the suspension of fundamental rights are only legitimate in exceptional situations; these situations are mainly due to internal commotion or external attack. The exceptionality of the situation has been qualified in several ways: 'serious and grave danger', 'real threat to the public order or the security of the State', etc. In several cases, the

---

[37] OAS, *The IACHR: Ten Years of Activities* (Washington, 1981), pp. 315 ff. For the doctrine on States of Emergency, see Ibid., pp. 336–8.

Commission has not accepted the declaration of emergency and the widespread suspension of human rights, because the circumstances did not warrant them.[38]

2. *Principle of temporariness.* The state of emergency is an institution 'essentially transitory in nature'; in other words, it cannot be established for an indefinite or prolonged period of time. This principle of temporariness seems to be a particular expression of the principle of proportionality.[39]

3. *Principle of non-derogability of fundamental rights.* In a proper state of emergency, fundamental human rights cannot be violated. Among the rights mentioned by the Commission as falling into this category are: the right to life, the right to be free from torture or other inhuman or degrading treatment or punishment, and the right to a fair trial. Among the violations of fundamental guarantees, it has referred to: the denial of the minimum guarantees of common article 3 of the 1949 Geneva Conventions, death sentences without due process of law before a competent tribunal, prolonged arbitrary detention without being charged and without fair trial and due process of law, deprivation of the freedom of the accused for a period longer than the maximum sentence he could receive, and explusion of nationals. As will be seen below, it seems that the Commission considers the principle of non-derogability of article 27(2) of the Convention to be applicable also to non-parties to the Convention.[40]

4. *Principle of proportionality.* All derogating measures which are excessive in respect to the threat, and therefore not strictly necessary, have to be considered inadmissible. The Commission also considers that states of emergency should not alter to any appreciable degree the independence of the various branches of the government or lead to the denial of the rule of law.

This doctrine has been considered by the Commission as the 'most admitted doctrine internationally', because it is the one which inspires the American Covention (art. 27), the ICCPR (art. 4), and the European Convention (art. 15). This doctrine and the principles referred to have also been applied by the IACHR to non-parties to the American Convention as a matter of general international law; thus, in the *Reports on Chile* in 1974 and 1985, *Paraguay* in 1978 and 1987, and *Uruguay* in 1978.[41]

---

[38] See above, Ch. 1.
[39] Ibid. See also Ch. 5, pp. 159 ff.
[40] See IACHR, *Report on Paraguay*, 1987. p. 16.
[41] This doctrine has been applied in the IACHR *Reports on Chile* in 1974 (see Buergenthal (ed.), *Protecting Human Rights*, pp. 362 ff.) and 1985 (pp. 43–45); in the IACHR *Reports on Paraguay*, 1978, p. 19 and 1987, pp. 15 ff.; in the IACHR *Report on Uruguay*, 1978, p. 19.

There is an even clearer formulation of these fundamental principles (with a greater emphasis on the principle of proportionality) in the chapter dedicated to the Commission's doctrine on terrorism and the limits of the repressive action of States.[42] These principles were equally applied in the case of non-parties to the Convention (e.g. Argentina in 1980).[43]

The legal basis for the extension of the main principles of the derogation clause to States non parties to the American Convention. This important question can be formulated as follows: on what grounds has the IACHR extended the application of the main principles of the derogation clause to States non-parties to the American Convention? Is it that the IACHR considers these principles to be part of customary international law and therefore applicable to all countries, or is it that the IACHR considers these principles to be regional custom? Not surprisingly, the IACHR has not given a complete legal justification for this extension; however, some important hints on the legal foundation of the extension of these principles appear in several cases.

The first case in which the IACHR addressed this question was in the 1974 *Report on Chile*.[44] In this case, which took place before the American Convention had entered into force, the Commission referred to the necessity of suspending human rights in emergencies, and said

Law, whether domestic or international, does not ignore such realities. It weighs them fairly and gives solution for dealing with them, while adequately evaluating the good that is endangered. With respect to American international law, which is the normative system that the Commission must take primarily into account, it must be understood that, in the absence of conventional standards in force in this area, the 'most accepted doctrine' is that which is set forth in the American Convention on Human Rights . . . which has been signed by twelve American countries (among them Chile) and whose ratification has already begun. The Convention contains an express provision in Article 27 establishing to what extent a state may restrict the protection of human rights in exceptional circumstances, such as war.[45]

Professor Meron has seen in this position a declaration by the IACHR that article 27 reflects a regional customary law norm for the American system.[46]

[42] OAS, *The IACHR: Ten Years of Activities*, pp. 339–42.
[43] IACHR, *Report on Argentina*, 1980, pp. 24–7. (The same principles were applied to a state party to the Convention: *Report on Colombia*, 1981, pp. 15–18.)
[44] IACHR, *Report on Chile*, 1974, pp. 2–4 (quoted in Buergenthal (ed.), *Protecting Human Rights* (3rd edn., 1990), pp. 362–3).
[45] Ibid. Chile ratified the Convention in 1990.
[46] Meron, *Human Rights and Humanitarian Norms*, p. 219 n. 262.

This position seems to have been reflected in other cases. Thus, in the *Report on Paraguay*, 1978, in which the Commission once again declared that this 'most accepted doctrine' inspired article 27 of the ACHR, the Commission added that the content of article 27 'is a reflection of convictions and beliefs that are firmly rooted in the minds of our people'.[47] Furthermore, in the 1987 *Report on Paraguay*, the Commission was aware of the fact that Paraguay had not ratified the American Convention and therefore was only bound by the American Declaration, which does not contemplate the possibility of derogating from human rights standards. However, the Commission considered that the criteria derived from article 27 'in its day embodied the hemisphere's thinking on the subject.' Following this reasoning, it considered that the principles of exceptional threat, temporariness, non-derogability of certain fundamental rights, and proportionality reflected 'unanimous doctrinary thinking'.[48]

Moreover, the Commission goes even further in its jurisprudence. It does not simply say that article 27 reflects the 'most accepted doctrine' in the American continent (what Meron qualifies as 'regional customary law'), but it seems to imply that this doctrine has a wider extension and can be defined as *'the most admitted doctrine internationally'*. In its 1974 *Annual Report*, which has been incorporated into the 'corpus' of the official Commission's doctrine, it said:

The Commission is not unaware of the reasons in favor of the attribution of special powers to the Executive Branch in exceptional situations, such as those which arise from internal commotion or external attack, but it takes into consideration the fact that the most admitted doctrine internationally, because it is which inspires the American Convention . . . (art. 27) as well as the UN ICCPR (art. 4) and the ECHR (art. 15), places precise limits on the use of special powers for the purpose of protecting human rights; and considers it necessary to harmonize the needs of a defence of a regularly established institutional order with the protection of the fundamental attributes of man.[49]

Furthermore, in the 1985 *Report on Chile*, the Commission confirmed this position by saying: 'Indeed, international law—as formalized in Article 4 of the ICCPR, to which Chile is a party, and article 27 of the ACHR, which embodies the most received doctrine on this subject—has imposed a series of requirements and exigencies if a State is to limit the exercise of the rights and freedoms that are internationally recognized.'[50]

The interpretation of the phrase 'the most accepted doctrine inter-

---

[47] IACHR, *Report on Paraguay*, 1978, p. 19.
[48] IACHR, *Report on Paraguay*, 1987, pp. 15–16.
[49] IACHR, *Annual Report*, 1974, p. 36. Included in IACHR, 'Doctrine', pp. 336–7. Also quoted in IACHR, *Report on Paraguay*, 1978, p. 19.
[50] IACHR, *Report on Chile*, 1985, p. 44.

nationally' presents some problems, mainly because the IACHR does not explain its meaning or provide any evidence for its claim.[51] The concept of 'the most accepted doctrine' is a common one in civil law systems and is built upon the writings of the most distinguished legal authors (it must not be forgotten that the original text of these findings of the IACHR is in Spanish and that most of the members of the Commission come from civil law systems). The concept serves to support judges in their task of applying the law and to fill possible gaps, by giving them the guidance of the 'most accepted doctrine.'[52] It seems that the position of the Inter-American Commission is similar in these cases, in so far as it has to apply certain principles on human rights in emergencies, but there is no possibility of a direct application of treaty law (derogation clause) because the States involved were not parties to the American Convention; therefore the Commission solves the problem by recurring to the 'most accepted doctrine'. In many cases, the most accepted doctrine is evidence of what the law is; in this respect, it would come close to the 'teachings of the most qualified publicists of the various nations', as a subsidiary means for the determination of the law. In the French text of article 38 of the Statute, this subsidiary source of law is called 'la doctrine' (similar to the Spanish text 'la doctrina'), which recalls the expression used by the IACHR when it referred to the 'most accepted doctrine'. However, the doctrine in international law can only be considered as evidence of the rules of law, and not as an independent source.[53] (To some extent, also the general principles of law in article 38 of the Statute 'were intended to equip the Court with the means of overcoming the deficiencies of international law resulting from its still imperfect development.'[54])

If this is the case, the 'most accepted doctrine internationally' will, in the jurisprudence of the IACHR, constitute evidence of general international law, in the same way as 'the most accepted doctrine in American international law' is evidence of what has been qualified as regional customary law. The position of the IACHR could be interpreted in the sense that it affirms that article 27 of the American Convention was declaratory at the time of regional customary law, and that similarly, article 15 of the European Convention and article 4 of the ICCPR were declaratory, at the

---

[51] However, in the 1985 *Report on Chile* the Commission refers to the writings of Higgins, Hartman, Questiaux, and Grossman on this topic, ibid., p. 44 n. 11. The IACHR uses a flexible language when it refers to this most accepted (received, admitted . . .) doctrine; hereafter the expression 'most accepted doctrine' will be used.

[52] See e.g. the Swiss Civil Code (art. 1), 'the most authoritative doctrine' (Williams, *The Swiss Civil Code: Sources of Law*, p. 61).

[53] A. Verdross, *Derecho international publico* (4th edn., Madrid, 1972), p. 99.

[54] H. Waldock, 'General Course', p. 56. See also on this point the remarks by Bin Cheng in his *General Principle of Law as Applied by International Courts and Tribunals* (London, 1953; repr. Cambridge, 1987), Introduction.

time of their signature, of customary international law. This interpretation needs close examination.

1. The contention that article 27 of the American Convention could at that time be deemed to constitute the crystallization of a rule of regional customary law is not free from difficulties. One may argue that in 1968 (just before the signature of the Convention), and as a consequence of the 'generating effect' created by the derogation clauses of the 1950 European Convention and the 1966 Covenant in the American continent, the 'most accepted doctrine' on the subject of 'human rights in states of emergencies' was precisely that expressed in the main principles of the derogation clause of those two treaties. This thinking was expressed in the extremely important 1968 IACHR Resolution on this topic. This resolution, which was the conclusion of a long period of study by the Commission on Human Rights in Emergencies, contains the main principles of the derogation clause.[55] Finally, following this argument, one may say that the crystallization of this process was article 27 of the American Convention, which at the time embodied the 'most accepted doctrine' (what Meron has qualified as regional customary law). Although this theory is not free from difficulties, that process could possibly have taken place by 1978. In fact, a certain homogeneity in the legal systems of the American States, in all of which the institution of a state of emergency has common characteristics, could have helped towards the creation of a regional custom.

2. Some aspects of the IACHR finding seems to imply that articles 15 of the ECHR and 4 of the ICCPR could at that time be considered to embody general international law; this contention is more controversial. It is hard to accept that the derogation clause of the European Convention, when it was signed in 1950, codified a rule of general international law ('the most accepted doctrine internationally'). It should not be forgotten that the ECHR was a regional treaty, which for the first time tried to regulate in general terms the issue of civil and political rights and to define with great precision the international obligations of States in this field. It was the first treaty on the matter and was signed a few years after the founding of the UN and less than two years after the Universal Declaration, which was not considered at the time to be obligatory as a matter of international law. Therefore, even if the drafting of the UN Covenant was following the same lines, it is hard to believe that article 15 of the ECHR could be considered as being declaratory of general international law.

It is also difficult to accept the contention that when the Covenant was signed in 1966, article 4 was declaratory of general international law.

---

[55] IACHR, 'Resolution on the Protection of Human Rights in Connection with the Suspension of Constitutional Guarantees or "State of Siege"' (1968) (included in Buergenthal and Norris (eds.), *The Inter-American System*, vol. iv, booklet 23, p. 8).

Although the possible 'generating effect' of the ECHR derogation clause could be seen as providing a possible means of creating a general rule, it is still difficult to have confidence in this view, especially if one takes into account that the Covenant was the first international treaty on human rights of a general character establishing clear legal obligations for States at a universal level. There is no evidence that by 1966 there existed a State practice as a matter of customary international law, in the sense that States behaved in accordance with the principles of the derogation clause.

The more plausible theory is this: the main principles of the derogation clause of the European Convention and the ICCPR as a provision of a norm-creating character have had a generating effect, first of all in the American continent, by creating a regional custom which was soon codified in the American Convention, and then by creating an emergent rule of general international law. The doctrine also provides evidence to show that these principles of the derogation clause are now emerging as general international law.[56]

In any case, whatever explanation one may find in the construction of the IACHR jurisprudence, the fact is that the Commission believes that the main principles of the derogation clause common to the three main treaties are not only part of 'American international law', but also form part of the 'most accepted doctrine internationally'. This goes some way towards proving that they are part of general international law. Accordingly, the IACHR has made this doctrine a part of its jurisprudence and has applied it to those States non-parties to the American Convention. Furthermore, no State, not even those affected by the application of these standards, has ever attacked the legal grounds of this position of the IACHR.

The jurisprudence of the IACHR on the application of the main principles of the derogation clause outside treaty-law has two important implications: (1) to affirm these principles within the American system as regional customary law; (2) to support the contention that these principles are also part of general international law, and could therefore also be applied by international bodies, especially those within the UN framework, when assessing human rights obligations in emergency situations for those States non-parties to the Covenant or to any other treaty with a derogation clause. This position seems to have been taken by the UN Special Rapporteur on States of Emergency.

(ii) *The Practice of the UN Organs.* In this section the practice of the UN organs on the question of human rights in emergencies will be analysed. This will be seen outside the context of the strict application by the HR Committee of article 4 of the Covenant as a matter of treaty law. As is

---

[56] See below, p. 260 n. 2.

well known, the Covenant established a special organ, the Human Rights Committee, to apply the standards of the Covenant. The jurisprudence of the Committee has been examined in great detail in previous chapters. What will now be analysed, in order to find evidence of the application of the principles of the derogation clause as general principles outside treaty law, is the practice of the UN organs in respect to this matter. Firstly, however, a brief account of the history of the involvement of UN organs with this topic will be of some value.

A brief historical outline. As early as 1977, the UN Sub-Commission on Prevention of Discrimination and Protection of Minorities expressed its deep concern at the manner in which certain countries applied the provisions relating to states of emergency. What the Sub-Commission underlined above all was the link between the numerous violations of human rights and states of emergency, and it paid special attention to the matter of the prolonged and often indefinite detention of unconvicted persons.[57] The Sub-Commission decided to undertake a comprehensive study of this question in order to advance the aim of the UN with respect to human rights. This project was finally entrusted to Mme Questiaux, who produced a valuable study.[58] In the first part of this report, Mme Questiaux outlines the main principles of the derogation clause of the three main treaties, the effect of emergencies on the rule of law and on detained persons. However, Mme Questiaux did not address the problem of whether some of the principles of the derogation clauses should be deemed to constitute general international law.[59] At the end of the report, she recommended *inter alia* that the Sub-Commission might include in its agenda a special item entitled 'implementation of the right of derogation provided for under article 4 of the ICCPR and violations of human rights', with the aim of drawing up a yearly list of States which proclaim or terminate states of emergency, and that the Sub-Commission should produce an annual report to the HR Commission analysing compliance with international standards of the legality of states of emergency; to this effect reference should be made to the principles of the derogation clause.[60]

This recommendation was accepted and put into effect by the UN Commission on Human Rights, which required the Sub-Commission to study further steps to ensure respect for human rights in emergencies

---

[57] Resolution 10 (XXX) of 31 Aug. 1977.

[58] E/CN. 4/Sub. 2/1982/15 (27 July 1982), *Study of the Implications for Human Rights of Recent Developments concerning Situations Known as States of Siege or Emergency.*

[59] None the less, see ibid., p. 19, in which she expresses the belief that the principle of non-derogability of fundamental rights is a principle of general international law and of *ius cogens.*

[60] Ibid., p. 44.

throughout the world, especially respect for those non-derogable rights contained in article 4(2) of the Covenant. The Commission gave urgent priority to this question due to its great importance.[61] In order to fulfil this task, the Sub-Commission appointed Mr Leandro Despouy as Special Rapporteur.[62]

Mr Despouy had to deal with an important problem which would influence the whole direction of his work: the problem was whether his study should be confined to those States parties to the Covenant (or other regional treaties with derogation clause), or should include all States. In the second case, a further question arose concerning the standards applicable to those States non-parties to any of those instruments. There were good reasons to hold that the work of the Special Rapporteur should be confined to States parties to those treaties (in fact, the wording of the item in the agenda of the Sub-Commission reads 'implementation of the right to derogate provided for under article 4 of the Covenant . . .', and moreover, Mme Questiaux's Report seems to refer only to treaty law). However, due to the far-reaching nature of Mr Despouy's mandate, the final decision was to include States non-parties as well.[63]

Having taken this position, the Special Rapporteur had to solve the further problem of the human rights standards in emergencies applicable to States non-parties to the Covenant. (It should not be forgotten that the mandate of the Special Rapporteur involved him in assessing the compliance of States with international standards). After explaining the principles of the derogation clause as developed by the supervisory organs under the treaties, Despouy had to consider to what extent 'these principles are customary in nature'. In an important statement, the Special Rapporteur said:

Having regard to the present state of international law, we believe that the answer should be that at least some of these principles are of the nature of customary law.[64]

The principles enumerated were: the principle of non-derogability, proclamation, legality, notification, exceptional threat, proportionality, temporariness, and non-discrimination. Referring to the Questiaux report, he confirmed that the national laws of most states are in conformity with these standards.[65] Despouy did not declare which principles

---

[61] UN Commission on HR, Resolution 1983/18.

[62] UN Commission on HR, Resolution 1985/23.

[63] See E/CN. 4/Sub. 2/1985/19, pp. 3–5. See also on this point the cautious position of Mr Al Khasawneh and the supporting opinion of the Int. Commiss. of Jurists in E/CN. 4/Sub. 2/1985/SR. 25/Add. 1, pp. 2–3 and 11.

[64] E/CN. 4/Sub. 2/1985/19 (17 June 1985). Explanatory paper prepared by Mr L. Despouy (Argentina).

[65] Ibid., p. 9., para. 24.

have become customary law; neither did he address this question in his first two annual reports, perhaps because, as he pointed out, during the first stage of his work as Special Rapporteur he concentrated on gathering information concerning states of emergency, leaving the question of the principles concerning human rights in emergencies in general international law for a later stage. Even though he does not deal with the point explicitly, his position as a matter of principle on the customary nature of some of these principles is very much in line with the position of the IACHR.

However, in one of his most recent reports,[66] the Special Rapporteur set up certain criteria and norms applicable in emergency situations; these criteria are part of an 'international legal framework deriving from prevailing international norms, the practice of international organizations and the internal law of States, which provides a frame of reference for the legality of states of emergency'.[67] These criteria are applicable to all States. Among the criteria identified by the Special Rapporteur the following were included: the principles of exceptional threat, proclamation, proportionality, limited geographic extent of the emergency, and the principle of non-derogability (he also stated his intention to draw up a list of non-derogable rights on the basis of international treaties and customary law).[68] Furthermore, Mr Despouy mentioned as a part of this legal framework: the need for a national monitoring mechanism (by Parliament or judicial review), the possible compensation for the victims of excessive emergency measures, and the need to maintain the power of non-military courts and the procedural guarantees of habeas corpus, *amparo*, and similar remedies.[69]

In order to produce his reports, the Special Rapporteur asked Governments through the UN Secretary-General to provide detailed information on states of emergency in order to assess their legality in the light of the main principles of the derogation clause. He therefore asked for detailed information on the following aspects: data of proclamation and termination, notification to international bodies (when applicable), legislative texts, reasons for the proclamation, the number and nature of rights suspended, the geographical extension of the emergency, persons affected, duration, etc.[70] It can easily be seen that the information referred to covers most of the aspects of the principles of the derogation clause.

[66] E/CN. 4/Sub. 2/1989/30/Add. 2/Rev. 1.
[67] Ibid., p. 2.
[68] Ibid. (paras. *a, b, c, d, e*).
[69] Ibid., pp. 2–3 (paras. *f, g*; and 4*b, c*).
[70] E/CN. 4/Sub. 2/1987/19 (18 Aug. 1987), p. 6, 'First Annual Report and List of States which since 1. 1. 1985 have Proclaimed, Extended or Terminated a State of Emergency, Presented by L. Despouy, Special Rapporteur'.

The response to this demand for information was quite impressive, especially if one takes into account not only the overall number of States which answered (68), but also the fact that twenty-six of these States were non-parties to the Covenant and many of them had been directly affected by a situation of emergency. The number of replies is especially significant when one remembers that a state of emergency is a highly sensitive matter and that the procedure of the Special Rapporteur implies a new development in this area. This information (together with that taken from other sources) has enabled the Special Rapporteur to produce four comprehensive reports on states of emergency in the world.[71]

No State has ever made a declaration against the competence of the Special Rapporteur to undertake these steps. Even a country like South Africa has recently sent information about the state of emergency declared there.[72] Moreover, the information received bears witness to the fact that even States non-parties to the relevant treaties are willing to fulfil the fundamental principles of the derogation clause. This is particularly true in respect to the non-derogable rights contained in article 4(2) of the Covenant; no State has ever officially derogated from these non-derogable rights. The answers received from States, in line with the position of the Special Rapporteur on the customary nature of some of the principles of the derogation clause, could provide further, although not conclusive, evidence concerning the *opinio iuris* of States on these principles as general international law, especially when it is taken into account that no State has objected to the work or to the principles laid out by the Special Rapporteur.[73]

A recent resolution of the UN Sub-Commission on Prevention of Discrimination and Protection of Minorities is also in line with the position of the Special Rapporteur on the importance of the principles of the derogation clause as general international law; the Sub-Commission reaffirmed their importance for all States; the principles mentioned are: the principle of non-derogability, official proclamation, exceptional threat, and temporariness. It also recognized the fundamental importance of enacting internal legislation in accordance with international norms for those States which have not yet done so.[74]

---

[71] For the first report see n. 70 above; for the second annual report: E/CN. 4/Sub. 2/1988/18 and Rev. 1 and Add. 1. For the third report: E/CN. 4/Sub. 2/1989/30, Add. 1 and Add. 2 Rev. 1. See also E/CN. 4/Sub. 2/1990/33 and Adds. 1, 2 (July–Aug. 1990). For the fourth annual report: E/CN. 4/Sub. 2/1991/28 (24 June 1991).

[72] E/CN. 4/Sub. 2/1989/30/Add. 1 (28 June 1989), E/CN. 4/Sub. 2/1990/33/Add 2, p. 3.

[73] The communications exchanged between the Special Rapporteur and the Government of Paraguay illustrate this new development within the UN organs, and the emerging obligations of States non-parties to the Covenant to comply with the principles of the derogation clause. See E/CN. 4/Sub. 2/1987/18 (22 May 1987).

[74] UN Sub-Commission Resolution 1988/24, adopted 1 Sept. 1988 without vote. 'Question of Human Rights and States of Emergency', in E/CN. 4/1989/3 (25 Oct. 1988). See also E/CN. 4/Sub. 2/1990/33, p. 2, para. 4.

The monitoring of human rights in emergencies by the UN Commission on Human Rights. As the International Law Association has pointed out: 'through the confidential procedure under ECOSOC Resolution 1503, *ad hoc* studies of country situations, special mechanisms to deal with particular grave human rights abuses, general debate on human rights conditions and the adoption of resolutions, the Commission has made itself into one of the most important bodies currently monitoring human rights abuses under states of emergency.'[75] After a detailed analysis of this monitoring function of the Commission in emergencies, some general conclusions can be drawn.

1. The contention that the situation of human rights in a country in a state of emergency belongs exclusively to the domestic jurisdiction of the State has systematically been rejected by the Commission and its *ad hoc* special committees or Special Rapporteurs when there is prima-facie evidence of gross violation of human rights; this has been a common contention by those States under scrutiny. The fact that there is a state of emergency does not prevent the exercise of the monitoring functions by the UN organs. These functions include at least the study of the situation through comprehensive reports and the making of recommendations.[76] Professor Brownile has pointed out that these powers of supervision by the UN political organs are exercised through discussions of human rights issues, publicity, fact-finding machinery, and other means under article 14 of the Charter; in practice, article 2(7) has not been interpreted in a restrictive way.[77] Therefore the contention that a state of emergency is an exclusive 'internal act of the state' has not been accepted by UN organs when gross violations of human rights were involved.[78]

2. The monitoring functions of these bodies arise from the UN Charter and the general competence of the UN Commission on Human Rights, and not from any special mechanism established by a particular human right treaty. Therefore the Commission studies those situations in respect both to parties and non-parties of human rights treaties; no distinction has been made as far as the right to monitoring these situations is concerned.

[75] ILA Seoul Report (1986), para. 32. For a survey of the supervision by the UN Commission on Human Rights in states of emergency, see ibid., paras. 32 ff.
[76] See E/CN. 4/1985/21 (19 Feb. 1985), p. 11 (where Ermacora quotes the UN Commission on South Africa which affirmed 'the universal right of study and recommendation' on general problems of human rights, A/2505).
[77] Brownlie, *Principles*, p. 571.
[78] See *Inter alia*: E/CN. 4/1310 (1979), Annexe *XXII*, 'Observations of the Government of Chile on the Report of the *Ad Hoc* Working Group.' See also the report itself, pp. 2 ff. Also A/31/253 (8 Oct. 1976), 'Report of the *Ad Hoc* Working Group on the Situation of Human Rights in Chile, para. 49; paras. 89 ff., E/CN. 4/1983/18 (21 Feb. 1983), 'Report on the Situation in Poland by the Under-Secretary Hugo Gobbi,' p. 2; E/CN. 4/1985/21 (19 Feb. 1985), 'Report on Afghanistan by Prof. Ermacora', p. 11.

3. In respect to parties to the Covenant, the Commission and other bodies refer to the derogation clause as the legal regime applicable, although the quality of the legal analysis and the evaluative statements depend on the different bodies and *ad hoc* working groups; in several of these reports, a thorough analysis of fulfilment with the strict conditions of the derogation clause has been made.[79] In the case of Poland, the UN Under-Secretary-General sent a very detailed questionnaire to the Polish Government asking for full information about the way in which the state of emergency declared in 1981 affected not only non-derogable rights but also those which can be suspended according to article 4, the assumption being that there was a disproportionate suspension of rights.[80]

4. A further step was taken by the *Ad Hoc* Working Group on Chile when it applied the derogation clause of article 4 to Chile even before the entry into force of the Covenant. Although it can be said that, according to article 18 of the Vienna Convention on the Law of Treaties, Chile, once it had signed and ratified the Covenant, was obliged to refrain from performing acts which would defeat the object and purpose of the treaty,[81] the reasoning of the *Ad Hoc* Working Group was different. It required Chile to comply with the main principles of the derogation clause as a matter of general international law. In its First Report in 1975, the Group said that 'these provisions (those of article 4) correspond to the generally recognised international standards of human rights in emergency situations'. The provisions explicitly mentioned by the Group were the main principles of the derogation clause, namely, the principles of exceptional threat, proportionality, consistency, non-derogability of fundamental rights (in particular arts. 6, 7, 8, 15, 18), and non-discrimination.[82]

In its Second Report, the *Ad Hoc* Working Group reaffirmed this doctrine and specified the legal basis for it; in an important statement the Group said: 'article 4 of the ICCPR, which has been ratified by Chile and is expected to enter into force on 23 March 1976, may be considered as reflecting the general international law of human rights on the subject of emergency situations and limitations of human rights'.[83] In order to support this contention, the Group made reference to the derogation clauses of the European and American Conventions, which contain the same main principles. This position is very much in line with the

---

[79] See the *Reports on Chile*, E/CN. 4/1310 (1979), pp. 14–16; A/31-253 (1976).

[80] E/CN. 4/1984/26, pp. 4–6. See also the UN HR Commission Resolution 1982/26 (arbitrary arrest, denial of the right of freedom of expression, peaceful assembly, right to form and join independent trade unions, etc.).

[81] E/CN. 4/SR. 1273 (1974) (Remarks of Cassesse of Italy), p. 85.

[82] A/10285 (7 Oct. 1975), p. 42, para. 106.

[83] E/CN. 4/1188 (4 Feb. 1976), p. 20, para. 59.

jurisprudence of the IACHR and that of the Special Rapporteur of the UN Sub-Commission.

5. Unfortunately, one of the characteristics of many reports of the UN Special Rapporteurs and *Ad Hoc* Working Groups in cases relating to States non-parties to the Covenant is a lack of analysis of the legal criteria used to assess human rights standards in states of emergency.[84] In the majority of cases, these bodies simply describe the gross violations of fundamental rights which have taken place. The legal basis for finding violations consists of the States legal obligations arising from the UN Charter and those parts of the UDHR which have become part of customary law.[85] The Special Rapporteurs and similar bodies have insisted above all on the principle of non-derogability of fundamental rights as a principle of general international law. Thus, the Special Rapporteur on Iran (Andrés Aguilar) maintained that the basic provisions of the UDHR have become customary international law and many of them norms of *ius cogens*.[86] In several cases, article 4 of the Covenant has been used as an 'authoritative interpretation' of the Charter and of the UDHR on this subject of human rights in states of emergency.[87]

A brief analysis of the case of South Africa will illustrate the general position of the UN organs in monitoring States' compliance with their human rights obligations in situations of emergency.

The Case of South-Africa. This case is interesting in so far as it refers to a State which is non-party to the Covenant, and which finds itself in a situation of emergency, and because it is a case which has received a good deal of attention from the international community and the UN organs.

In a recent article, a leading South-African constitutional lawyer has pointed out that 'in recent years . . . South Africa's security laws have received almost as much attention from the international community as her race laws'.[88] In fact, at least since 1967, the UN organs have been concerned with the situation of human rights in South-Africa under the security laws. As the opposition to apartheid from the black population grew, the South-African authorities had to resort to all kinds of measures to overcome the unrest. It is therefore interesting to see how the UN organs have reacted and to find out on which principles they have founded their analysis of human rights standards in emergencies.

---

[84] See the remarks of the ILA Report to the Seoul Conference, 1986, in respect of the cases of Bolivia, Guatemala, and Equatorial Guinea, pp. 19–20.

[85] See e.g. the *Report on Iran*, E/CN. 4/1985/20, pp. 8–9; and the *Report on Afghanistan* E/CN. 4/1985/21, p. 36.

[86] *Report on Iran*, E/CN. 4/1985/20, pp. 19–20.

[87] See the positions of Aguilar and Gross Espiel in E/CN. 4/1985/20, pp. 8–9, and E/CN. 4/1500, p. 52, respectively.

[88] J. Dugard, 'The Conflict between International Law and South-Africa Law', *South-Africa Journal of Human Rights*, 2 (1986), 16.

As in many other cases in which UN organs try to assess compliance with human rights obligations, States under scrutiny usually resort to the defence of domestic jurisdiction. South Africa is no exception, having pointed out that measures employed by a State to maintain national security fall exclusively within the domestic jurisdiction of that State.[89] Professor Dugard has shown that this position, which seems to be sound in principle, has two exceptions: emergency laws may in fact become a matter of international concern either when they become so arbitrary and oppressive that they violate human rights norms of customary international law, or when they are used by one racial group to suppress another in a manner which violates the non-discriminatory norms.[90] This position of Professor Dugard is in line with the present state of general international law on human rights.[91] It is clear that States cannot use the defence of emergencies to justify gross violations of human rights.[92]

Moreover, there are some rights which have already acquired the status not only of customary law but also of *ius cogens*; these rights are non-derogable in general international law even in states of emergency; they include: the right to life, the right to be free from slavery, torture, or other inhuman or degrading treatment or punishment, systematic racial discrimination, and prolonged arbitrary detention.[93] Therefore, even States non-parties to the Covenant are obliged to comply with these standards in states of emergency as a matter of customary international law.

The reports of the *Ad Hoc* Working Group on South Africa have over the years contained a description of gross violations of most of these fundamental rights which have taken place in South-Africa.[94] As is well-known, South Africa declared a formal state of emergency on 21 July 1985.[95] However, due to the strict security laws in force before that date,

---

[89] See the South Africa's reponse to Resolution 418 (1977) in its letter of 4 Nov. 1977 to the UN Secretary-General (quoted in Dugard, 'Conflict', p. 16).

[90] Dugard, 'Conflict', p. 16.

[91] See L. Henkin, 'Human Rights and Domestic Jurisdiction', in T. Buergenthal (ed.), *Human Rights, International Law and the Helsinki Accord*, (New York, 1977), pp. 21 ff. See also F. Ermacora, 'Human Rights and Domestic Jurisdiction,' *Recueil*, 124. 2 (1968), 371–451.

[92] Several UN General Assembly Resolutions have insisted over and over again on the need 'to cease using the state of emergency or siege for the purpose of violating human rights' (e.g. UNGA Resolution 31/124 of 15 Dec. 1979).

[93] See above, pp. 215 ff. (US *Restatement*). See below, pp. 251–2.

[94] See *inter alia* E/CN. 4/950, (1967); E/CN. 4/984 and Add: 1–18 (1969); E/CN. 4/1020 (1970); E/CN. 4/1050 (1971); E/CN. 4/1076 (1972); E/CN. 4/1111 (1973); E/CN. 4/1135 (1974); E/CN. 4/1159 (1975); E/CN. 4/1222 (1977); E/CN. 4/1270 (1978); E/CN. 4/1485 (1982); E/CN. 4/1985/8 (1985); E/CN. 4/1986/8 (1986); E/CN. 4/1987/8 (1987); E/CN. 4/1988/8 (1988).

[95] The state of emergency has been renewed annually, the last time on 8 June 1989. However, on 8 June 1990, President De Klerk announced the end of the State of Emergency (with the exception of the territory of Natal). See E/CN. 4/Sub. 2/1990/33, Add. 2 (17 Aug. 1990).

'black South-Africa may be said to have always lived under a permanent state of emergency'.[96] In fact, there is little difference between the standards applied by the Group before and after the formal declaration of emergency, mainly because these reports refer to the violation of those rights from which there can be no derogation. A brief analysis of these two phases will establish this point clearly.

Before the formal declaration of emergency in 1985: Before the declaration, the Working Group found several violations of some fundamental human rights. Thus, the Group attested the following practices against the *right to life* for which the Government was held responsible: deaths in prison, disappearances, killings by law enforcement officials as a consequence of using disproportionate force against unarmed demonstrators, and the imposition of capital punishment against generally accepted international standards, namely, for minor offences, in summary trials without full guarantees of fair trial and due process of law, with discrimination against black people in the majority of the death sentences, and retroactive application of the death penalty.[97]

In violation of the right to be *free from torture* and inhuman or degrading treatment or punishment, the Group found that there existed in South Africa an 'administrative practice' of torture and ill-treatment of prisoners and detainees, including children. The Group also found that the practice of *prolonged arbitrary detention* was against basic international standards; these detentions could last up to six months, with the detainee usually being kept incommunicado, without any contact with a counsel, and without being brought before a judge. Finally, the area of violations of the *principle of non-discrimination* has been illustrated in all the reports. Leaving aside the whole system of apartheid as a system based on discriminatory grounds,[98] the application of emergency laws have also been found to be discriminatory in several areas, e.g. in the application of capital punishment, in the restriction of the right to freedom of movement and residence, in the adscription to labour camps, etc.

Since the imposition of a formal state of emergency: Since the imposition of the state of emergency in 1985, the UN Commission on Human Rights has condemned 'the dramatic escalation of violation of human rights'.[99] In its preliminary evaluation of the state of emergency, the *Ad Hoc* Working Group has pointed out some practices against

[96] E/CN. 4/1986/6 Annexe (3 Sept. 1985), p. 1, 'Violations of Human Rights in Southern Africa: Report of the *Ad Hoc* Working Group of Experts'.

[97] The US *Restatement* also considers that capital punishment for minor offences is against customary international law when 'grossly disproportionate to the crime', ii. 164.

[98] For a good analysis of the problem, see V. Jaichand, 'International Human Rights Law, South Africa and Racial Discrimination', *Journal of Legislation*, 15 (Notre Dame Law School, 1988), 29–44.

[99] UN Commission on HR, 'Resolution 1988/9.

international standards along the same lines as those mentioned before.[100]
The legal standards which these practices violate were contained, according to the Group, in the UDHR, the ICCPR, and other international
instruments, as declaring general principles of law. Thus, the detention
laws, due to their arbitrary character and the lack of safeguards against
possible abuses, were in violation of article 9 of the UDHR and of
general principles of law; the provisions on capital punishment, discrimination, the practice of torture, and the use of excessive and disproportionate force by law enforcement officials were also found to be in
violation of other articles of the UDHR. Even if there is no general
agreement on the fact that the whole UDHR has already become general
international law, or that it is reflecting general principles of law, nobody
doubts that the standards which refer to those fundamental rights
mentioned by the Group have become general international law, and
some of them norms of *ius cogens*.

In the preliminary evaluation of the state of emergency, the UN Group
of Experts considered that part of the South-African legislation violated
the principle of proportionality, in the light of the standards of article 4 of
the Covenant.[101] Thus, the principle of proportionality is considered to be
a general principle of law when assessing the legality of restrictions on
human rights in emergencies also outside treaty law. Moreover, the use of
article 4 as declaratory of international standards of human rights in
emergencies for all States has also been adopted in a recent study of the
International Commission of Jurists; of all the principles of article 4,
this study especially underlines that of non-derogability.[102] Interestingly
enough, the Commission used the derogation clause of the Covenant
as the legal criterion for assessing the legality of human rights in emergencies after pointing out that South Africa does not deny that it has
international obligations to respect human rights, but had tried to
justify its non-compliance with them with the plea of necessity of *force
majeure*.[103] The International Commission of Jurists, in line with the
position of other international bodies, submits the legality of the plea of
necessity to the strict conditions of the main principles of the derogation
clause as the legal regime applicable in general international law.
According to this report, the gross violations of non-derogable rights
cannot be excused or justified on the basis of any supposed national
emergency.

---

[100] See E/CN. 4/1986/6, Appendix, 'Violations of Human Rights in Southern Africa:
Report of the *Ad Hoc* Working Group: Preliminary Evaluation of the State of Emergency,'
pp. 1–3.
[101] Ibid., p. 2, para. 7.
[102] G. Bindman, ed., *South-Africa: Human Rights and the Rule of Law.* (London, 1988)
[103] Ibid., pp. 148–9. For another excuse of the Government of South Africa, see below.
p. 254.

The case of South Africa, and the reaction of the UN organs, brings out several important points:

1. All States have international obligations to respect basic human rights and fundamental freedoms, even if they are not parties to any general international treaty on human rights;

2. Although the question of derogating measures from human rights standards taken in a state of emergency or on grounds of national security is in principle within the exclusive domestic jurisdiction of the State, when these measures result in gross violations of fundamental human rights they are also a matter of international concern;

3. The violation of some fundamental rights, such as those mentioned by the UN *Ad Hoc* Working Group, are never permitted, not even in order to deal with the emergency: these standards are part of customary international law (as the USA *Restatement* has pointed out) or of the general principles of law recognized by civilized nations (as the Working Group has indicated several times);

4. The main principles of the derogation clause, especially the principle of proportionality, constitute important guide-lines in the assessment, as a matter of general international law, of the legality of restrictions on human rights even for those States non-parties to the Covenant, when these States try to excuse their non-compliance with human rights obligations under customary law with the plea of necessity.

(d) The Acceptance of these Principles by States Non-Parties to the Human Rights Treaties

Finally, another effective way by which treaty-norms may become customary international law is through the recognition of those standards by non-parties to human rights treaties. Due to the quasi-universal character of some of these-multilateral treaties (e.g. the ICCPR) containing fundamemntal principles on human rights, it is very unlikely that States non-parties to these treaties will explicitly reject these norms of a general character; this fact will lead to an explicit recognition by States of the binding character of these norms. The evidence examined in this chapter as far as some of the principles of the derogation clause are concerned confirms this position. Within the Inter-American system, no State, not even those affected by their application, has ever attacked on legal grounds the application by the IACHR of the principles of the derogation clause to States non-parties to the American Convention.[104] Likewise, within the UN system no State has ever objected to the work and the position of principle taken by the Special Rapporteur on the

[104] See above, p. 242.

application of the principles of the derogation clause to States non-parties to the Covenant. The same can be said in respect to the application of the same principles by the UN Commission on Human Rights and other specialized organs.[105]

A particular expression of implied acceptance in building customary international law on human rights has been mentioned by Schachter when commenting on the reaction of States to international criticism because of violations of human rights. He has said:

> It is important as well to consider whether the conduct criticised is defended by the perpetrators as legitimate, or, as is often the case, denied on factual grounds. In the latter event, one may plausibly infer that the State accepts the principles involved while denying its application in the particular circumstances.[106]

This, for instance, seems to be the case of South Africa, which did not contest the standards invoked by the Special Rapporteur but denied that the conduct complained of ever took place.[107] This position undoubtedly strengthens the legal force of the standards of the derogation clause invoked by the Special Rapporteur.

5. THE APPLICATION OF THE PRINCIPLES OF THE DEROGATION CLAUSE TO OTHER AREAS OF INTERNATIONAL LAW

(a) State Responsibility for Injury to Aliens (Human Rights of Aliens in States of Emergency in General International Law)

State responsibility for injury to nationals of other States is a well-known area of international law which preceded in time the international protection of the human rights of individuals in general. There is no doubt that nowadays there is a certain overlap and cross-fertilization between these two branches of international law, because both ultimately try to protect individual rights against possible abuses by the State. However, they also maintain a certain autonomy. In addition to those rights that a State is obliged to respect for all persons subject to its jurisdiction, under the law of State responsibility for injury to aliens there are some kinds of injuries to aliens that have not been recognized as violating human rights.[108] It is interesting in this context to look at which principles regulate the human rights of aliens in situations of emergency according to this branch of international law. The recent US *Restatement* has tried

---

[105] See above, p. 246.

[106] Schachter, 'General Course', p. 338.

[107] E/CN. 4/Sub. 2/1989/30/Add. 1, Reply of the Government of the Republic of South Africa, pp. 6–13.

[108] See the recent formulation of these obligations in the USA *Restatement*, ii. 184 ff.

to answer these questions according to general principles of international law governing State responsibility.[109]

After recalling how there are some rights in contemporary international law which are deemed non-derogable and binding on all States, the *Restatement* provides a detailed explanation of the legal regime of the derogation clause of the Covenant (art. 4), and adds: 'the derogations permissible in emergency under the Covenant are presumably permissible also under this section (State responsibility for injury of aliens) in relation to nationals of other States as a matter of customary law'. Thus, subject to the principle of proportionality contained in article 4 of the Covenant, 'a State could lawfully seize or regulate property and detain or regulate the activities of persons, whether nationals or aliens'.[110] In other words, the conclusion of the US *Restatement* is that the principles of the derogation clause of the Covenant, especially the principle of proportionality, are prima facie the legal regime applicable to derogations of human rights of aliens in emergencies as a matter of customary international law. This position gives further evidence that the main principles of the derogation clause are emerging as principles of general international law, and thus works in the same direction as the findings of the IACHR, the ILO organs, and the Special Rapporteur of the UN HR Commission.

This position of the US *Restatement* also has a supplementary importance. At the beginning of this chapter, it was shown that the legal regime of the derogation clause is an application and adaptation of the doctrine of State necessity to human rights in states of emergency. This doctrine in international law is envisaged as a justification for a non-performance of a legal obligation when certain strict conditions are met, and it therefore excludes State responsibility. The position of the ILO organs on the legal regime applicable to those ILO Conventions dealing with human rights issues, but without a derogation clause, has been to apply to that situation the doctrine of necessity in general international law, although, as has been seen, the formulation of some of those principles was borrowed from the derogation clause of the Covenant.[111] The position of the US *Restatement*, on the other hand, goes further, in the sense that when it is necessary to apply general principles of international law to this subject (human rights of aliens in emergencies), the *Restatement*, instead of applying the doctrine of necessity, which would seem to be prima facie the obvious choice, applies article 4 of the Covenant (which contains similar principles) as a matter of general international law. This position further strengthens the use of the main principles of the

---

[109] Ibid., p. 185, comment *a*.
[110] Ibid., p. 189, comment *h*.
[111] For the doctrine of the ILO organs, esp. in the *Greek* and *Polish cases*, see above, Ch. 9.

derogation clause as general international law, the assumption being that these principles contain the appropriate adaptation of the doctrine of necessity to human rights in emergencies.

(b)   The Law of Belligerent Occupation: The Case of Israel

The main principles of the derogation clause could play an important role in the application of human rights treaties to occupied territories. In fact, the flexibility of the derogation clause (mainly through the principle of proportionality) could complement the Fourth Geneva Convention standards, and balance the two paramount principles of the general regime of occupation under international law, namely, the military imperatives of the occupation and the well-being of the population.[112] The case of the Israeli occupied territories could illustrate the application in general international law of this sound position.

The situation in the occupied territories is a special one under international law. First of all, because the main international instrument applicable, as far as human rights are concerned, is the Fourth Geneva Convention of 1949 which deals with the law of occupation. Secondly, because, even if the situation in these territories can be classified under article 4 of the ICCPR as a 'public emergency threatening the life of the nation', due to the continuous unrest, Israel has not ratified the Covenant, and therefore the derogation clause could not be applied as treaty law. Thirdly, because although the Fourth Geneva Convention contains detailed legislation on the human rights of civilians in occupied territories, it is obvious that there is a need for this to be widened to include other human rights as defined in other international instruments like the UDHR or the ICCPR, especially because of the prolongation of the military occupation. Thus, Cohen considers that

The UDHR and the International Covenants, as interpretations of the UD, could be used to supplement the traditional law of belligerent occupation in order to ensure for the civilian population the maximum of human rights protection in occupied territory during prolonged belligerent occupation. Although the UD is not binding and the Covenants are binding only on those States parties to them, nevertheless the application of the law of human rights as expressed in these instruments could serve as a guide to the occupying power in the treatment of occupied populations under a situation of prolonged belligerent occupation.[113]

According to the same view, a government is bound in situations of occupation by three fundamental principles: the principle of non-

---

[112] L. Oppenheim, *International Law*, ii. *Disputes, War and Neutrality*, 7th edn., ed. H. Lauterpacht (London, 1952), 430.

[113] E. Cohen, *Human Rights in the Israeli-Occupied Territories, 1967–1982* (Manchester, 1986), p. 9.

derogability of fundamental rights, the principle of consistency (with reference to its international obligations under the Hague and Geneva Conventions, and other possible international instruments), and the principle of non-discrimination. Moreover, the principle of proportionality is also relevant in the sense that 'the closer an armed conflict situation grows into a peacetime situation, the more human rights contained in the Universal Declaration and the Covenants apply.'[114]

On the other hand, Meron has pointed out that

a State in the security situation of Israel is, no doubt, entitled to invoke reasons of security in order to derogate from various provisions of the Universal Declaration, but her purpose should be to achieve the maximum recognition of human rights as may be consistent with necessary and legitimate security considerations.[115]

Meron agrees with the view that derogations to human rights obligations are acceptable only if events make them necessary and if they are proportionate to the dangers that those events represent.[116]

These principles of necessity and proportionality are precisely among the most important principles of the derogation clause of the three treaties. Thus, this position confirms the general thesis of this section on the applicability of some of the fundamental principles of the derogation clause as a general international law which is therefore binding on States non-parties to the Covenant. Interestingly enough, Amnesty International has also referred to article 4 of the Covenant in order to assess the rights of the population under belligerent occupation.[117]

6. THE PRINCIPLES OF THE DEROGATION CLAUSE AS 'GENERAL PRINCIPLES OF LAW'

Another complementary piece of evidence reinforces the position that the main principles of the derogation clause can be deemed to constitute general international law; these principles could be considered to be general principles of law recognized by most of the world legal systems according to article 38 of the ICJ Statute. In fact, Meron has pointed out that 'it is surprising that the "general principles of law recognized by civilized nations". . . have not received greater attention as a method for

---

[114] Ibid., p. 65.
[115] T. Meron, 'West Bank and Gaza: Human Rights and Humanitarian Law in the Period of Transition,' *Israel YBHR* 9 (1979), 113.
[116] Ibid.
[117] Amnesty International, *Report and Recommendations of an Amnesty International Mission to the Government of the State of Israel (3–7 June 1979) including the Government's Response and A. I. Comments* (London, 1980), p. 47.

obtaining greater legal recognition for the principles of the Universal Declaration and other human rights instruments.' He adds: 'as human rights norms stated in international instruments come to be reflected in national laws, . . . article 38(1)(c) will increasingly become one of the principal methods for the maturation of such standards into the mainstream of international law.' Thus, the distinction between customary law and general principles will eventually become blurred.[118] The recent US *Restatement* also considers one of the sources of international obligations on human rights the 'general principles of law common to the major legal systems of the world.'[119]

In the Introduction, it was contended that the institution of states of emergency and its encroachment upon individual rights is a very old institution in public law and can be traced back to Roman times. The institution came into being in exceptional situations of a temporary character when the life of the nation was threatened. Modern public law has recognized the institution and has incorporated it. Several writers have confirmed that most modern national legislation contains the institution, and that it has similar characteristics to those of the derogation clause. Thus, Questiaux affirms that the guarantees in international law (through the principles of the derogation clause) are a reflection of those contained in municipal systems.[120] As has been pointed out in reference to the European Convention, the derogation clause is a transposition to that Convention of the same principles recognized in the legislation of the European States. The same can be said in respect of the American Convention and in relation to the legislations of American States.[121] Moreover, a recent study has contended that there is a striking convergence between the principles evolved in public international law and the principles codified in constitutions or judicially evolved in different legal systems.[122] At the same time, Despouy considers that international law has put together and systematized guarantees that are already recognized in municipal legislation; by doing that, international law, through the

---

[118] Meron, *Human Rights and Humanitarian Norms*, pp. 88–9.

[119] US *Restatement*, ii. 152. See the interesting reflections of Merrills on the role played by some general principles of law in the jurisprudence of the European Court of HR, 'Development', pp. 160–83.

[120] Questiaux, *Study*, p. 20, para. 73.

[121] Ergec, 'Les Droits de l'homme', pp. 56–100; Ganshof van der Meersch, 'La Protection des droits de l'homme en droit constitutional comparé', in *Rapports generaux du IXᵉ Congrès international de droit comparé (Teheran 1974)* (Brussels, 1977), pp. 659 ff. For the American System see Norris and Desio, 'The Suspension of Guarantees'. The regime on emergencies found in the new constitutions of African States is in line with the principles of the derogation clause; see B. D. Nwabueze, *Constitutionalism in the Emergent States*, (London, 1973), pp. 174–353.

[122] A. M. Singhvi, 'The Law of Emergency Powers: A Comparative Study', Ph. D. thesis (1986, Cambridge), p. 638.

derogation clause of the Covenant, has filled some gaps in domestic legislation.[123] Other writers have seen in the derogation clause of the Covenant, which reflects the doctrine of suspension of human rights in emergencies, a general principle of law recognized by the major legal systems.[124]

It is true that considerable caution is required in transposing the principles of municipal law to the plane of international law, even if these principles are contained in most legal systems; in international law this procedure has been used in a restrictive way, its main purpose being to fill the gaps left by the other sources of international law (treaty and custom). However, general principles seem to have had a great importance, especially in this area of the relation between the State and individuals.[125] If this is true, recourse could be had to general principles of law, in the case of human rights in emergencies, in order to fill the gaps in the rules applicable for States non-parties to human rights treaties with a derogation clause, something which has in fact been done by the IACHR and the ILO organs. The fact of the inclusion of principles of municipal law on human rights in emergencies, reflected in the derogation clause of the three main treaties, proves that these principles are appropriate for application at the international level.[126] In this respect, Schachter believes that national law principles are often suitable for international application in the area of human rights.[127]

---

[123] Despouy, 'Etats d'exception', p. 4.

[124] Kaufman and Mosher, 'General Principles of Law and the UN Covenant on Civil and Political Rights', *ICLQ* 27 (1978), 610–13.

[125] See the interesting reflections of Schachter on this point in 'General Course,' pp. 74 ff.

[126] Schachter finds that the most important limitation on the use of municipal law principles in the international arena arises from the requirement that these principles must be appropriate for application on the international level, ibid., p. 78.

[127] Ibid., p. 79.

Conclusion of the Inquiry: The Principles which Constitute Emergent Principles of General International Law Governing Human Rights in States of Emergency

In the light of all the evidence adduced in the last part of this study, both in respect of the principles of the doctrine of necessity in general international law, and in respect of the principles of the derogation clause, it would seem that there are some principles which constitute emergent principles of general international law.[1] These principles are, first of all, those common to the doctrine of necessity and the derogation clause, namely, the principle of exceptional threat, the principle of non-derogability of fundamental rights, and the principle of proportionality. As well as these common principles, one should add the principle of non-discrimination which also seems to be in a very advanced state of crystallization.[2] Moreover, it could be said that some of these substantive principles are in fact already principles of general international law. This seems to be the case with the principles of proportionality and non-discrimination, and, at least as far as the four common non-derogable rights are concerned, the case with the principle of non-derogability. The same could probably be said in respect of the principle of exceptional threat. Although some writers have mentioned the principle of proclamation among the emergent principles, it seems that, for the reasons mentioned above, it is not as close to becoming customary law as the

---

[1] The concept of 'emergent rules (or principles) of customary international law' was used by the ICJ in the *North Sea Continental Shelf cases*, see paras. 61, 62, 63, 69. See also dissenting opinion of Sorensen, p. 244.

[2] The contention that these principles are emerging as principles of general international law has been tentatively pointed out by some writers. Thus, one of the commentators of the 'Siracusa Principles' has noted that 'compliance with these principles is becoming part of the customary international law of human rights': O'Donnell in *HR Quarterly* 7 (1985), p. 34 (see also Hartman's commentary, p. 131). For the 'Siracusa Principles', see principle No. 75. See also O'Donnell, *States of Exception*, *Int. CJ Review*, 21 (1978), 53, in which he considered the principles of exceptional threat, proportionality, and non-derogability as having 'universal validity'. The Int. Commiss. of Jurists has pointed out that some of the requirements of article 4(1) of the ICCPR suggest 'the emergence of rules of customary international law', in *State of Siege or Emergency and their Effects on Human Rights: Observations and Recommendations of the ICJ* (Aug. 1981) (repr. in E/CN. 4/Sub. 2/NGO. 93 (1981)). The 'Oslo Statement' considers art. 4 as containing international law norms of universal validity (pp. 2 ff.). See also ILA Montreal Report (1982), p. 94.

other four.[3] These four substantive principles are the most important and would provide both States and international tribunals with fundamental criteria with which to judge the conduct of States in respect of human rights in emergencies according to general international law. There follows a brief comment on these principles.

1. THE PRINCIPLE OF EXCEPTIONAL THREAT

According to the three main treaties, the derogation from human rights standards in emergencies is legitimate only in the case of 'war or public emergency threatening the life of the nation'. On the other hand, according to general international law, the plea of emergency (or necessity), is accepted as a justification for non-compliance with international obligations only in 'exceptional circumstances' and when there is no other alternative course of action. Of course, it is theoretically possible for these two standards, the one arising from treaty law and the other from general international law, to be different, for example, in the case of States which have agreed in a treaty to declare states of emergency and suspend human rights in conditions which are stricter than those required in general international law.[4] However, if one looks at the qualifications placed upon this standard by general international law, this does not in fact seem to be the case. Thus, Ago refers to the existence of an 'extremely grave and imminent peril', and the ILO Special Commission, in the case of Poland, refers to 'circumstances of extreme gravity'.[5] Also the UN Special Rapporteur on States of Emergencies, when trying to establish a general standard of emergency, speaks of 'exceptional circumstances . . . involving a serious and imminent threat to a country'.[6] Therefore it does not seem that these two standards are prima-facie different.

Of course, concepts such as 'exceptional circumstances' and 'grave danger' are difficult to define in abstract terms and to apply in particular cases. However, there is another important reason in favour of the application in general international law of a standard similar to that carefully worked out by the jurisprudence of the international bodies entrusted with the application of the main treaties on human rights. These institutions have developed an important jurisprudence in the

---

[3] See above, pp. 232–3.
[4] See the position of Graefrath in the UN HR Committee: CCPR/C/SR. 349, p. 10, para. 37. See also Meron, *Human Rights and Humanitarian Norms*, p. 218.
[5] 'Ago Report', p. 20, para. 19. For the *Polish case* under ILO see *Polish case*, p. 127. See also Bin Cheng, *General Principles*, p. 31, 'under very exceptional circumstances'.
[6] E/CN. 4/Sub. 2/1985/19, p. 3.

interpretation of the concept of public emergency. This jurisprudence, which can be deemed to be quite similar in the work of the European organs, the UN Human Rights Committee, and the Inter-American Court and Commission, has found a careful balance between the need of the State to derogate in order to protect the nation, and the enjoyment of individual rights. Seen in this light, the concept of 'public emergency threatening the life of the nation' is in fact a very workable and realistic one, and there is no doubt that it can be of assistance to other international organs when applying the concept of 'exceptional circumstances' in general international law. In fact, some of these international bodies have already, not surprisingly, made reference to that jurisprudence, and in particular to the concept of public emergency as elaborated by the European organs in almost forty years of case-law.[7]

In the *Greek case*, for instance, the ILO organs, applying general international law, reached the same conclusion as the European Commission when it applied the derogation clause, in the sense that nothing emerged from the evidence before the ILO organs to allow them to conclude that in Greece in 1967 there existed a state of emergency or exceptional circumstances which would justify temporary non-compliance with the ILO Conventions.[8] This finding supports the position that an emergency must attain a certain degree of gravity in order to justify a temporary non-compliance with human rights obligations.

2. THE PRINCIPLE OF PROPORTIONALITY

Similar considerations can be made about the principle of proportionality. As has been continually seen in the present 'inquiry', this principle appears in almost all the findings of judicial or quasi-judicial bodies as one of the main principles used in order to assess the legality of measures of derogation. The doctrine has also insisted on the importance of this principle in derogations under general international law. Thus, Higgins has pointed out that 'derogations to human rights obligations are acceptable only if events make them necessary and if they are proportionate to the danger that those events represent'.[9] Moreover, this principle has been regarded as a general principle of international law whose special

[7] See e.g. the reference of the IACHR in the *Miskitos case*, p. 115; See also A/10285 (1275), p. 47, para. 122 (UN *Ad Hoc* Working Group on Chile).

[8] ILO, *Greek case*, p. 26. For the finding of the European Commission on this point, see *YBECHR* (1969), p. 76.

[9] R. Higgins, 'Derogations under Human Rights Treaties' *BYBIL* 48 (1976–7), 282–3. See also Meron, 'West Bank and Gaza', p. 113; and McDougal *et al.*, 'Human Rights and World Public Order: A Framework for Policy-Oriented Enquiry', *AJIL* 63 (1969), 267.

relation with limitations on human rights has been underlined.[10] The principle of proportionality refers not only to the nature of the measures taken, in the sense that they must be proportionate to the threat, but also includes what the IACHR has called 'the principle of temporariness' (which means that they cannot last longer than the emergency itself), and the limitation that they must be extended in geographical terms only to those places affected by the emergency. These limitations are the logical consequences of the principle of proportionality and would therefore seem to be applicable in general international law as well. In the interpretation of the principle of proportionality, the rich jurisprudence of the European organs could be of great assistance for other international bodies when applying the principle in general international law.

3. THE PRINCIPLE OF NON-DISCRIMINATION

There is no doubt that the principle of non-discrimination occupies an important position in the field of human rights; moreover, the UN Charter contains several references to the enjoyment of human rights without any discrimination.[11] Non-discrimination on racial grounds, and also on religious or sexual grounds, is considered a rule of customary international law.[12] Due to the fundamental importance of this principle in international law, and to the fact that it is also mentioned as a condition for the lawful derogation of rights in the UN Covenant and the American Convention, it seems that it should also be considered as a principle applicable in general international law relating to human rights in states of emergency. In any case, if a State wishes to take measures which establish differences between several ethnic or religious groups of the population, it has to prove that it is necessary to take these measures in order to overcome the emergency, or, in other words, that the measures have an objective and legitimate justification, and that they are also proportionate to the threat. This standard established by international bodies when applying treaty law should be extended to general international law.[13]

---

[10] See Ch. 5, Sect. 1.
[11] For the importance of the principle in the human rights field, see Jimenez de Arechaga, 'International Law', pp. 174–7. See also McKean, *Equality and Discrimination*, pp. 277–87.
[12] Brownlie, *Principles*, pp. 598–601; US *Restatement*, pp. 165–6; Lillich, 'Civil Rights', pp. 132–3.
[13] The UN Special Rapporteur on Emergencies has suggested that all States should include in their domestic legislation the principle of non-discrimination as a non-derogable principle in emergencies: E/CN. 4/Sub. 2/1990/33, p. 5.

4. THE PRINCIPLE OF NON-DEROGABILITY OF FUNDAMENTAL RIGHTS

This is another principle of crucial importance in the legal regime of human rights in emergencies. The principle *in se* has doubtless already emerged as a principle of customary international law; this means that even in situations of emergency there are some fundamental rights which cannot be derogated from. This principle has been considered as a general principle of law recognized by civilized nations, according to article 38 of the ICJ Statute.[14] The problem, of course, is to determine which rights are to be considered non-derogable. As has been seen, the three treaties establish different lists of non-derogable rights. And yet, at least the four common non-derogable rights can be assumed to constitute norms of *ius cogens* which are therefore non-derogable even for States non-parties to these treaties.[15] Furthermore, the ILA Montreal Report considers the list of non-derogable rights of the three treaties to be 'indicative of some emerging customary norms in the area of non-suspendable human rights in states of emergency'.[16]

Although this list is a useful indicator of non-derogable rights, it is not immediately clear that all those rights are in fact non-derogable in general international law. In fact, in Chapter 4, two objections to the lists of non-derogable rights were put forward. The first objection was that the lists do not contain some fundamental rights which are indispensable for the protection of human beings and very much at risk in emergencies (special mention was made of some minimum guarantees against arbitrary detention and some others concerning fair trial). The second objection was that the list of non-derogable rights in the ICCPR and in the ACHR contain some rights which are not so indispensable and at risk in emergencies.[17]

Although, as Meron has pointed out, the international community as a whole has not established a uniform list of non-derogable rights and there is no immediate prospect of a consensus reaching beyond the four common non-derogable rights,[18] some remarks were made about the

---

[14] Questiaux, *Study*, p. 19. See also Kaufman and Mosher, 'General Principles', p. 612; Despouy, 'L'État d'exception', p. 10.

[15] According to the US *Restatement*, 'prolonged arbitrary detention' would also be against *ius cogens* norms; however, if the conditions of the derogation clause are met in a state of emergency, the detention would not presumably be arbitrary (pp. 174–5).

[16] ILA Montreal Report (p. 20). On the other hand, the ILA, Paris Report contains 16 non-derogable rights which are believed to be general principles of law (p. 70). See also ILA Warsaw Report, p. 39.

[17] See above, pp. 97–8.

[18] Meron, 'On a Hierarchy of International Human Rights', *AJIL* 80 (1986), 16. The UN Special Rapporteur on Emergencies has announced his intention of drawing a list of non-derogable rights: E/CN. 4/Sub. 2/1989/30/Add. 2, Rev. 1, p. 2.

criteria that should govern a possible list. First, the list of non-derogable rights should be short and should include only those rights which must be protected because they are so fundamental and in danger of being violated in emergencies. According to this criterion, the four common non-derogable rights should be included in the list. Secondly, minimum guarantees against arbitrary detention and others relating to fair trial should also be included. This is for two reasons: first, because they are so fundamental that the derogation of these guarantees would put at risk the right to life and the right to be free from torture, and secondly, because no public emergency could justify the derogation of these minimum guarantees.[19] There is no reason why these same fundamental rights and guarantees should be derogated from by States non-parties to the three treaties, as a matter of general international law. The findings of some international organs applying human rights standards according to general international law support this position.[20]

In any case, the principle of proportionality, which is the main criterion for derogation in general international law, provides a strong safeguard against possible doubts in concrete cases concerning the non-derogable character of certain rights. In fact, a State which wants to derogate from some fundamental human rights in a situation of emergency according to the doctrine of necessity must prove that the derogation is necessary and proportionate to the threat.[21]

---

[19] See above, Ch. 4.

[20] See e.g. the position of the IACHR (above, pp. 237 f.). For the position of the UN Commission on HR and other 'Ad hoc' bodies, see above, pp. 247–9.

[21] See the interesting remarks by Meron in 'On a Hierarchy of International Human Rights', p. 20.

Conclusions of Part II

1. The importance of identifying general principles governing human rights in states of emergency in the context of general international law has been shown at the beginning of this part of the study. There are two main reasons why it is important to identify them: first, because there are some human rights treaties which have no derogation clauses (i.e. the African Charter and some ILO Conventions) and which therefore include no explicit indication of the legal regime applicable in emergencies; and secondly, because there are many States which are not yet parties to human rights treaties but which declare states of emergency and derogate from human rights standards.

2. There is no doubt that at this stage in the evolution of international law, all States have obligations to respect human rights arising from customary international law. However, one of the problems is to agree upon a list of customary norms. Nevertheless, the object of this study is not the elaboration of a complete list of customary norms, but rather the identification of those principles which could justify non-compliance with these obligations in a state of emergency.

3. In order to identify these principles, two main lines of inquiry have been followed. The first has focused on the doctrine of State necessity, which it has found to be the doctrine which best corresponds to situations of emergency. The doctrine of necessity is a well-known circumstance excluding responsibility for wrongfulness in international law, and it has been recognized by the ILC in the draft articles on the law of State responsibility. However, this doctrine is submitted in its application to strict conditions in order to avoid abuses. These conditions can be expressed in the terminology used to describe some of the principles contained in the derogation clause: the doctrine of necessity must respect three fundamental principles, namely, the principles of exceptional threat, proportionality, and non-derogability of peremptory norms. An illustration of the application of this doctrine of necessity in several cases relating to human rights in emergencies by the ILO organs has been examined. In the absence of any derogation clause in some ILO Conventions, the ILO organs have in the *Greek* and *Polish cases* identified the plea of emergency with that of necessity in general international law, and have applied the principle of exceptional threat and that of proportionality as principles of general international law governing this

matter. In other cases, the ILO organs have also made reference to the principles of temporariness and non-discrimination.

4. The second line of inquiry (the object of the study in Chapter 10) has analysed the contention that some of the principles of the derogation clause are emerging as principles of general international law. This contention is in part based on the fact that the derogation clause has been seen as a particular application (or adaptation) of the doctrine of necessity to the subject of human rights in states of emergency in the main treaties. However, a close examination of the derogation clause has shown that it contains three types of principles. The first type of principle is a clear reflection of the principles of the doctrine of necessity (the principles of exceptional threat, proportionality, and non-derogability of fundamental rights). The second type of principle, that is, the principle of non-discrimination, is an important principle in the field of human rights, although it is not explicitly mentioned among the principles governing the doctrine of necessity in general international law. Finally, the third type of principle is that of a 'procedural' character, namely, the principles of proclamation and notification. Even if these latter principles seem to be more suitable for application within the framework of treaty law, they could in theory become customary norms in the future. However, the principles which appear to be emerging as strong candidates for becoming general principles of international law are the principles belonging to the first two types, in other words, the principle of exceptional threat, the principle of proportionality, the principle of non-derogability, and the principle of non-discrimination.

5. A detailed examination of all the existing evidence supports the contention that these principles are emerging as customary norms. This is so, *first of all*, not only because of the fundamental importance of the UN Covenant on Civil and Political Rights as a general multilateral treaty of a quasi-universal character, but also because of the repetition of these principles of the derogation clause in the other regional treaties dealing with the same rights. Thus, the European and the American Convention, the two most important regional treaties on the subject, contain similar principles in their derogation clauses. *Secondly*, the decisions of international organs in this field have confirmed the applicability of these principles of the derogation clause as norms of general international law. Thus, the IACHR has applied these principles of the derogation clause outside treaty law, in the sense that they have been applied to States non-parties to the American Convention. In its constant jurisprudence, the IACHR has held that these principles of the derogation clause constitute not only the 'most accepted doctrine' in the American Continent and therefore regional customary law, but also the 'most accepted doctrine internationally', as it is embodied in the derogation clauses of the three

treaties under consideration. This position supports the contention that these principles can be deemed to constitute principles of general international law.

6. The recent developments within the organs of the United Nations also support this contention. Thus, after a period in which the UN Commission and Sub-Commission on Human Rights studied the question of human rights in emergencies within the framework of treaty law, the recently appointed Special Rapporteur on States of Emergency had to find some international standards applicable to States non-parties to the Covenant. This need arose from the fact that his mandate includes the study and assessment of the legality of states of emergency and the suspension of human rights of all States, and not only those parties to the Covenant which are subjected to the derogation clause as a matter of treaty law. Following the same line as the IACHR, the Special Rapporteur has considered some of the principles of the derogation clause to be principles of general international law. Moreover, many States have sent the information requested by the Special Rapporteur following the fundamental principles of the derogation clause. The Commission and the special organs, such as the *Ad Hoc* Working Groups and Special Rapporteurs, have in particular cases applied some of these principles to States non-parties to the Covenant.

7. The acceptance by States of this process of formation of customary norms has been shown by the fact that no State has rejected the position of principle taken by the international monitoring organs on the application of these principles of the derogation clause outside treaty law. This has been seen within the Inter-American system, and in respect to the new developments occurring within the UN Commission on Human Rights through the Special Rapporteur. The same could be said with respect to the application by the ILO organs of some of these principles arising from the doctrine of necessity in similar cases.

8. Further evidence also supports the contention that the main principles of the derogation clause are emerging as principles of general international law: the principles have also been considered to be '*general principles of law*' recognized by civilised nations, according to article 38(1)(c) of the ICJ Statute. It is well-established that the institution of 'state of emergency' permitting derogations from human rights is present in most legal systems; moreover, the principles regulating the institution in human rights treaties is considered to be a reflection of the legal regime of national legislations.

9. In the light of all this evidence, it seems that at least the principles of exceptional threat, the principle of proportionality, the principle of non-derogability of fundamental rights, and the principle of non-discrimination can be considered as emergent principles of general

international law in a very advanced state of crystallization. Moreover, it could be said that some of these principles are already principles of general international law. This seems to be true in the case of the principles of proportionality and non-discrimination, and of the principle of non-derogability of the four common rights recognized in the three main treaties; in fact, this last principle can be considered not only as a norm of customary law but also of *ius cogens*. The principle of proclamation is another candidate to become a customary principle in this field, but its crystallization seems at the present time to be more remote than that of the other principles.

Concluding Remarks

Due to the fact that detailed conclusions have been included at the end of each chapter, there is no need for them to be repeated here; there follow, therefore, just a few concluding remarks.

In the first part of the study, it was noted that there is a strong similarity between the derogation clauses of the three main treaties; in fact, they contain the same main principles, with a very similar wording. This fact is not surprising if one takes into account the history of the *travaux préparatoires* of the three treaties and their mutual influence. These principles establish a legal regime governing the human rights obligations of States in emergencies; these principles put precise and strict limits on the conduct of States in emergencies. These principles are: the principles of exceptional threat, proclamation, notification, non-derogability of fundamental rights, proportionality, non-discrimination, and consistency.

Moreover, the jurisprudence of the international bodies entrusted with the interpretation and application of the derogation clause under the three treaties has been similar. In fact, the jurisprudence of the European Commission and Court, which has the longest case-law concerning emergencies, has had a remarkable influence on the construction of the derogation clause by the other more recent international bodies. In other words, although the interpretation of the derogation clause could in theory have been different (due to the differences between the European Convention and the other two treaties), the practice of the UN Human Rights Committee and that of the Inter-American Commission and Court show a similar construction of the principles of the derogation clause. This is an important and welcome development, because the European organs have produced a jurisprudence which is very successful in balancing the need of States to overcome the emergency and the respect for fundamental human rights. The fact that, over the last decade and in fulfilment of their task of monitoring States' compliance with human rights obligations, the UN Human Rights Committee and the Inter-American Commission of Human Rights have strongly insisted on the application of the principles of the derogation clause has strengthened the importance of these principles.

One of the main objectives stated at the beginning of our inquiry, that is, the analysis of the principles governing human rights in states of emergency in general international law, has been carried out in the final part of this study. Four principles have been identified as constituting emergent principles of general international law in an advanced state of crystallization; moreover, it could probably be said that some of them are already principles of general international law. This means that, in emergencies, even States non-parties to international treaties with a derogation clause have to respect some principles and cannot use the plea of emergency as an excuse for non-compliance with these principles. Among the principles which emerge as customary international law, the principle of exceptional threat, the principle of non-derogability of fundamental rights, the principle of proportionality, and, finally, the principle of non-discrimination have special importance. These principles establish precise and strict limits on the conduct of States in situations of emergencies.

Thus, States can use derogations from human rights obligations arising from customary law only in exceptional situations, that is, when the emergency reaches a certain level of gravity. Minor political crises or disturbances do not justify the declaration of emergency and the derogation from human rights. Secondly, there are some fundamental human rights from which no derogation whatsoever is permitted in emergencies. This principle of non-derogability of fundamental rights has already become customary international law, at least as far as the four common non-derogable rights are concerned and probably also in respect of some minimum guarantees both against arbitrary detention and ensuring due process of law. Thirdly, the derogation of rights in emergencies is limited by two important conditions: one is the principle of proportionality and the other is the principle of non-discrimination. Accordingly, derogating measures must be strictly required by the exigencies of the situation, and cannot involve arbitrary discrimination.

The application of these principles as general international law has been shown by means of the analysis of the practice of the ILO organs, the IACHR, the new procedure taken under the UN Special Rapporteur on States of Emergency, and the UN Commission on Human Rights. The practice of these bodies in a few important cases is the first evidence of the emergence of these principles as general international law. Of special importance has been the application by the IACHR of the main principles of the derogation clause to those States members of the OAS but not parties to the American Convention, as principles reflecting 'the most accepted doctrine internationally'. This position on the emergent character of these principles as customary norms can also be seen in the new developments taken by the UN Special Rapporteur. At the same time, the potential significance of these principles has been underlined by their being applied to two other areas of international law: State responsibility for injury to aliens (human rights of aliens in states of emergencies), and the law of belligerent occupation.

Furthermore, the international organs under the three treaties have produced an important jurisprudence concerning the interpretation of these principles; this jurisprudence will undoubtedly have a notable influence on the way in which these principles crystallize as principles of customary international law. Concepts such as the kind of emergency which justifies derogations ('a public emergency threatening the life of the nation'), the construction of the principle of proportionality of the measures of derogation, the concept of the margin of appreciation, etc., will assist international monitoring organs in the task of assessing standards of human rights in emergencies according to general international law.

In conclusion, this study has shown the existence of precise limits on States' derogations from human rights standards in situations of emergency not only in treaty law but also according to general international law. The identification of these standards in general international law is extremely important insofar as many States are not yet parties to the UN Covenant on Civil and Political Rights. Moreover, international monitoring bodies have sometimes been unaware of the importance of these human rights obligations in emergencies and have therefore not exercised their functions with full effectiveness.

Select Bibliography

Ago, R., 'The International Wrongful Act of the State, Source of International Responsibility', *YBILC* 2. 1 (1980) ('Ago report').

Alexander, G., 'The Illusory Protection of Human Rights by National Courts during Periods of Emergency', *Human Rights Law Journal*, 5, (1984), 1–67.

Alston, P., 'Conjuring up New Human Rights: A Proposal for Quality Control', *AJIL* 78 (1984), 607–21.

Alston P., and Simma, B. 'First Session of the UN Committee on Economic, Social and Cultural Rights', *AJIL* 81 (1987), 747–56.

—— 'Second Session of the UN Committee on Economic, Social and Cultural Rights', *AJIL* 82 (1988), 603–15.

Amnesty International, *Argentina: The Military Junta and Human Rights: Report of the Trial of the Former Junta Members* (London, 1987).

—— *Report and Recommendations of an Amnesty International Mission to the Government of the State of Israel (3–7 Jure 1979) including the Government's Response and A.I. Comments* (London, 1980).

—— *Report on Torture* (2nd edn., London, 1975).

—— *Torture in the Eighties* (London, 1984).

—— USA: *The Death Penalty* (London, 1987).

Andrews, J. A., (ed.), *Human Rights in Criminal Procedure* (The Hague, 1982).

'Atti San Remo, 1970': see under International Institute of Humanitarian Law.

Balladore Pallieri, G., 'Les Transferts Internationaux des Populations', *Annuaire de l'Institut de droit international*, 44 (1952), 138–99.

Barboza, J., 'Necessity (Revisited) in International Law', in *Essays in Honour of Judge Lachs*, ed. Makarczyk (The Hague, 1984), pp. 27–43.

Baxter, R., 'Human Rights and Humanitarian Law: Confluence or Conflict', *Australian YBIL*, 9 (1985), 94–105.

—— 'Multilateral Treaties as Evidence of Customary International Law', *BYBIL* 41 (1965–6), 125 ff.

—— 'Treaties and Custom', *Recueil*, 129 (1970), vol. 1.

Bello, E., 'The African Charter on Human and Peoples' Rights', *Recueil*, 194. 5 (1985), 9–268.

Bernhardt, R. (ed.), *International Enforcement of Human Rights: Report Submitted to the Colloquium of the International Association of Legal Science (Heidelberg, 28–30 August 1985)* (Heidelberg, 1987).

Bin Cheng, *General Principles of Law as Applied by International Courts and Tribunals* (London, 1953; repr. Cambridge, 1987).

Bindman, G. (ed.), *South-Africa: Human Rights and the Rule of Law* (London, 1988).

Bloed A., and van Dijk, P. (eds.), *Essays on Human Rights in the Helsinki Process* (The Hague, 1985).

BLOED, A., and VAN HOOF, G. J. H., 'Some Aspects of the Socialist View of Human Rights', in Bloed and van Dijk (eds.), *Essays on Human Rights in the Helsinki Process*, pp. 29–55.

BONNER, D., *Emergency Powers in Peacetime* (London, 1985).

BOSSUYT, M., *Guide to the* Travaux préparatoires *of the International Covenant on Civil and Political Rights* (The Hague, 1987).

—— 'La Distinction juridique entre les droits civiles et politiques et les droits économiques, sociaux et culturels', *HRJ* 8 (1975), 783.

BOYLE, K., 'The Concept of Arbitrary Deprivation of Life', in Ramcharan (ed.), *The Right to Life in International Law*, pp. 221–44.

—— 'Human Rights and the Northern Ireland Emergency', in Andrews (ed.), *Human Rights in Criminal Procedure*, pp. 144–61.

—— 'Human Rights and Political Resolution in Northern Ireland', *Yale Journal of World Public Order*, 9 (1982), 156–77.

BROWNLIE, I., *Basic Documents in International Law* (3rd edn., Oxford, 1983).

—— *Basic Documents on Human Rights* (2nd edn., Oxford, 1981).

—— *Brownlie's Law of Public Order and National Security*, ed. M. Supperstone (London, 1981).

—— *International Law and the Use of Force by States* (Oxford, 1963).

—— *The Law relating to Public Order* (London, 1968).

—— *Principles of Public International Law* (4th edn., Oxford, 1990).

—— 'Humanitarian Intervention', in J. N. Moore (ed.), *Law and Civil War in the Modern World* (London, 1974).

—— 'Interrogation in Depth: The Compton and Parker Reports', *Modern Law Review*, 35 (1972), 501–7.

—— 'The Rights of Peoples in Modern International Law', *Bulletin of the Australian Society of Legal Philosophy*, 9. 33 (1985), 104–19.

BUERGENTHAL, T., 'The Advisory Practice of the Inter-American Human Rights Commission', *AJIL* 73 (1979), 1–27.

—— 'The Inter-American Court, Human Rights and the OAS', *HRLJ* 7 (1986), 157.

—— 'Judicial Interpretation of the American Human Rights Convention', in OAS, *Human Rights in the Americas*, pp. 253–60.

—— 'Proceedings against Greece under the European Convention of Human Rights', *AJIL* 62 (1968), 441–50.

—— 'To Respect and to Ensure: State Obligations and Permissible Derogations', in Henkin, *The International Bill of Rights*, pp. 72–91.

—— (ed.), *Contemporary Issues in International Law: Essays: Honour of L. B. Sohn* (Kehl, 1984).

—— (ed.), *Human Rights, International Law and the Helsinki Accord* (New York, 1978).

—— and NORRIS, R. E. (eds.), *The Inter-American System* (4 vols.; New York, 1984).

—— NORRIS, R. E., and D. SHELTON (eds.), *Protecting Human Rights in the Americas: Selected Problems* (3rd edn., Kehl-Strasbourg, 1990).

CALOGEROPOULOS-STRATIS, A., *Droit humanitaire et droits de l'homme: La Protection de la personne en période de conflit armé* (Geneva, 1980).

CAMARGO, P. P., 'The American Convention on Human Rights', *HRJ* 3 (1970), 333 ff.

CAMERON, I., 'Turkey and Article 25 of the ECHR', *ICLQ* 37 (1988), 887–925.

CAMPBELL, T. D. (ed.), *Human Rights from Rhetoric to Reality* (Oxford, 1986).

CAPOTORTI, F., 'L'Extinction et la suspension des traités', in *Recueil*, 134 (1971), 417–588.

CARTY, A., 'Human Rights in a State of Exception: The ILA and the Third World', in Campbell (ed.), *Human Rights from Rhetoric to Reality*, pp. 60–79.

CASTBERG, F., *The European Convention on Human Rights* (Leiden, 1974).

CERNA, C., 'Human Rights in Conflict with the Principle of Non-Intervention: The Case of Nicaragua before the 17th Meeting of Consultation of Ministers of Foreign Affairs', in OAS, *Human Rights in the Americas*, pp. 93–107.

CHOWDHURY, S. R., *Rule of Law in a State of Emergency: The Paris Minimum Standards of Human Rights Norms in a State of Emergency* (London, 1989).

CHURCHILL, R. R., and LOWE, A. V., *The Law of the Sea* (2nd rev. edn., Manchester, 1988).

COHEN, E., *Human Rights in the Israeli-Occupied Territories, 1967–1982* (Manchester, 1986).

COHEN-JONATHAN, G., *La Convention européenne des droits de l'homme* (Paris, 1989).

—— 'Les Réserves à la Convention européenne de droits de l'homme: A propos de l'arrêt Belilos du 29 avril 1988', *Revue generale de droit international public* (1989), pp. 273–315.

Council of Europe, *Collected Texts of the ECHR* (1986).

—— *Digest of Strasbourg. Case-Law Relating to the ECHR*, (6 vols.; 1985).

—— 'Problems Arising from the Coexistence of the UN Covenants on Human Rights and the ECHR', Report of the Committee of Experts, Strasbourg, 1970, doc. H(70) 7.

COUSSIRAT-COUSTÈRE, V., 'La Réserve française à l'article 15 de la Convention européenne des droits de l'homme', *HRJ* 2 (1975), 269–75.

—— 'La Réserve française à l'article 15 de la Convention européenne des droits de l'homme', *Journal du droit international*, 2 (1975), 269–75.

CRUZ VILLALON, P., *Estados Exceptionales y Suspension de Garantias* (Madrid, 1984).

DAES, A., 'Restrictions and Limitations on Human Rights', in R. Cassin, *Amicorum Discipulorumque* (4 vols.; Paris, 1969–72), iii. 80–93.

—— 'Study of the Individual's Duties to the Community and the Limitations on Human Rights and Freedoms under Article 29 of the Universal Declaration of Human Rights (Part III: Protection of Human Rights in Time of Public Emergency)', in E/CN. 4/Sub. 2/432/Add. 7 (1980).

DELBRUCK, J., 'Proportionality', in Bernhardt (ed.), *Encyclopedia of Public International Law* (1984), vii. 396–400.

DESPOUY, L., 'États d'exception en Europe continentale et en Amerique latine', *Recueil des Cours*, International Institute of Human Rights (Strasbourg) 14th Session, 1983, pp. 1–25.

DIMITREJEVIC, V., *The Roles of the Human Rights Committee* (Europa Institut der

Universität des Saarlandes, Saarbrucken, 1985).

DINSTEIN, Y., 'Human Rights in Armed Conflicts: International Humanitarian Law', in Meron (ed.), *Human Rights in International Law*, ii. 345–68.

—— 'The Right to Life, Physical Integrity and Liberty', In Henkin (ed.), *The International Bill of Rights*, pp. 114–38.

DRAPER, G. I. A. D., 'The Relationship between the Human Rights Regime and the Law of Armed Conflicts', in 'Atti San Remo, 1970', pp. 141 ff.

DRZEMCZEWSKI, A. Z., *European Human Rights Convention in Domestic Law: A Comparative Study* (Oxford, 1983).

—— *The Applicability of Customary International Human Rights Law in the English Legal System*, HRJ 8 (1975), 71–83.

DUGARD, J., *Human Rights and the South-Africa Legal Order* (New Jersey, 1978).

—— 'The Conflict between International Law and South-Africa Law: Another Divisive Factor in Society', *South-Africa Journal of Human Rights*, (1986), 1–28.

ELAGAB, O. Y., *The legality of Non-Forcible Counter-Measures in International Law* (Oxford, 1988).

EL KOUHENE, M., *Les Garanties fondamentales de la personne en droit humanitaire et droits de l'homme* (The Hague, 1985).

ERGEC, R., *Les Droits de l'homme a l'épreuve des circonstances exceptionnelles: Étude sur l'article 15 de la Convention europeenne des droits de l'homme* (Brussels, 1987).

ERMACORA, F., 'Human Rights and Domestic Jurisdiction', *Recueil*, 124. 2 (1968), 371–451.

European Convention of Human Rights, *Collected Edition of the* Travaux préparatoires *of the ECHR* (8 vols.; The Hague, 1975–85).

FARER, T., 'Elections, Democracy and Human Rights: Toward Union', *HR Quarterly*, 11 (1989), 504–21.

—— 'Human Rights during Internal Armed Conflicts: The Law applicable to Western Hemisphere States', in OAS, *Human Rights in the Americas*, pp. 146–51.

FAWCETT, J., *The Application of the European Convention on Human Rights* (2nd edn., Oxford, 1987).

—— 'The Role of the United Nations in the Protection of Human Rights: Is it misconceived?', in A. Eide and A. Schou (eds.), *International Protection of Human Rights* (1968), pp. 95 ff.

FERNANDEZ de SOTO, G., 'La Desaparicion forzada de personas: Un crimen de Lesa Humanidad', in OAS, *Human Rights in the Americas*, pp. 152–63.

FISCHER, D., 'Reporting under the UNCCPR: The First Five Years of the Human Rights Committee', *AJIL* 76 (1982), 142–53.

FITZMAURICE, G., 'The General Principles of International Law considered from the Standpoint of the Rule of Law', *Recueil*, 92. 2 (1957), 1–227.

—— 'Some Reflections on the ECHR and on Human Rights', in *Völkerrecht als Rechtsordnung Internationale Gerichtsbarkeit Menschenrechte*, Festschrift for H. Mösler (Berlin, 1983), pp. 203–19.

FRANCK, T., 'Preventive Detention and Other Emergency Powers in Africa and Asia', in Franck (ed.), *Human Rights in Third World Perspective* (New York, 1982), pp. 49–252.

Gaja, G., 'A "New" Vienna Convention on Treaties between States and International Organizations', *BYBIL* 58 (1987), 253–69.

Ganshof van der Meersch, W. J., *Organisations européenes* (Brussels, 1966).

—— 'La Protection des droits de l'homme en droit constitutional compare, in *Rapports generaux du IXᵉ Congrès international de droit comparé (Teheran 1974)* (Brussels, 1977), pp. 659 ff.

Garcia Amador, F. V., 'Atribuciones de la CIADH (IACHR) en relacion con los Estados Miembros de la OEA (OAS) que no son partes en la Convention de 1969', in OAS, *Human Rights in the Americas*, pp. 177–87.

Garcia-Bauer, C., *La Conferencia Inter-Americana de Rio de Janeiro y su importancia para la proteccion de los Derechos Humanos*, in OAS, *Human Rights in the Americas*, pp. 62–79.

Garro, A., 'The Role of the Argentine Judiciary in Controlling Governmental Action under the State of Siege', *Human Rights Law Journal*, 4 (1983), 283–344.

Ghandhi, P. R., 'The Human Rights Committee and Derogations in Public Emergencies', *German YBIL* 32 (1989), 321–61.

Green, L. C., 'Human Rights and the General Principles of Law', in Green (ed.), *Law and Society* (Leiden, 1975), pp. 283–320.

Greenspan, M., 'The Protection of Human Rights in Time of Warfare', *Israel YBHR* 1 (1971), 228–45.

Grossman, C., 'Algunas consideraciones sobre el regimen de situaciones de excepcion bajo la Convencion Americana de Derechos Humanos', in OAS, *Human Rights in the Americas*, pp. 121–34.

Groves, H. E., 'Emergency Powers', *Int. CJ Review*, 3 (1961), 1 ff.

Gurria Lacroix, J., 'Antecedentes de la suspension de garantias', *Boletin Mexicano de Derecho Comparado*, 31–2 (1978), 69–88.

Haba, E., *Tratado basico de Derechos Humanos* (2 vols.; San José, Costa Rica, 1986).

Harris, D. J., *The European Social Charter* (New York, 1984).

Hartman, J. F., 'Derogation from Human Rights Treaties in Public Emergencies: A Critique of Implementation by the European Commission and Court of Human Rights and the Human Rights Committee of the United Nations', *Harvard International Law Journal*, 22 (1981), 1–52.

—— 'Working Paper for the Committee of Experts on the Article 4 Derogation Provision', *HR Quarterly* 7 (1985), 89–131.

Henkin, L., 'Human Rights and Domestic Jurisdiction', in Buergenthal (ed.), *Human Rights, International Law and the Helsinki Accord*, pp. 21–39.

—— (ed.), *The International Bill of Rights: The ICCPR* (New York, 1981).

Herczegh, G., 'État d'exception et droit humanitaire: Sur l'article 75 du protocol I de 1977', *Recueil des Cours*, International Institute of Human Rights (Strasbourg), 14th Session, 1983.

Higgins, R., 'Derogations under Human Rights Treaties', *BYBIL* 48 (1976–7), 281–320.

—— 'The European Convention on Human Rights', in Meron (ed.), *Human Rights in International Law*, pp. 495–549.

—— 'Reality and Hope in International Human Rights', *Hofstra Law Review*, 9 (1981), 1485.

HOLLAND, D. C., 'Emergency Legislation in the Commonwealth', *CLP* 13 (1960), 138.

HOWARD, R., 'Evaluating Human Rights in Africa: Some Problems of Implicit Comparisons', *HR Quarterly*, 6 (1984), 161–79.

HUMAN RIGHTS COMMITTEE, *Selected Decisions under the Optional Protocol*, i. *Second to Sixteenth Sessions, 1977–1982*, UN (New York, 1985).

—— *Selected Decisions under the Optional Protocol*, ii. *Seventeenth to Thirty-Second Sessions, October 1982–1988*, UN (New York, 1990).

Human Rights: Status of International Instruments (up to 1 Sept. 1989), UN (New York, 1989).

HUMPHREY, J. P., 'The UDHR: Its History, Impact and Judicial Character', in Ramcharan (ed.), *Thirty Years after the UDHR*, pp. 21–40.

IMBERT, P. H., *Les Réserves aux traites multilateraux: Évolution du droit et de la practique depuis l'avis consultative donné par la Court international de justice le 28 Mars 1951* (Paris, 1979).

—— 'Reservations and Human Rights Conventions', in Maier (ed.), *Protection of Human Rights in Europe*, pp. 87–121.

—— 'Reservations to the ECHR before the Strasbourg Commission: The Temeltasch Case', *ICLQ* 33 (1984), 558–95.

International Commission of Jurists, *Basic Facts* (Geneva, 1962).

—— *Development, Human Rights and the Rule of Law*, Report of the Conference held in the Hague on 27 Apr.–1 May 1981 (Oxford, 1981).

—— *The Dynamic Aspects of the Rule of Law in the Modern Age*, Report on the Proceedings on the South-East Asian and Pacific Conference of Jurists (Bangkok, 1965).

—— *Executive Action and the Rule of Law*, A Report on the Proceedings of the International Congress of Jurists, Rio de Janeiro, 1962 (Geneva, 1962).

—— *Human and Peoples' Rights in Africa and the African Charter*, Nairobi Conference, 1985 (Geneva, 1986).

—— *Human Rights in a One-Party-State: Seminar on Human Rights, their Protection and the Rule of Law in a One-Party-State*, Dar-es-Salaam, 1976 (London, 1978).

—— *Report of the International Congress of Jurists (Athens 1955)* (The Hague, 1956).

—— *The Rule of Law and Human Rights, Principles and Definitions as Elaborated at the Congresses and Conferences held under the Auspices of the International Commission of Jurists, 1955–66* (Geneva, 1966).

—— *The Rule of Law in a Free Society*, A Report of the International Congress of Jurists, New Delhi, 1959 (Geneva, 1962).

—— *States of Emergency: Their Impact on Human Rights* (Geneva, 1983).

—— *States of Siege or Emergency and their Effects on Human Rights: Observations and Recommendations* (1981; repr. in E/CN. 4/Sub, 2/NGO. 93 (1981).

—— 'The Committee's Role regarding States of Emergency', *ICJ Review* (1983), 41 ff.

—— 'Lagos Conference on the Rule of Law 1961', in Brownlie, *Basic Documents on Human Rights*, pp. 426 ff.

International Institute of Humanitarian Law, *Human Rights and Humanitarian Law: Proceedings of the International Conference on Humanitarian Law*, San Remo, 1970 ('Atti San Remo, 1970').

International Law Association, *Human Rights and States of Emergency*, Report of the Committee on the Enforcement of Human Rights Law to the 62nd Conference, Seoul, 1986 (London, 1988). pp. 108–97. (ILA Seoul Report).

—— *Human Rights in States of Emergency*, Second Interim Report of the Committee of Enforcement of Human Rights Law to the 62nd Conference, Warsaw, 1988 (forthcoming) (ILA Warsaw Report).

—— *Human Rights in a State of Emergency*, Report of the Sub-Committee on Human Rights to the 59th Conference, Belgrade, 1980 (London, 1982), pp. 90–145 (ILA Belgrade Report).

—— *Minimum Standards of Human Rights Norms in a State of Exception*, Report of the Committee of the Enforcement of Human Rights Law to the 61st Conferences, Paris, 1984 (London, 1986), pp. 56–96. (ILA Paris Report).

—— *Problems in the Implementation of Human Rights: Minimum Standards of Human Rights Norms in a State of Exception*, Report of the Sub-Committee on Human Rights to the 60th Conference, Montreal, 1982 (London, 1984), pp. 88 ff. (ILA Montreal Report).

—— *Report of the Committee on Human Rights to the 58th Conference (Manila 1978)* (London, 1980), pp. 79–158 (ILA Manila Report).

JACOBS, F. G., *The European Convention on Human Rights* (Oxford, 1975).

—— 'International Law and Human Rights: State Jurisdiction and State Responsibility', in *International Law Conference on Cyprus, 1979, Organized by the Cyprus Bar Council, Nicosia April 30–May 3* (1979), pp. 74–108.

JAGOTA, S. P., 'State Responsibility: Circumstances Precluding Wrongfulness', *NYIL* 16 (1985), 266–71.

JAICHAND, V., 'International Human Rights Law, South Africa and Racial Discrimination', *Journal of Legislation*, 15 (Notre Dame Law School, 1988), pp. 29–44.

JHABVALA, F., 'The Soviet-Bloc's View of the Implementation of Human Rights Accords', *HR Quarterly*, 7 (1985), 461–91.

JIMENEZ de ARECHAGA, E., 'International Law in the Past Third of a Century', *Recueil*, 159. 1 (1978), 1–344.

—— International Responsibility', in M. Sorensen (ed.), *Manual of Public International Law* (London, 1968), pp. 531–603.

KARTASHKIN, V. A., 'Covenants on Human Rights and Soviet Legislation', *HR Journal*, 10 (1977), 97–115.

KAUFMAN, N., and MOSHER, S., 'General Principles of Law and the UN Covenant on Civil and Political Rights', *ICLQ* 27 (1978), 596–613.

'KINGSTON SEMINAR', *UN Seminar on the Effective Realization of Civil and Political Rights at the National Level, Kingston, 1967*, ST/TAO/HR/29.

KISS, A., 'Les Fonctions du secretaire-general du Conseil de l'Europe comme dépositaire des conventions européennes', 2 *AFDI* (1956), 680–8.

—— 'Permissible Limitations and Derogations to Human Rights Conventions', *Recueil des Cours*, International Institute of Human Rights (Strasbourg), 14th Session, 1983, pp. 1 ff.

—— 'Permissible Limitations on Rights', in Henkin (ed.), *The International Bill of Rights*, pp. 290–310.

LATTANZI, F., *Garanzie dei diritti dell 'uomo nel diritto internazionale generale* (Milan, 1983).

LAUTERPACHT, H., *International Law and Human Rights* (London, 1950).

LILLICH, R., 'Civil Rights', in Meron (ed.), *Human Rights in International Law*, pp. 115–70.

—— 'The Paris Minimum Standards of Human Rights in a State of Emergency', *AJIL* 79 (1985), 1072–81.

McBRIDE, S., 'The Inter-Relationship between Humanitarian Law and the Law of Human Rights', in 'Atti San Remo 1970, pp. 83 ff.

McDOUGAL, M., *et al.*, *Human Rights and World Public Order: The Basic Policies of an International Law of Human Rights* (New York, 1980).

—— *et al.*, 'Human Rights and World Public Order: A Framework for Policy-Oriented Enquiry', *AJIL* 63 (1969), 237–69.

McKEAN, W., *Equality and Discrimination under International Law* (Oxford, 1983).

McRAE, D. M., 'Proportionality and the Gulf of Maine Boundary Dispute', *Can. YBIL* 19 (1981), 287–301.

MAIER, I. (ed.), *Protection of Human Rights in Europe: Limits and Effects*, Proceedings of the 5th International Colloquy about the European Convention on Human Rights, Frankfurt-Main, 1980 (Heidelberg, 1982).

MANGAN, B., 'Protecting Human Rights in National Emergencies: Shortcomings in the European System and a Proposal for Reform, *HR Quarterly*, 10 (1988), 372–94.

MANOS, J., *La Clause de derogation dans la Convention européenne des droits de l'homme (article 15)* (Geneva, 1974).

MARIE, J. M., 'International Instruments Relating to Human Rights, Classification and Chart Showing Ratifications as of 1 January 1990', *HRLJ* 11 (1990), 175–202.

MARKS, S., 'Emerging Human Rights: A New Generation for the 1980s?', *Rutgers Law Review*, 33 (1981), 435–52.

—— 'La Notion de periode d'exception en matière des drotis de l'homme', *HRJ* 8 (1975), 821–58.

—— 'Principles and Norms of Human Rights Applicable in Emergency Situations: Underdevelopment, Catastrophes and Armed Conflicts', in Vasak (ed.), *International Dimensions of Human Rights*, pp. 175–213.

MARTINS, D., *The Protection of Human Rights in Connection with the Suspension of Guarantees or 'State of Siege'*, OAS Ser. L/V/II. 15, doc. 12 (1966) (repr. in OAS, *The Organization of American States and Human Rights*, pp. 122–54).

—— 'La Suspension de garantias o Estado de Sitio ante el Derecho Constitucional Internacional de los Estados Americanos, in *Revista de la Facultad de Derecho y Ciencias Sociales* (Montevideo, 1966), pp. 453–518.

MATHEWS, A., Freedom, State Security and the Rule of Law, Dilemmas of the Apartheid Society (Cape Town, 1986).

MEDINA, C., *The Battle of Human Rights: Gross, Systematic Violations and the Inter-American System* (The Hague, 1988).

MENDEZ, J., 'La Participacion de la CIADH (IACHR) en los conflictos entre los Miskitos y el Gobierno de Nicaragua', in OAS, *Human Rights in the Americas*, pp. 306–18.

MERON, T., *Human Rights and Humanitarian Norms as Customary Law* (Oxford, 1989).

—— *Human Rights in Internal Strife: Their International Protection* (Cambridge, 1987).

—— *Human Rights Law-Making in the United Nations: A Critique of Instruments and Process* (Oxford, 1986).

—— 'Applicability of Multilateral Conventions to Occupied Territories', *AJIL* 72 (1978), 542–57.

—— 'Draft Model Declaration on Internal Strife', *International Review of the Red Cross*, 262 (1988), 59–76.

—— 'The Geneva Conventions as Customary Law', *AJIL* 81 (1987), 348–70.

—— 'Human Rights in Time of Peace and in Time of Armed Strife: Selected Problems', in Buergenthal (ed.), *Contemporary Issues in International Law*, pp. 1–21.

—— 'The International Convention on the Elimination of All Forms of Racial Discrimination and the Golan Heights', *Israel YBHR* 8 (1978), 222–39.

—— 'On a Hierarchy of International Human Rights', *AJIL* 80 (1986), 1–23.

—— 'On the Inadequate Reach of Humanitarian and Human Rights Law and the Need for a New Instrument', *AJIL* 77 (1983), 589.

—— 'Towards a Humanitarian Declaration on Internal Strife', *AJIL* 78 (1984), 859–69.

—— 'West Bank and Gaza: Human Rights and Humanitarian Law in the Period of Transition', *Israel YBHR* 9 (1979), 106–20.

—— (ed.), *Human Rights in International Law: Legal and Policy Issues* (Oxford, 1984).

MERRILLS, J. G., *The Development of International Law by the European Court of Human Rights* (Manchester, 1988).

'MEXICO SEMINAR', *UN Mexico Seminar on Amparo, Habeas Corpus and Other Similar Remedies (15–28 August 1961)*, ST/TAO/HR/12.

MONTEALEGRE, H., *La Seguridad del Estado y los Derechos Humanos* (Santiago de Chile, 1979).

MORRISON, C., 'Margin of Appreciation in European Human Rights Law', *Human Rights Journal*, 6 (1973), 263–86.

MOSLER, H., 'The International Society as a Legal Community', *Recueil* 140. 4 (1974), 1–320.

MOVCHAN, A., *Human Rights and International Relations* (Moscow, 1988).

MOYER, C., and PADILLA, D., 'Executions in Guatemala as Decreed by the Courts of Special Jurisdiction in 1982–3, in OAS, *Human Rights in the Americas*, pp. 280–9.

NIKKEN, P., *La Proteccion international de los Derechos Humanos: Su desarrollo progresivo* (Madrid, 1987).

NORRIS, R. E., and DESIO, P., 'The Suspension of Guarantees: A Comparative Analysis of the American Convention on Human Rights and the Constitutions of the States Parties,' *The American University Law Review*, 30 (1980), 189–223.

NOWAK, M., 'The Effectiveness of the ICCPR: Stocktaking after the First Eleven Sessions of the UN Human Rights Committee', 1 *HRLJ* (1980) 136–70.

—— 'UN Human Rights Committee: Survey of Decisions given up till July 1984', *HRLJ* 5 (1984), 199–219.

—— 'Un Human Rights Committee: Survey of Decisions given up till July 1986', *HRLJ* 7 (1986), 287–307.

—— 'UN Human Rights Committee: Survey of Decisions given up till July 1989', *HRLJ* 11 (1990), 139–56.

NWABUEZE, B. D., *Constitutionalism in the Emergent States* (London, 1973).

—— *Presidentialism in Commonwealth Africa* (London, 1974).

OAS, *Handbook of Existing Rules Pertaining to Human Rights: the Inter-American System* (Washington, 1985).

—— *Human Rights in the Americas: Homage to the Memory of Carlos A. Dunshee de Abranches* (Washington, 1984).

—— *The IACHR: Ten Years of Activities (1971–1981)* (Washington, 1982).

—— *The Organization of American States and Human Rights, 1960–1967* (Washington, 1972).

O'BOYLE, M., 'Emergency Situations and the Protection of Human Rights: A Model Derogation Provision for a Northern Ireland Bill of Rights', *Northern Ireland Legal Quarterly*, 28 (1977), 160–87.

—— 'Torture and Emergency Powers under the European Convention on Human Rights: Ireland v U.K.', *AJIL* 71 (1977), 686 ff.

O'DONNELL, D., 'Commentary (to the "Siracusa Principles") by the Rapporteur on Derogation', *HR Quarterly*, 7 (1985), 23–34.

—— 'Legitimidad de los Estados de Exception, a la luz de los instrumentos de Derechos Humanos', in Haba, *Tratado basico de Derechos Humanos*, ii. 679 ff.

—— 'States of Exception', *Int CJ Review* 21, (1978), 52–60.

OKERE, O., 'The Protection of Human Rights in Africa and the African Charter on Human and Peoples' Rights: A Comparative Analysis with the European and American Systems', *HR Quarterly*, 6 (1984), 141–60.

OPSAHL, T., 'Emergency Derogation from Human Rights', *Nordic Journal on Human Rights*, 3 (1987), 4–6.

OSAKWE, C., 'Soviet Human Rights Law under the USSR Constitution of 1977: Theories, Realities and Trends', *Tulane Law Review*, 56 (1981), 249–93.

'Oslo Statement', 'Osle Statement on Norms and Procedures in Times of Public Emergency or Internal Violence (17 June 1987)', *Nordic Journal on Human Rights* 3 (1987), 2–3 (repr. in E/CN. 4/Sub. 2/1987/31).

PACHOTA, V., 'The Development of the ICCPR', in Henkin (ed.), *The International Bill of Rights*, 32–71.

PELLET, A., 'La Ratification de la Convention européenne des droits de l'homme', *Revue de droti public* (1975), 1319–79.

Personal Justice Denied: Report of the Commission on Wartime Relocation and Internment of Civilians (Washington, 1982).

PILLITU, P. A., *Lo stato di necessità nel diritto internazionale* (Perugia, 1981).

QUESTIAUX, N., *Study of the Implications for Human Rights of Recent Developments concerning Situations Known as State of Siege or Emergency* (E/CN, 4/Sub, 2/1982/15).

—— 'La Convention européenne des droits de l'homme et l'article 16 de la constitution du 4 octobre 1958', *HRJ* 3 (1970), 651–63.

RAMCHARAN, B. A., 'The International Law Commission and Human Rights', *HRJ* 8 (1975), 9–20.

—— (ed.), *The Right to Life in International Law* (The Hague, 1985).

—— (ed.), *Thirty Years after the UDHR* (The Hague, 1979).

ROBERTS, A., 'The Applicability of Human Rights Law during Military Occupation', *Review of International Studies*, 13 (1987), 39–48.

—— and GUELFF, R., (ed.), *Documents on the Laws of War* (2nd edn., Oxford, 1989).

ROBERTSON, A. H., *Human Rights in the World: An Introduction to the Study of the International Protection of Human Rights* (3rd edn., Machester, 1989).

—— 'Human Rights as the Basis of International Humanitarian Law', in 'Atti San Remo 1970', pp. 55 ff.

—— 'The Implementation System: International Measures', in Henkin (ed.), *The International Bill of Rights*, pp. 332–69.

—— 'The Lawless Case', *BYBIL* 37 (1961), 536 ff.

—— 'The Political Background and Historical Development of the ECHR', British Institute of International and Comparative Law (London, 1965), pp. 25–38.

—— (ed.), *Privacy and Human Rights*, Manchester 1973.

RODLEY, N., *The Treatment of Prisoners under International Law* (Oxford, 1987).

—— 'Human Rights and Humanitarian Intervention: The Case-Law of the World Court', *ICLQ* 38 (1989), 321–33.

—— 'UN Action and Procedures against Disappearances, Summary or Arbitrary Executions, and Torture', *HR Quarterly* 8 (1986), 700–30.

ROSENNE, S., 'The Depositary of International Treaties', *AJIL* 61 (1967), 923–45.

ROUSSEAU, C., *Le Droit des conflicts armés* (Paris, 1983).

SALMON, J. A., 'Faut-il codifier l'état de nécessité en droit international?', in Makarczyk (ed.), *Essays in Honour of Judge Lachs* (The Hague 1984), pp. 236–70.

SALVIA, M., 'La Notion de proportionalité dans la jurisprudence de la Commission et de la Cour européenne de droits de l'homme', *Diritto communitario e degli scambi internazionali*, 17 (1978), 463–93.

SAPIENZA, R., 'International Legal Standards on Capital Punishment', in Ramcharan (ed.), *The Right to Life in International Law*, pp. 284–96.

SCHACHTER, O., 'General Course in Public International Law: International Law in Theory and in Practice', *Recueil*, 178. 5 (1982), 9–396.

SCHWELB, E., 'The Law of Treaties and Human Rights', *Archiv des Völkerrechts* 16 (1974–5), 1–27.

SHELTON, D., 'Application of Death Penalty on Juveniles in the US: Violation of Human Rights Obligations within the Inter-American System', *HRLJ* 8 (1987), 345–61.

—— 'State Practice on Reservations on Human Rights Treaties', *Annuaire canadien des droits de l'homme* (1983), 205–34.

SINCLAIR, I., *The Vienna Convention on the Law of Treaties* (2nd edn., Manchester, 1984).

SINGHVI, A. M., 'The Law of Emergency Powers: A Comparative Study', Ph.D. thesis (Cambridge, 1986).

'Siracusa Principles', 'The Siracusa Principles on the Limitation and Derogation Provisions in the ICCPR', *HR Quarterly* 7 (1985), 3–130.

SMITH, G. A., 'The European Convention on Human Rights and the Right of Derogation: A Solution to the Problem of Domestic Jurisdiction', *Howard Law Journal* 11 (1965), 594–606.

—— (ed.), *International Protection of Human Rights* (New York, 1973).

STEIN, T., 'Derogations from Guarantees Laid down in Human Rights Instruments', in Maier (ed.), *Protection of Human Rights in Europe*, pp. 123–33.

SUPPERSTONE, M., *Brownlie's Law of Public Order and National Security* (London, 1981).

SUY, E., 'Droits des traités et droits de l'homme', *Völkerrecht als Rechtsordnung Internationale Gerichtsbarkeit Menschenrechte*, Festschrift for H. Mösler (Berlin, 1983), pp. 935–47.

TUMANOV, V. A., 'International Protection of Human Rights: Soviet Report', in Bernhardt (ed.), *International Enforcement of Human Rights*, pp. 21–4.

TUNKIN, G. I., 'General International Law in the International System', Recueil, 147. 4 (1975), 1–218.

—— *Theory of International Law* (London, 1974).

UMOZURIKE, U. O., 'The African Charter on Human and Peoples' Rights', *AJIL* 77 (1983), 902–12.

UN, *Multilateral Treaties Deposited with the Secretary-General*, Status as at 31 Dec. 1986 (New York, 1987).

US *Restatement*: US *Rest*. 3rd.: *Restatement of the Foreign Relations Law of the USA*, American Law Institute (Washington, 1987).

VAN DIJK, P., and VAN HOOF, G. J. H., *Theory and Practice of the ECHR* (Deventer, 1984).

VAN HOOF, F., 'The Protection of Human Rights and the Impact of Emergency Situations under International Law with Special Reference to the Present Situation in Chile', *HRJ* 10 (1977), 213–48.

VARGAS CARRENO, E., 'Las Observaciones. *in loco* practicadas por la Comision Americana de Derechos Humanos', in OAS, *Human Rights in the Americas*, pp. 290–305.

VASAK, K. (ed.), *International Dimensions of Human Rights* (2 vols.; Westport, Conn., 1982).

—— 'L'Histoire des problèmes de la ratification de la Convention par la France', *HRJ* 3 (1970), 558–66.

VELU, J., 'The European Convention on Human Rights and the Right to Respect for Private Life, the Home and Communications', in Robertson (ed.), *Privacy and Human Rights*, pp. 12–95.

VERDROSS, A., *Derecho internacional publico* (4th edn., Madrid, 1972).

VILLEVIEILLE, J. F., 'La Ratification par la France de la Convention européenne des drotis de l'homme', *AFDI* 29 (1973), 922–7.

WALDOCK, H., 'General Course on Public International Law,' *Recueil*, 106 (1962), 1–250.

—— 'Human Rights in Contemporary International Law and the Significance of the European Convention', British Institute of International and Comparative Law (London, 1965), pp. 1–23.

WALKATE, J. A., 'The Human Rights Committee and Public Emergencies', *Yale Journal of World Public Order*, 9 (1982), 133–46.

WARBRICK, C., *'The Protection of Human Rights in National Emergencies'*, in F. E. Dowrick (ed.), *Human Rights Problems, Perspectives and Texts* (Farnborough, 1979), pp. 89–106.

WEISSBRODT, D., 'Protecting the Right to Life: International Measures against Arbitrary Killing or Summary Killings by Governments', in Ramcharan (ed.), *The Right to Life in International Law*, pp. 297–314.

—— 'The Three "Theme" Special Rapporteurs of the UN Commission on Human Rights', *AJIL* 80 (1986), 685–95.

ZANGHI, C., 'La Declaration de la Turquie relative à l'article 25 de la Convention européenne des droits de l'homme', *Revue general de droit international public* 93 (1989), pp. 69–85.

ZAYAS, A., *et al.*, 'Application of the International Covenant on Civil and Political Rights, under the Optional Protocol by the Human Rights Committee', *German YBIL* 28 (1985), 9–64.

Index

Afghanistan 155, 196–7
African Charter 209–10
Ago, R. 220, 228, 261
arbitrary detention
  minimum guarantees 108–14
  incommunicado 109–10
Argentina 122, 137–8, 161

Barbados 134–5
Bolivia 23, 52, 69–70, 163, 187
Brogan case 109
Brownlie, I. 140 n. 2, 214 n. 1, 247
Buergenthal, Th. 129
burden of proof, the problem of 47

Chile:
  and the principle of exceptional threat
    21, 26
  and the proclamation of emergency 36–7
  judicial control 41–2
  and the reporting procedure 48–9
  deprivation of nationality 98–9
  and the rights of detainees 114
  and proportionality 154, 156, 167
  and non-discrimination 183
  and customary international law 238–9,
    248
Colombia 25–6, 61, 120–1, 158–9
Congo 133
consistency, the principle of 190–1
Cuba 186
Cyprus case:
  and the principle of proclamation 37–8
  margin of appreciation 46
  and notification 67–9, 79
  and non-discrimination 182
  and the principle of consistency 194–6,
    199–200, 204

De Becker case 151
derogation, notice of 60–4
derogation clause
  origin 7–8
  legal nature 227–8
  and customary international law 229–33
Despouy, L. 244–6, 258
'dictator' (in Roman times) 7
due process rights 115–6

Ecuador 74–5
El Salvador 156, 186
emergency:
  definition 11–16
  interpretation of the concept 16–27
  characteristics 27–30
  circumstances 30–1
Ermacora, F. 29, 146
Evans, Sir V. 8 n. 6, 121, 146
exceptional threat, *see* emergency

Fawcett, Sir J. 45, 71–2, 200
force majeure 220–1, 224–5
France . . . v. Turkey 64

Genocide case 129
Germany, Federal Republic of 136–7
good faith, the principle of 44–5
Greek case, The:
  and the concept of emergency 18–20,
    27–8, 36
  and international control of emergencies
    43
  and the notice of derogation 60–1, 71
  and articles 17–18 of the ECHR 104–5
  and military courts 119
  and proportionality 149, 156
  and non-discrimination 181–2
  and the principle of consistency 193–4
  ILO Report 223–5, 262
Guatemala 122, 127–9

habeas corpus 111–13
Hartman, J. F. 94
Higgins, R. 132, 262
Human Rights Committee:
  reporting procedure 48–9
  Optional Protocol 51

IA Court HR 112, 121, 127–9, 159–60
IACHR, control of emergencies 51–5
International Commission of Jurists 108–9,
  113–14
International Law Association (ILA) 30–1,
  41
Ireland v. *U.K.*
  and the concept of emergency 28–9
  and the notice of derogation 61–2, 77

and proportionality 144–50
and non-discrimination 177–81
and the principle of consistency 193
Israel 256–7
ius cogens 96, 229, 249–50, 264, 269

Jacobs, F. G. 130
Jimenez de Arechaga, E. 114, 230
judicial control by domestic courts 40–2

Landinelli case 21, 70, 75–6, 81, 157
Lawless case:
  and the concept of emergency 16–8
  characteristics of the emergency 27–9
  and proclamation 38–9
  and the principle of good faith 44
  the margin of appreciation 45–7
  and the notice of derogation 60, 63
  and the notification requirement 65–7, 71
  and article 18 of the ECHR 105
  and extrajudicial guarantees 113
  and due process rights 118
  and proportionality 144–6, 148–9
  and the principle of consistency 193
limitation clauses 8–10, 141

Macias case 99–100
Malta 134
margin of appreciation 45–7
Meron, Th. 215–18, 238–9, 257, 264
Mexico 130–1, 136, 138

ne bis in idem 235
Nicaragua
  and the concept of emergency 24–5
  and non-derogable rights 101
  and proportionality 146, 154–5, 162–6
  and non-discrimination 182, 187
Nicaragua case (v. *US*), *The* 217–18
Nicaragua–Miskitos case
  and the concept of emergency 24
  and proclamation 38
  and international control 52–3
  and the notice of derogation 61
  and proportionality 161–2
  and non-discrimination 174, 176, 187–9
  and the principle of consistency 197–9
non-derogability, the principle of 87–8
non-derogable rights,
  the list 94–6
  the four common 96–7
  reservations to 127–39

non-discrimination, the principle of 170–7
notification:
  the rationale of the principle 58–9
  see also derogation, notice of
North Sea Continental Shelf cases 230

Paraguay 30, 167, 239
Peru 155
Poland 248, 261
Polish case, The. ILO Report. 225–6
proclamation, the requirement of 34–7
proportionality, the principle of 140–2

Questiaux, N. 28, 110, 116, 243–4, 258

reservations 127–39

Schachter, O. 254, 259
self-defence 220–1, 224–5
South Africa 211, 249–54
state necessity 221–6
state of emergency, *see* emergency
Suriname 79, 168, 187
Syria 121

Trinidad and Tobago 131, 136

United Kingdom (U.K.):
  and the origin of the derogation clause
    7–9
  and the concept of emergency 12
  and non-derogability 87–8, 91
  and proportionality 153–5
  and non-discrimination 173–4
  and the Diplock Courts 183–4
  and consistency 191–2
  see also *Ireland* v. *U.K.*
UN Commission on Human Rights 247–9
Uruguay 74, 102, 119–20, 139, 153–4,
    157–8
USA 175, 177, 191–2
US Restatement 215–17, 231 n. 19, 234,
    253–5

Velu, J. 105

Waldock, Sir H. 17, 42, 45
war 12–14
  laws of 201–5
  belligerent occupation 256–7

Weinberger case 184–5